Advances in Industrial Crystallization

Advances in Industrial Crystallization

Editors

Heike Lorenz
Erik Temmel

MDPI • Basel • Beijing • Wuhan • Barcelona • Belgrade • Manchester • Tokyo • Cluj • Tianjin

Editors
Heike Lorenz
Max Planck Institute for Dynamics of Complex Technical Systems
Germany

Erik Temmel
Sulzer Chemtech Ltd.
Switzerland

Editorial Office
MDPI
St. Alban-Anlage 66
4052 Basel, Switzerland

This is a reprint of articles from the Special Issue published online in the open access journal *Crystals* (ISSN 2073-4352) (available at: https://www.mdpi.com/journal/crystals/special_issues/Industrial_Crystallization).

For citation purposes, cite each article independently as indicated on the article page online and as indicated below:

LastName, A.A.; LastName, B.B.; LastName, C.C. Article Title. *Journal Name* **Year**, *Volume Number*, Page Range.

ISBN 978-3-0365-0330-1 (Hbk)
ISBN 978-3-0365-0331-8 (PDF)

Cover image courtesy of Heike Lorenz.

© 2021 by the authors. Articles in this book are Open Access and distributed under the Creative Commons Attribution (CC BY) license, which allows users to download, copy and build upon published articles, as long as the author and publisher are properly credited, which ensures maximum dissemination and a wider impact of our publications.

The book as a whole is distributed by MDPI under the terms and conditions of the Creative Commons license CC BY-NC-ND.

Contents

About the Editors vii

Preface to "Advances in Industrial Crystallization" ix

Erik Temmel and Heike Lorenz
Advances in Industrial Crystallization
Reprinted from: *Crystals* 2020, *10*, 997, doi:10.3390/cryst10110997 1

Elena Kotelnikova, Roman Sadovnichii, Lyudmila Kryuchkova and Heike Lorenz
Limits of Solid Solutions and Thermal Deformations in the L-Alanine–L-Serine Amino Acid System
Reprinted from: *Crystals* 2020, *10*, 618, doi:10.3390/cryst10070618 5

Alexander A. Bredikhin, Dmitry V. Zakharychev, Zemfira A. Bredikhina, Alexey V. Kurenkov, Aida I. Samigullina and Aidar T. Gubaidullin
Stereoselective Crystallization of Chiral 3,4-Dimethylphenyl Glycerol Ether Complicated by Plurality of Crystalline Modifications
Reprinted from: *Crystals* 2020, *10*, 201, doi:10.3390/cryst10030201 19

Maxim Oshchepkov, Vladimir Golovesov, Anastasia Ryabova, Anatoly Redchuk, Sergey Tkachenko, Alexei Pervov and Konstantin Popov
Gypsum Crystallization during Reverse Osmosis Desalination of Water with High Sulfate Content in Presence of a Novel Fluorescent-Tagged Polyacrylate
Reprinted from: *Crystals* 2020, *10*, 309, doi:10.3390/cryst10040309 37

Elena Horosanskaia, Lina Yuan, Andreas Seidel-Morgenstern and Heike Lorenz
Purification of Curcumin from Ternary Extract-Similar Mixtures of Curcuminoids in a Single Crystallization Step
Reprinted from: *Crystals* 2020, *10*, 206, doi:10.3390/cryst10030206 53

Nahla Osmanbegovic, Lina Yuan, Heike Lorenz and Marjatta Louhi-Kultanen
Freeze Concentration of Aqueous [DBNH][OAc] Ionic Liquid Solution
Reprinted from: *Crystals* 2020, *10*, 147, doi:10.3390/cryst10030147 69

Ying-Wai Cheong, Ka-Lun Wong, Boon Seng Ooi, Tau Chuan Ling, Fitri Khoerunnisa and Eng-Poh Ng
Effects of Synthesis Parameters on Crystallization Behavior of K-MER Zeolite and Its Morphological Properties on Catalytic Cyanoethylation Reaction
Reprinted from: *Crystals* 2020, *10*, 64, doi:10.3390/cryst10020064 83

Dan Zhu, Shihao Zhang, Pingping Cui, Chang Wang, Jiayu Dai, Ling Zhou, Yaohui Huang, Baohong Hou, Hongxun Hao, Lina Zhou and Qiuxiang Yin
Solvent Effects on Catechol Crystal Habits and Aspect Ratios: A Combination of Experiments and Molecular Dynamics Simulation Study
Reprinted from: *Crystals* 2020, *10*, 316, doi:10.3390/cryst10040316 99

Yu Bai, Yingming Wang, Shijie Zhang, Qi Wang and Ri Li
Numerical Model Study of Multiple Dendrite Motion Behavior in Melt Based on LBM-CA Method
Reprinted from: *Crystals* 2020, *10*, 70, doi:10.3390/cryst10020070 115

Dominic Wirz, Marc Hofmann, Heike Lorenz, Hans-Jörg Bart, Andreas Seidel-Morgenstern and Erik Temmel
A Novel Shadowgraphic Inline Measurement Technique for Image-Based Crystal Size Distribution Analysis
Reprinted from: *Crystals* **2020**, *10*, 740, doi:10.3390/cryst10090740 131

Miao Liang, Zhijun Wang, Hai Wu, Li Yu, Bo Sun, Huan Zhou, Feng Yu, Qisheng Wang and Jianhua He
Microplates for Crystal Growth and in situ Data Collection at a Synchrotron Beamline
Reprinted from: *Crystals* **2020**, *10*, 798, doi:10.3390/cryst10090798 157

Christian Melches, Hermann Plate, Jürgen Schürhoff and Robert Buchfink
The Steps from Batchwise to Continuous Crystallization for a Fine Chemical: A Case Study
Reprinted from: *Crystals* **2020**, *10*, 542, doi:10.3390/cryst10060542 173

Erik Temmel, Jonathan Gänsch, Andreas Seidel-Morgenstern and Heike Lorenz
Systematic Investigations on Continuous Fluidized Bed Crystallization for Chiral Separation
Reprinted from: *Crystals* **2020**, *10*, 394, doi:10.3390/cryst10050394 187

Dennis Hülsewede, Erik Temmel, Peter Kumm and Jan von Langermann
Concept Study for an Integrated Reactor-Crystallizer Process for the Continuous Biocatalytic Synthesis of (S)-1-(3-Methoxyphenyl)ethylamine
Reprinted from: *Crystals* **2020**, *10*, 345, doi:10.3390/cryst10050345 203

Ming Li, Lulu Gu, Tao Li, Shiji Hao, Furui Tan, Deliang Chen, Deliang Zhu, Yongjun Xu, Chenghua Sun and Zhenyu Yang
TiO_2-Seeded Hydrothermal Growth of Spherical $BaTiO_3$ Nanocrystals for Capacitor Energy-Storage Application
Reprinted from: *Crystals* **2020**, *10*, 202, doi:10.3390/cryst10030202 217

Hatem Abushammala and Jia Mao
Impact of the Surface Properties of Cellulose Nanocrystals on the Crystallization Kinetics of Poly(Butylene Succinate)
Reprinted from: *Crystals* **2020**, *10*, 196, doi:10.3390/cryst10030196 233

About the Editors

Heike Lorenz studied Chemistry at Freiberg University of Mining and Technology and received her Ph.D. from Otto von Guericke University of Magdeburg. After research periods at the Otto von Guericke University and the Max Planck Institute for Dynamics of Complex Technical Systems in Magdeburg she obtained her habilitation with a thesis on heterogeneous processes on the example of combustion of solids and crystallization from solution. Since 1998 she is senior researcher at the mentioned Max Planck Institute in Magdeburg and leads a crystallization research team. She was visiting scientist at the International Flame Research Foundation, Ijmuiden/The Netherlands and invited lecturer at Rouen University. Since 2010 she is Adjunct Professor (apl. Prof.) at Otto von Guericke University. Heike Lorenz authored or co-authored more than 160 papers, 10 book contributions, and has edited 2 international conference books. She is a member of the European Federation of Chemical Engineering's (EFCE) Working Party on Crystallization and the Crystallization Board of DECHEMA/VDI-GVC's ProcessNet in Germany. Her research interests include crystallization for separation and product design, process monitoring and purification of fine chemicals, large-scale industrial and natural products. Since 2019 she is a member of the Editorial Board of *Crystals* and Editor-in-Chief of its *Industrial Crystallization* Section.

Erik Temmel studied chemical process engineering at the Otto von Guericke University and received his Ph.D. for research on continuous crystallization at the Max Planck Institute in Magdeburg. After several post-doctoral fellowships, e.g. at DSM (Switzerland), he is now leading the R&D for crystallization at Sulzer Chemtech Ltd (Switzerland).

Preface to "Advances in Industrial Crystallization"

Crystallization plays an important role in many manufacturing industries for the production of food, drugs, biopharmaceuticals, flavors, dyestuffs and pigments and agrochemicals, as well as ceramics and metals. Crystallization process design in industry is generally based on key performance indicators (KPIs) which include costs, process parameters, such as yield, productivity and process simplicity, and product properties, such as purity, solid-state form, crystal size and morphology. These KPIs are usually specific for each application. In relation to this connection, the book compiles fifteen articles collected for a first Special Issue of the recently established Industrial Crystallization Section of *Crystals*.

Relating to several of the abovementioned industry sectors, the contributions comprise different current facets of crystallization research, from the very fundamentals up to industrial application. Many of the topics covered may be classified as follows: (i) from the crystallization basics to processes, (ii) crystal shape development, (iii) measurement techniques, (iv) continuous crystallization, (v) process intensification, (vi) melt crystallization and (vii) nanoparticles in crystallization. Most articles are strongly based on challenging experimental work and also involve novel techniques.

The individual articles are introduced in the *Editorial* of the book in the context of the main topics mentioned above.

We thank all authors whose contributions are included in this Special Issue for their excellent work, and their inspiring articles.

Heike Lorenz, Erik Temmel
Editors

Editorial

Advances in Industrial Crystallization

Erik Temmel [1,2] and Heike Lorenz [2,*]

1. Sulzer Chemtech Ltd., Gewerbestrasse 28, 4123 Allschwil, Switzerland; erik.temmel@sulzer.com
2. Max Planck Institute for Dynamics of Complex Technical Systems, Sandtorstrasse 1, 39106 Magdeburg, Germany
* Correspondence: lorenz@mpi-magdeburg.mpg.de

Received: 29 October 2020; Accepted: 31 October 2020; Published: 3 November 2020

Dear colleagues,

We are pleased to present the Crystals' Special Issue on "Advances in Industrial Crystallization". We are grateful for the large quantity of submitted manuscripts and we would like to thank all authors of the selected 15 publications for their efforts. In the following, a brief introduction describing the idea for creating this Special Issue and an overview of the related topics covered by the contributions is given.

As a general perspective from the industry, each process choice and design is individualized to each of the innumerous applications. For example, knowledge of basic thermodynamics yields the overall "map" for the crystallization process. The general solid–liquid equilibrium (SLE) type defines the separability of a mixture, depending on the operation conditions (i.e., composition, temperature, etc.) and determines the route to solidify the demanded solid-state form. Together with the kinetics of mass transfer and crystallization, which give insights into the expected crystal size distribution and shape development, an educated evaluation of various processes is feasible [1–5].

A decision between the reasonable options is taken afterward, based on the expected process of key performance indicators or constraints related to the specific application. For expensive fine chemicals, like enantiomers and other APIs, product purity and yield are commonly the focus, while for bulk chemicals, like inorganic salts or monomers (e.g., acrylic acid), productivity and process simplicity is mostly decisive. Subsequently, first lab-scale trials are usually conducted to evaluate the process choice and confirm the initial expectations. Suitable upscaling strategies are applied, afterward, to "level" the process up to the final plant, which can have capacities between a few kilograms or several hundred thousand tons per year. A careful monitoring of the critical parameters, like the liquid phase composition and temperature, but also the crystal shape and size distribution, exploiting suitable measurement techniques, is crucial at this stage of process development [6].

However, mostly empirical upscaling strategies exist in industry today and the above-mentioned detailed fundamental information are commonly not available and cannot be measured in the typical time of an industrial project. Hence, industry is inevitably dependent on academic research, which tirelessly helps to clarify the required essential issues.

Reflecting this successful relationship, all together 15 publications are summarized in this Special Issue. They comprise several contemporary aspects of crystallization and simultaneously give a comprehensive overview of industrially relevant topics in the field. The main subjects covered include (i) from fundamentals towards crystallization processes, (ii) crystal shape development, (iii) measurement techniques, (iv) continuous crystallization, (v) process intensification, (vi) melt crystallization, and (vii) nanoparticles in crystallization. In the following we briefly introduce the respective papers.

i. **From fundamentals towards crystallization processes:** As stated above, sophisticated crystallization process design relies on fundamentals like the present phase equilibria (e.g., solid–liquid, solid–solid) and the prevailing crystallization kinetics (like nucleation, growth, agglomeration, and breakage) [3,7–13]. The diversity of the individual properties in the solid state [9,12] or the solidification process [10] of different products is one major challenge for each process engineer. Hence, continuously novel processes [11,13] and process combinations [10] are discovered to deal with these various issues to fulfill the desired task, like product purity or productivity enhancement. The materials studied from the authors comprise biologically active components or their precursors as amino acids, curcumin from a plant extract, or a particular chiral dimethylphenyl glycerol ether [9,11,12]. The application of specific antiscalants to inhibit gypsum scaling in RO desalination, and utilization of freeze concentration for recycling of an ionic liquid from its aqueous solution are introduced in articles [10,13].

ii. Beside purity and solid-state form, the crystal size distribution and the crystal shape are often target properties of a product and also decide on the overall process performance [4,5]. Although the crystal morphology of a particular compound is primarily determined by its crystal lattice, the individual growth rates of the different crystallographic faces can lead to alternative crystal geometries. The latter might be caused by lattice defects at the crystal surface and crystallization conditions like temperature, supersaturation, impurities present, stirrer energy introduced, and hydrodynamic conditions in general [2,4,5,14]. Hence, there is a need for prediction or empirical study of **crystal shape development**, an issue that is considered in the Special Issue articles [15–17]. It can be shown for numerous substances that the crystal environment is decisive for the morphology, after the process [15], but the fundamental mechanisms [16] are still not completely clarified and there is a great demand for predictive tools [17] for the process design.

iii. In this connection, sophisticated **measurement techniques** are of particular importance [5,6,18]. To date, there is still a need for suitable measurement techniques to efficiently study and control crystallization processes. New methods for monitoring the process at the point of crystal formation (i.e., in situ), shed light on the evolution of the crystal size distribution [19] or the solid-state formation [20], for example. In [19], the authors introduce a new inline probe for image-based measurement of crystal size distribution, which is also applicable at larger scales. Whereas, in [20], equipment for the in situ small-scale determination of the crystal structure on the example of lysozyme is developed.

iv. **Continuous crystallization** as a common operation mode for most bulk chemical manufacturing processes has recently gained much interest from the pharmaceutical industry, and is thus, again focus of academic research [4,18,21–24]. Even though the basics were developed until the 80s, there is still a great need for upscaling procedures [22], general process development [23], and new ideas for the efficient upstream and downstream combination [24]. The examples studied cover a broad range of applications as an industrially relevant fine chemical produced in a Draft Tube Baffled (DTB) crystallizer, at targeted particle-related properties [22], a continuous fine chemical enantiomer separation in coupled fluidized bed crystallizers [23], and the continuous synthesis of a chiral drug intermediate via an integrated biocatalysis-crystallization concept [24].

v. A novel field of interest is related to **process intensification** by integration of different unit operations, including crystallization [10,18,23,24]. The related Special Issue contributions demonstrate that combining crystallization and crystal size classification in one apparatus can lead to highly efficient processes that still meet the demanding purity requirements for pure enantiomers [23], and that combination with membrane concentration [10] can give rise to completely new fields of research. Additionally, the integration of a biocatalytic reaction with crystallization to overcome unfavorable chemical reaction equilibria [24] has attracted much interest, lately.

vi. Even though the focus of this Special Issue is mainly crystallization from solution, the ongoing activities in the field of industrial **melt crystallization** [4,5,18,25] are represented as well [13,17]. Especially, in the common domains of freeze crystallization [5,13,18] and metallurgy [17],

vii. Beside the purification or separation of enantiomers, fine chemicals or multi-component mixtures, where the isolation of a pure target product is of interest [11,12,23], the specific production of solids with defined characteristics is the main task for **nanoparticles** [15,26–29]. In the recent past, they gained increasing interest in industry and also in medical applications, due to their beneficial properties, which could be individualized for a specific duty [26,27]. The examples included in the Special Issue refer to new applications like capacitor energy-storage [28] and composite materials utilizing cellulose nanocrystals to reinforce biodegradable PBS polymers [29] as well as efficient catalysts in the cyanoethylation of methanol [15].

new application fields are discovered as, for example, the recycling of ionic liquids [13], and novel, simulation tools are applied to elucidate the basic mechanisms of solid formation [17].

The range of fundamental and application-oriented aspects addressed in the present Special Issue on Advances in Industrial Crystallization highlights the progress and future directions of research in our field. We hope you will enjoy and appreciate the authors contributions, which might inspire new and fruitful projects.

Heike Lorenz and Erik Temmel

Magdeburg/Germany and Allschwil/Switzerland, October 2020

References

1. Tung, H.-H.; Paul, E.L.; Midler, M.; McCauley, J.A. *Crystallization of Organic Compounds*, 1st ed.; John Wiley & Sons, Inc.: Hoboken, NJ, USA, 2009.
2. Myerson, A.S. *Handbook of Industrial Crystallization*, 2nd ed.; Butterworth-Heinemann: Boston, MA, USA, 2001.
3. Nývlt, J.; Söhnel, O.; Matuchová, M.; Boul, M. *The Kinetics of Industrial Crystallization*; Elsevier Science Ltd.: Amsterdam, The Netherlands, 1985.
4. Beckmann, W. *Crystallization-Basic Concepts and Industrial Applications*, 1st ed.; Wiley-VCH: Weinheim, Germany, 2013.
5. Lewis, A.; Seckler, M.; Kramer, H.; van Rosmalen, G. *Industrial Crystallization-Fundamentals and Applications*; Cambridge University Press: Cambridge, UK, 2015.
6. Chianese, A.; Kramer, H.J.M. *Industrial Crystallization Process Monitoring and Control*; Wiley-VCH: Weinheim, Germany, 2012.
7. Mullin, J.W. *Crystallization*, 4th ed.; Butterworth-Heinemann: Oxford, UK, 2001.
8. Lorenz, H. Solubility and Solution Equilibria in Crystallization. In *Crystallization*, 1st ed.; Wiley-VCH: Weinheim, Germany, 2013; pp. 35–74.
9. Kotelnikova, E.; Sadovnichii, R.; Kryuchkova, L.; Lorenz, H. Limits of Solid Solutions and Thermal Deformations in the L-Alanine–L-Serine Amino Acid System. *Crystals* **2020**, *10*, 618. [CrossRef]
10. Oshchepkov, M.; Golovesov, V.; Ryabova, A.; Redchuk, A.; Tkachenko, S.; Pervov, A.; Popov, K. Gypsum Crystallization during Reverse Osmosis Desalination of Water with High Sulfate Content in Presence of a Novel Fluorescent-Tagged Polyacrylate. *Crystals* **2020**, *10*, 309. [CrossRef]
11. Horosanskaia, E.; Yuan, L.; Seidel-Morgenstern, A.; Lorenz, H. Purification of Curcumin from Ternary Extract-Similar Mixtures of Curcuminoids in a Single Crystallization Step. *Crystals* **2020**, *10*, 206. [CrossRef]
12. Bredikhin, A.A.; Zakharychev, D.V.; Bredikhina, Z.A.; Kurenkov, A.V.; Samigullina, A.I.; Gubaidullin, A. Stereoselective Crystallization of Chiral 3,4-Dimethylphenyl Glycerol Ether Complicated by Plurality of Crystalline Modifications. *Crystals* **2020**, *10*, 201. [CrossRef]
13. Osmanbegovic, N.; Yuan, L.; Lorenz, H.; Louhi-Kultanen, M. Freeze Concentration of Aqueous [DBNH][OAc] Ionic Liquid Solution. *Crystals* **2020**, *10*, 147. [CrossRef]
14. Nývlt, J.; Ulrich, J. *Admixtures in Crystallization*, 1st ed.; Wiley-VCH: Weinheim, Germany, 1995.
15. Cheong, Y.-W.; Wong, K.-L.; Ooi, B.; Ling, T.C.; Khoerunnisa, F.; Ng, E.-P. Effects of Synthesis Parameters on Crystallization Behavior of K-MER Zeolite and Its Morphological Properties on Catalytic Cyanoethylation Reaction. *Crystals* **2020**, *10*, 64. [CrossRef]

16. Zhu, D.; Zhang, S.; Cui, P.; Wang, C.; Dai, J.; Zhou, L.; Huang, Y.; Hou, B.; Hao, H.; Zhou, L.; et al. Solvent Effects on Catechol Crystal Habits and Aspect Ratios: A Combination of Experiments and Molecular Dynamics Simulation Study. *Crystals* **2020**, *10*, 316. [CrossRef]
17. Bai, Y.; Wang, Y.; Zhang, S.; Wang, Q.; Li, R. Numerical Model Study of Multiple Dendrite Motion Behavior in Melt Based on LBM-CA Method. *Crystals* **2020**, *10*, 70. [CrossRef]
18. Yazdanpanah, N.; Nagy, Z.K. (Eds.) *The Handbook of Continuous Crystallization*, 1st ed.; Royal Society of Chemistry: Cambridge, UK, 2020.
19. Wirz, D.; Hofmann, M.; Lorenz, H.; Bart, H.-J.; Seidel-Morgenstern, A.; Temmel, E. A Novel Shadowgraphic Inline Measurement Technique for Image-Based Crystal Size Distribution Analysis. *Crystals* **2020**, *10*, 740. [CrossRef]
20. Liang, M.; Wang, Z.; Wu, H.; Yu, L.; Sun, B.; Zhou, H.; Yu, F.; Wang, Q.; He, J. Microplates for Crystal Growth and in situ Data Collection at a Synchrotron Beamline. *Crystals* **2020**, *10*, 798. [CrossRef]
21. Temmel, E. Design of Continuous Crystallization Processes. Ph.D. Thesis, Otto von Guericke University, Magdeburg, Germany, 2016.
22. Melches, C.; Plate, H.; Schürhoff, J.; Buchfink, R. The Steps from Batchwise to Continuous Crystallization for a Fine Chemical: A Case Study. *Crystals* **2020**, *10*, 542. [CrossRef]
23. Temmel, E.; Gänsch, J.; Seidel-Morgenstern, A.; Lorenz, H. Systematic Investigations on Continuous Fluidized Bed Crystallization for Chiral Separation. *Crystals* **2020**, *10*, 394. [CrossRef]
24. Hülsewede, D.; Temmel, E.; Kumm, P.; Von Langermann, J. Concept Study for an Integrated Reactor-Crystallizer Process for the Continuous Biocatalytic Synthesis of (S)-1-(3-Methoxyphenyl)ethylamine. *Crystals* **2020**, *10*, 345. [CrossRef]
25. Ulrich, J.; Glade, H. *Melt Crystallization: Fundamentals, Equipment and Applications*, 1st ed.; Shaker Verlag: Aachen, Germany, 2003.
26. Bhushan, B. *Springer Handbook of Nanotechnology*, 4th ed.; Springer: Berlin, Germany, 2017.
27. Nivethaa, E.A.K.; Martin, C.A.; Frank-Kamenetskaya, O.V.; Kalkura, S.N. Chitosan and Chitosan Based Nanocomposites for Applications as a Drug Delivery Carrier A Review. In *Processes and Phenomena on the Boundary Between Biogenic and Abiogenic Nature. Lecture Notes in Earth System Sciences*; Frank-Kamenetskaya, O., Vlasov, D., Panova, E., Lessovaia, S., Eds.; Springer: Cham, Switzerland, 2019; pp. 23–37. [CrossRef]
28. Li, M.; Gu, L.; Li, T.; Hao, S.; Tan, F.; Chen, D.; Zhu, D.-L.; Xu, Y.; Sun, C.; Yang, Z. TiO2-Seeded Hydrothermal Growth of Spherical BaTiO3 Nanocrystals for Capacitor Energy-Storage Application. *Crystals* **2020**, *10*, 202. [CrossRef]
29. Abushammala, H.; Mao, J. Impact of the Surface Properties of Cellulose Nanocrystals on the Crystallization Kinetics of Poly(Butylene Succinate). *Crystals* **2020**, *10*, 196. [CrossRef]

Publisher's Note: MDPI stays neutral with regard to jurisdictional claims in published maps and institutional affiliations.

© 2020 by the authors. Licensee MDPI, Basel, Switzerland. This article is an open access article distributed under the terms and conditions of the Creative Commons Attribution (CC BY) license (http://creativecommons.org/licenses/by/4.0/).

Article

Limits of Solid Solutions and Thermal Deformations in the L-Alanine–L-Serine Amino Acid System

Elena Kotelnikova [1], Roman Sadovnichii [1], Lyudmila Kryuchkova [1] and Heike Lorenz [2,*]

[1] Department of Crystallography, St. Petersburg State University, Universitetskaya emb. 7/9, 199034 St. Petersburg, Russia; kotelnikova.45@mail.ru (E.K.); rsadovnichii@gmail.com (R.S.); kryuchkova.2106@gmail.com (L.K.)
[2] Max Planck Institute for Dynamics of Complex Technical Systems, Sandtorstrasse 1, 39106 Magdeburg, Germany
* Correspondence: lorenz@mpi-magdeburg.mpg.de

Received: 4 June 2020; Accepted: 13 July 2020; Published: 16 July 2020

Abstract: The limits of solid solutions and thermal deformations in the L-alanine–L-serine (L-ala–L-ser) amino acid system have been determined. Thirteen amino acid mixtures with various proportions of the components L-ser/L-ala were studied using powder X-ray diffraction techniques. It was found that the regions of solid solutions in the system are rather limited and cover less than 10 mol. % from each component side. The thermal behavior of the components L-ser and L-ala and the composition L-ser/L-ala = 90/10 were studied by temperature-resolved powder X-ray diffraction. The heating of L-ser and L-ala only causes thermal deformations, while two-phase mixtures with the 90/10 L-ser/L-ala ratio form solid solutions at elevated temperatures. Additionally, the parameters of the thermal deformation tensor for L-ser and L-ala were calculated, and the figures of their thermal expansion coefficients were plotted and analyzed. The study conducted is of high applicability, since amino acids are active components of various biological, geological, and technological processes, including those at elevated temperatures, and have numerous applications in life-science industries.

Keywords: L-serine; L-alanine; enantiomers; isomorphic miscibility; thermal expansion; PXRD; TRPXRD

1. Introduction

Amino acids are typical representatives of molecular crystals with chiral molecules. They are considered as some of the most actively synthesized organic compounds, since they are widely used—for example, in the pharmaceutical and food industries—and, consequently, participate in a variety of biological and technological processes taking place at different temperatures [1–3]. Amino acids are abundant in geological media, as, in contrast to proteins, they are able to survive relatively high temperatures. Being typical representatives of chiral organic compounds, they are used for determining the age of sedimentary rocks [4–6]. This method of dating is based on the capability of L- and D-enantiomers of amino acids to undergo mutual transformations leading to racemization in the scale of geological time. The above-mentioned underlines the significance of the present work, which aims to solve several fundamental and applied problems related to discovering the limits of solid solutions and the thermal behavior of the components of amino acid systems.

In the review of B. Saha [7] and our recent work [8], it was already mentioned that the number of publications reporting thermal deformations of organic crystal structures is scarce compared to the number of related investigations on inorganic compounds. At the same time, publications on thermal deformations of amino acids or chiral substances, which play a particularly important role in living matter, are even less numerous. Examples are the works from B. Nicolaï et al. [9] and ourselves [8,10–12]. We investigated the thermal deformations of crystal structures in the following systems: the components and two solid solutions formed in the L-threonine–L-*allo*-threonine diastereomer system [10], the components in the L-malic acid—D-malic acid system [8,11], and the components formed in the L-valine–L-isoleucine amino acid system [12].

L-Amino acid enantiomers are known to more frequently occur in nature and have more practical applications. This fact motivated our interest in binary systems of different amino acids with the same chirality. The reported studies of such systems are rather scarce and, according to our knowledge, include the publications [13,14] and our works [15–19].

On the contrary, rather numerous are investigations into systems of different amino acids which also have different chiralities—i.e., where the amino acid molecules are D- and L-enantiomers (see, for example, the review [16]). Here, it is worth mentioning a great contribution to the investigation of racemic compositions of the above type systems made by B. Dalhus and C. Görbitz [20–23].

The present work continues our study of systems consisting of L-enantiomers of different amino acids. In the following, we report the results of (1) the limits of solid solutions formed in the L-alanine–L-serine system, and (2) the thermal deformations of crystal structures of L-alanine (L-ala) and L-serine (L-ser). The work aims at a deeper understanding of the structure-property relationships of amino acids as chiral compounds.

2. Materials and Methods

2.1. Materials

Alanine, $C_3H_7NO_2$, and serine, $C_3H_7NO_3$, (Figure 1) are aliphatic proteinogenic amino acids and are found in many naturally occurring proteins. Both amino acids have one chiral center and, therefore, can exist as L- and D-enantiomers. In the serine molecule (Figure 1b), the methyl (CH_3) end group in alanine (Figure 1a) is replaced by a methylene (CH_2) moiety with an OH group.

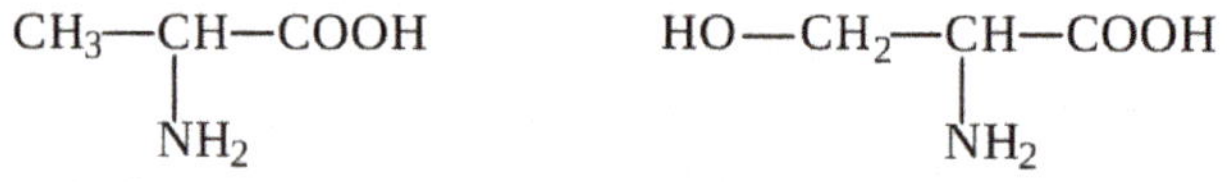

$$CH_3-CH-COOH \qquad HO-CH_2-CH-COOH$$
$$\quad\ \ \ |\qquad\qquad\qquad\qquad\qquad\quad\ \ |$$
$$\quad\ \ NH_2\qquad\qquad\qquad\qquad\qquad\ NH_2$$

(a) \qquad\qquad\qquad\qquad (b)

Figure 1. Structural formulae of alanine (**a**) and serine (**b**) molecules.

The known crystal structures of both amino acids are characterized by the same space group $P2_12_12_1$. The first results of deciphering the crystal structures of alanine and serine were reported several decades earlier [24,25]. Later, these results were refined and high-pressure data were added [26,27]. Moreover, numerous investigations included studies of the mechanical, optic, magnetic, and electron properties of alanine and serine crystals [28–32].

L-alanine and L-serine (99% purity) were obtained from Merck, Zug, Switzerland, and were used as obtained. As a solvent, deionized water was applied.

2.2. Methods

Crystals in the L-alanine–L-serine system were obtained by spontaneous crystallization from the aqueous solutions of 13 different L-ala/L-ser mixtures by reducing the temperature with the subsequent evaporation of the solvent. Mixtures with different contents of the components (mol.%) of L-ser/L-ala = 0/100, 7/93, 10/90, 15/85, 25/75, 35/65, 50/50, 65/35, 75/25, 85/15, 90/10, 93/7, and 100/0 were weighed on a Shimadzu AX200 scale (accuracy 0.0001 g). The solutions were prepared for a saturation temperature of 50 °C with the constant stirring of the solution using a magnetic stirrer, and afterwards were filtered through a microporous filter (pore size 1.2 μm). The solution was poured into a Petri dish and kept at room temperature for 5–7 days. The evaporation rate of the solvent was regulated by changing the position of the lid on the Petri dish. The resulting crystalline precipitate was removed from the solution and quickly dried on filter paper.

All the 13 samples obtained were investigated by means of Powder X-ray Diffraction (PXRD) using a Rigaku MiniFlex II diffractometer (Rigaku Co., Tokyo, Japan) with $Cu_{K\alpha}$ and $Co_{K\alpha}$ irradiation and the 2θ range of 5–60°. Three samples (L-ala, L-ser, and a sample of composition L-ser/L-ala = 90/10 mol. %) were studied by Temperature-Resolved Powder X-ray Diffraction (TRPXRD). These experiments were conducted in the atmospheric air using a Rigaku Ultima IV diffractometer (Rigaku Co., Tokyo, Japan) equipped with a high-temperature accessory and the following settings: $Co_{K\alpha}$ irradiation, the 2θ range of 5–60°, and the temperature range of 23–200 °C with the temperature pitch of 10 °C. The X-ray patterns were processed and the unit cell parameters in the whole temperature range were calculated using the PDXL 2.7 (Rigaku Co., Tokyo, Japan) and Topaz (Bruker AXS GmbH, Karlsruhe, Germany) software.

The temperature dependences of the orthorhombic unit cell parameters were used to calculate the thermal expansion coefficients (CTE) of the L-ala and L-ser crystal structures (α, 10^{-6} °C^{-1}) along the three orthogonal axes of the thermal deformation tensor (α_{11}, α_{22}, α_{33}). The values obtained were utilized to plot the projections of the CTE figures onto the *ab*, *ac*, and *bc* planes of the corresponding orthorhombic crystal structures. In the orthorhombic syngony, the directions of the tensor axes coincide with those of the crystallographic axes. The TEV software (Thomas Langreiter and Volker Kahlenberg, Institute of Mineralogy and Petrography, Innsbruck, Austria) was used to calculate the thermal deformation tensor and plot the CTE figures.

3. Results and Discussion

3.1. Limits of Solid Solutions

Relatively similar sizes and shapes of the alanine and serine molecules (Figure 2a,b) would suggest isomorphic miscibility or, in other words, the formation of solid solutions in their mixtures.

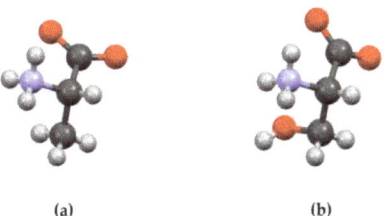

Figure 2. Molecules of alanine (**a**) and serine (**b**).

This hypothesis was totally valid in the case of the L-threonine–L-*allo*-threonine diastereomers with the same chirality (L referring to C_α), where the authors observed the phenomenon of the continuous isomorphic miscibility of the components, which is only rarely found in chiral systems [10]. On the other hand, in the system of the L-threonine–D-threonine enantiomers [33] and the system of the D-threonine–L-*allo*-threonine diastereomers [34], both containing components with different chiralities, despite the identical molecular compositions and very close sizes and shapes of the molecules, the regions of solid solutions appeared to be rather limited. Three other investigated systems formed by the L-enantiomers of different amino acids—i.e., L-valine–L-isoleucine [15,16], L-valine–L-leucine [17], and L-isoleucine–L-leucine [18]—are characterized by the presence of non-equimolar discrete heterocompounds (V_2I (valine/isoleucine = 2:1), V_3L (valine/leucine = 3:1), and I_3L (isoleucine/leucine = 3:1), respectively) and limited solid solutions in the vicinities of the heterocompounds and pure components.

As seen in Figure 3 from the X-ray patterns of the co-precipitated mixtures of L-ala and L-ser, in the L-alanine–L-serine system neither discrete heterocompounds nor solid solutions were found within the studied composition range. The X-ray patterns shown contain two sets of peaks, one corresponding to L-ala and the other to L-ser. The relative intensities of each peak set vary with the changes in the mixture composition, which proves that the mixtures studied are conglomerates or just mechanical mixtures.

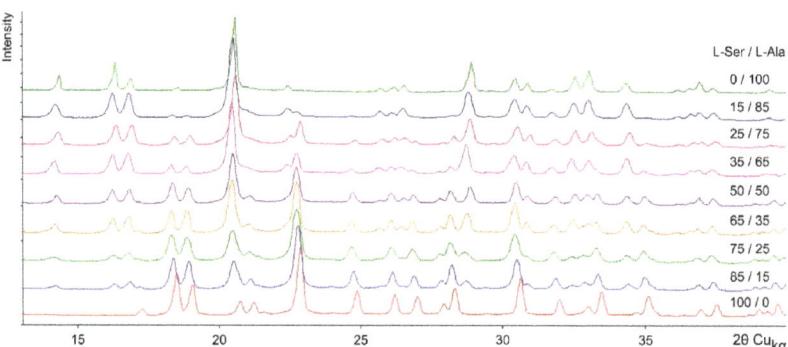

Figure 3. X-ray patterns (2θ Cu$_{K\alpha}$) of co-precipitated mixtures (mol. %): L-ser/L-ala = 0/100, 15/85, 25/75, 35/65, 50/50, 65/35, 75/25, 85/15, and 100/0.

In order to more precisely delineate the potential limits of the solid solutions in the L-alanine–L-serine system, several co-precipitated mixtures with compositions close to the pure components were prepared and investigated. The obtained X-ray patterns are shown in Figure 4. These experiments succeeded in revealing a limited isomorphic miscibility in the vicinity of the components. This is proved by the absence of the admixture phase peaks and the altered peak positions in the samples with L-ser/L-ala ratios of 7/93 and 93/7 compared to the corresponding positions of the L-ser and L-ala peaks. This is also reflected in the nature of changes in the respective calculated parameters and volume of the orthorhombic cells given in Table 1. A monocrystal obtained at the L-ser/L-ala ratio of 93/7 was studied by means of Single Crystal X-ray Diffractometry (SCXRD), but revealed only the presence of the L-ser phase. This fact can be explained by the low sensibility of the SCXRD method in the diagnostics of solid solutions with low contents of light molecules. The samples with L-ser/L-ala compositions of 10/90 and 90/10 (Figure 4) show the additional presence of the other solid solution and thus biphasic behavior.

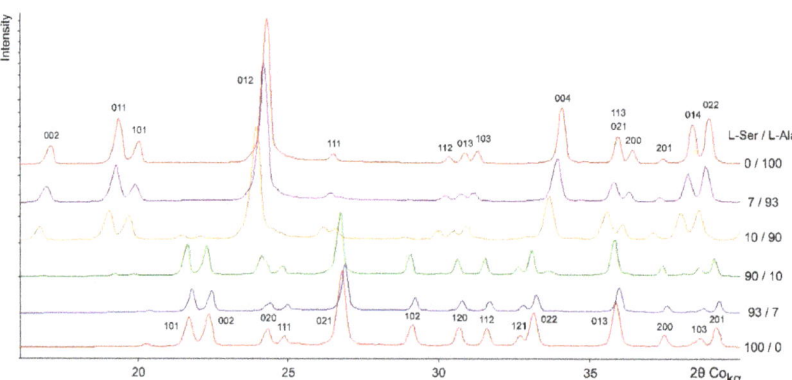

Figure 4. X-ray patterns (2θ Co$_{K\alpha}$) of the co-precipitated mixtures (mol. %): L-ser/L-ala = 0/100, 7/93, 10/90, 90/10, 93/7, and 100/0.

Table 1. Parameters and volume of the orthorhombic cell for L-ser, L-ala, and their solid solutions with L-ser/L-ala ratios of 7/93 and 93/7.

L-Ser/L-Ala	a, Å	b, Å	c, Å	V, Å^3
0/100	5.7859(5)	6.0340(4)	12.3441(6)	430.96 (5)
7/93	5.7908(6)	6.0346(5)	12.3717(8)	432.34 (7)
100/0	5.6129(3)	8.5910(5)	9.3456(4)	450.66 (4)
93/7	5.6096(2)	8.5824(4)	9.3397(3)	449.65 (3)

3.2. Thermal Deformations

Figures 5 and 6 show the X-ray patterns of L-ala and L-ser, respectively, registered at various temperatures using the TRPXRD method. According to the data obtained, in the temperature range of 23–200 °C, L-serine and L-alanine do not undergo any polymorph transformations. Both components are exposed to thermal deformations manifested as various shifts of the peaks either towards low or high 2θ values depending on the hkl indices of the peaks. It should be noted that a small splitting of the unambiguously indexed peak 020 in Figure 6, which is present at lower temperatures, is most likely caused by texture effects.

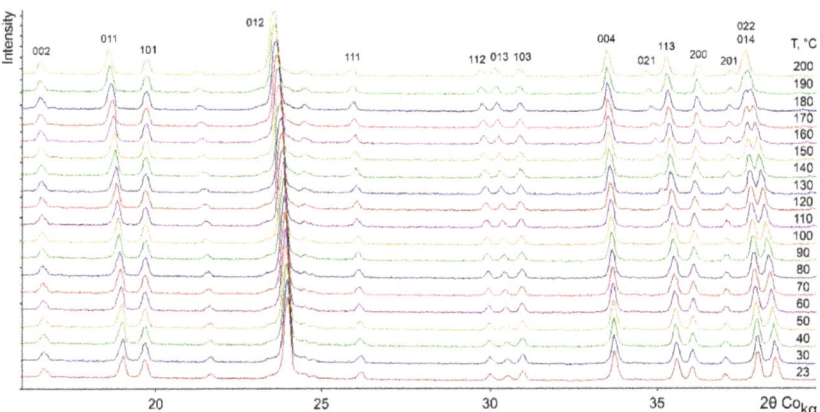

Figure 5. X-ray patterns (2θ Co$_{K\alpha}$) of an L-ala sample obtained at various temperatures.

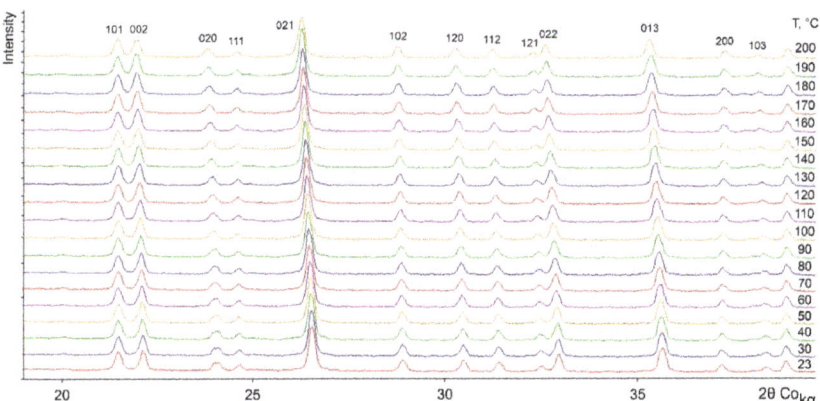

Figure 6. X-ray patterns (2θ Co$_{K\alpha}$) of an L-ser sample obtained at various temperatures.

Figure 7 presents a cutout of the temperature-resolved X-ray patterns of a sample with the L-ser/L-ala ratio of 90/10 studied in a temperature range between 25 and 210 °C. Initially, the sample was clearly a physical mixture of L-ser- and L-ala-rich solid solutions (as already mentioned in Figure 4). Despite a very low L-ala content, the existence of a two-phase mixture is obvious in the X-ray pattern by the presence of the 012 peak, as the most intensive peak of the L-ala phase. This peak is located close to the 020 peak of the L-ser phase (see the arrows in Figure 7). As the temperature rises, both peaks shift towards low 2θ values, with the shift of the 012 Ala peak being greater than that of the 020 Ser peak. Furthermore, the intensity of the 012 Ala peak gradually decreases until it disappears at 175 °C, while the intensity of the 020 Ser peak slightly increases, despite the fact that the sublimation point of L-ala (315 °C) significantly exceeds that of L-ser (228 °C). This could be a result of an increase in the isomorphic miscibility of L-ser and L-ala molecules when the temperature is close to 175 °C. This deems probable if one takes into account the following considerations: (1) the mixture with a similar composition of L-ser/L-ala = 93/07 was shown to form a solid solution at room temperature (see Figure 4), and (2) elevated temperatures usually cause the limits of solid solutions to widen.

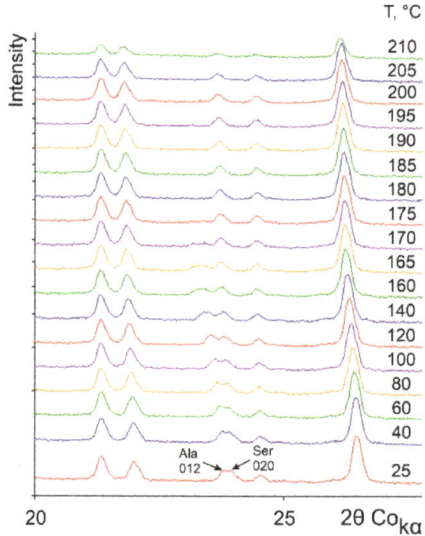

Figure 7. Fragments of the X-ray pattern (2θ Co$_{K\alpha}$ = 20–28°) of a sample with the L-ser/L-ala ratio of 90/10 registered at different temperatures.

Figure 8 shows changes in the orthorhombic cell parameters and volume V of L-ala and L-ser versus temperature. As the temperature increases, the a parameter decreases, while the b and c parameters, as well as the V volume of the orthorhombic cells of L-ala and L-ser, increase. Thereby, the changes of both parameters and volume are more pronounced in the unit cell of L-ala than for L-ser. The functions plotted allowed to estimate the thermal expansion coefficients (CTE) of the volumes (α_V) of the corresponding orthorhombic cells. It is to be mentioned that the $\alpha_V = 170.8 \times 10^{-6}$ °C^{-1} obtained for L-ala almost by half exceeds the corresponding value calculated for L-ser, $\alpha_V = 113.2 \times 10^{-6}$ °C^{-1}.

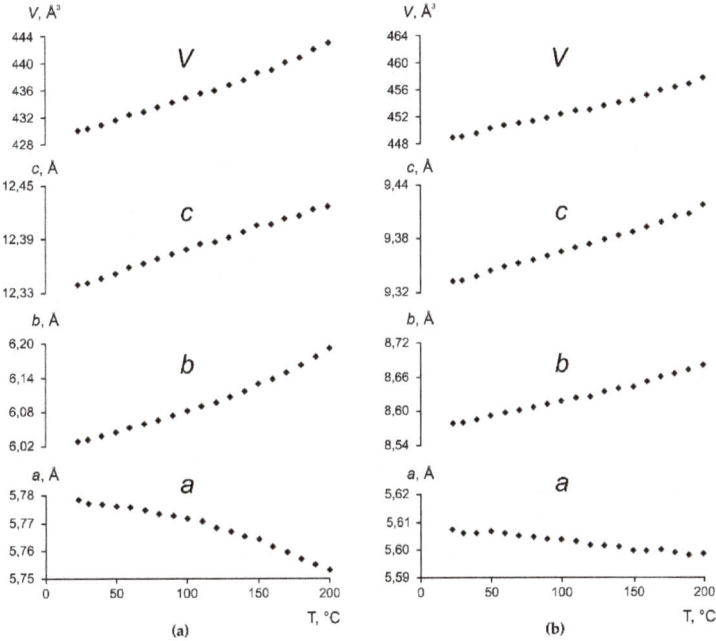

Figure 8. Changes in orthorhombic cell parameters a, b, and c (Å) and volume V (Å^3) versus the temperature for (**a**) L-ala and (**b**) L-ser.

The temperature dependences of the orthorhombic cell parameters and volume V were approximated by polynomials of the first and second order. These data were used to calculate the parameters of the thermal deformation tensor and the coefficients of thermal expansion along the crystallographic axes of L-ala and L-ser, which are summarized in Table 2.

Table 2. Thermal expansion coefficients ($\alpha \times 10^{-6}$ °C^{-1}) of the L-ala and L-ser orthorhombic crystal structures along the axes of the thermal deformation tensor: $\alpha_{11} = \alpha_a$, $\alpha_{22} = \alpha_b$, and $\alpha_{33} = \alpha_c$.

Amino Acid	23 °C			100 °C			200 °C		
	α_a	α_b	α_c	α_a	α_b	α_c	α_a	α_b	α_c
L-alanine	−6.7	79.3	41	−22.1	139.2	40.9	−42.2	213.9	40.7
L-serine	−9.6	52.6	40.7	−9.6	64.1	48.2	−9.6	78.7	57.7

The data represented in this table, in turn, were used for plotting the figures of the thermal expansion coefficients (CTE) for L-ala (Figure 9) and L-ser (Figure 10). For better understanding, the figures also show projections of the figures onto the ab, ac, and bc planes of the corresponding crystal structures.

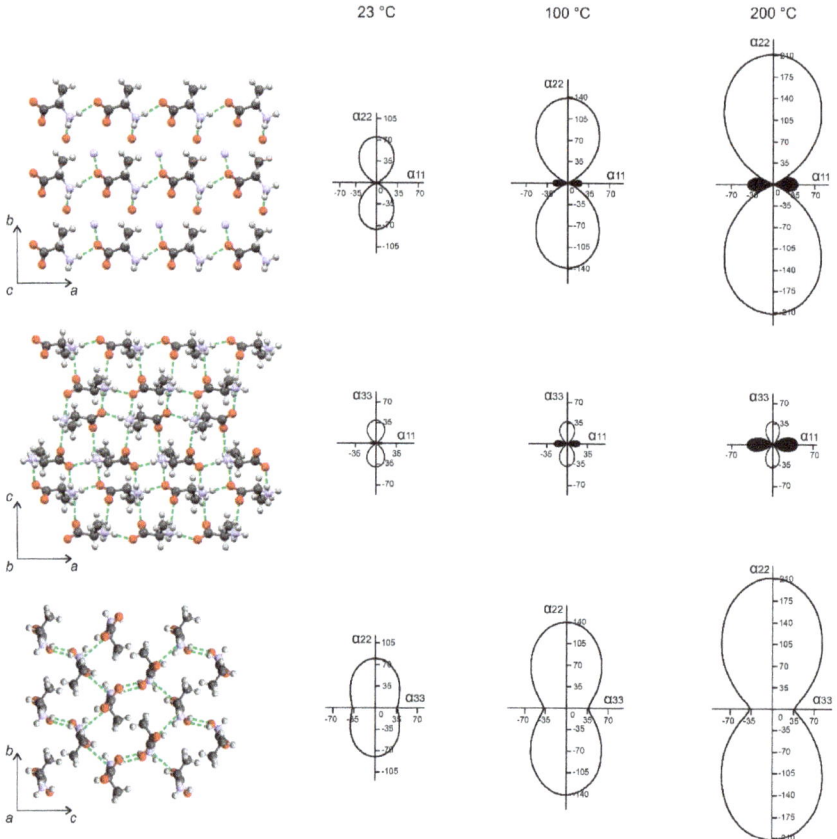

Figure 9. Projections of the figures of the thermal expansion coefficients (CTE) onto the *ab*, *ac*, *and bc* planes of the L-ala orthorhombic cell. The CTE figures are plotted for the temperatures of 23, 100, and 200 °C. Hydrogen bonds are shown as dashed lines. The projections of the orthorhombic cell are plotted using the structural data from CSD (identifier LALANINE54) [35].

An examination of the CTE figures for three different temperature conditions reveals that the thermal deformations of the crystal structures of both L-ala and L-ser are distinctly anisotropic. Both the crystal structures demonstrate a noticeable thermal expansion along the crystallographic axes *b* and *c* and a very significant negative (anomalous) thermal expansion (more precisely, contraction) in the direction of the *a* axis. The crystal structures of the amino acids L-ala and L-ser can be considered as frame structures containing relatively large cavities bound by hydrogen and Van der Waals contacts. When heated, the so-called "hinge mechanism" is realized [36,37]. Its most important feature is the synchronous change of two linear parameters of the unit cell in opposite directions with the "neutral" behavior of the third parameter and the volume. In the present case, both amino acids show a multidirectional synchronous change in the parameters *a* and *b*, as the hinges can be considered the "frames" of molecules connected by hydrogen and Van der Waals bonds. When heated, some of the atoms move away from each other, which leads to the convergence of the other part of the atoms with each other. In this case, one or another specific anisotropy is caused by the different geometries and concentrations of the hydrogen bonds of the N–H... O and O–H... O type in the crystal structures of the acids. At the same time, despite some similarities, the thermal deformations of L-ala and L-ser have some individual patterns. In this connection, the differences of the two amino acid molecules with the different end groups (L-ala: CH$_3$ group; L-ser: CH$_2$OH group) should be mentioned.

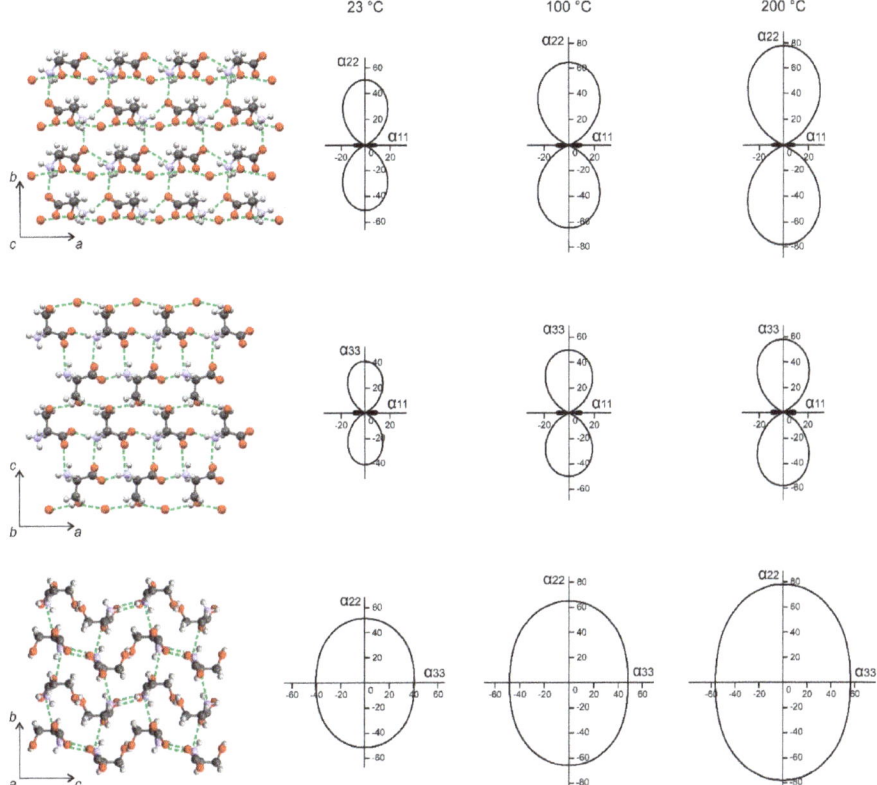

Figure 10. Projections of the figures of the thermal expansion coefficients (CTE) onto the *ab*, *ac*, and *bc* planes of the L-ser orthorhombic cell. The CTE figures are plotted for the temperatures of 23, 100, and 200 °C. Hydrogen bonds are shown as dashed lines. The projections of the orthorhombic cell are plotted using the structural data from CSD (identifier LSERIN41) [35].

L-alanine (Figure 9, Table 2): For this amino acid, the rise of temperature results in a very considerable increase in the thermal expansion coefficients in the direction of the *b* axis, while they only slightly change in the *c* direction. In the *a* direction, the negative thermal expansion (contraction) coefficient increases noticeably. As a whole, the thermal deformation anisotropy increases with the elevation of temperature, but this process follows different patterns depending on the projection onto the *ab*, *ac*, and *bc* crystal planes.

In the projection onto the *ab* plane, the L-ala molecules are interconnected with the N–H ... O bonds to form chains positioned along the *a* direction. In the opposite direction (in the direction of the *b* axis), the contacts are much weaker due to their Van der Waals nature. Consequently, the maximum thermal expansion is observed along the *b* axis and the negative thermal expansion takes place along the *a* axis. As seen from the projection onto the *ac* plane, the hydrogen bonds of the N–H ... O type exist in the direction of both *a* and *c* axes, but their geometries and concentrations differ depending on the particular direction and, therefore, the hydrogen contacts have different strengths. In a virtual "competition" between the hydrogen bonds, those positioned along the *c* axis are the weakest, and so this direction is characterized by a relatively low thermal expansion, while along the *a* axis the structure contracts. The same reasoning can be followed to examine thermal expansion anisotropy in the projection onto the *bc* plane.

L-serine (Figure 10, Table 2): The elevation of temperature results in an increase in the thermal expansion coefficients in the direction of the *b* axis, but to a lesser extent in comparison with the L-ala molecule. Along the *c* axis, the thermal expansion coefficient becomes noticeably greater, while the negative thermal expansion coefficient in the direction of the *a* axis does not change. This is the principal difference between the thermal deformations in L-ser and L-ala. The anisotropy of the thermal deformations in L-ser can be observed to have various manifestations in the projections onto the *ab*, *ac*, and *bc* crystal planes.

In the direction of the *a* axis, the thermal expansion is negative (anomalous) due to the presence of strong hydrogen bonds of the N–H . . . O and O–H . . . O types along this direction. In the *bc* plane, the influence of the N–H . . . O hydrogen bonds is reinforced by the Van der Waals contacts made by the methylene (CH_2) groups. In the neighboring molecules, these groups are directed towards each other. As a result, the crystal structure undergoes thermal expansion in the directions of the crystallographic axes *b* and *c*. It is interesting to note that in this plane the thermal deformation anisotropy was not observed and, consequently, was not affected by alterations of temperature.

3.3. Discussion

In one of our works [18] we reviewed some particularities of crystal structures of several (but not all) proteinogenic amino acids to estimate their abilities to form heteromolecular discrete compounds consisting of molecules with different and the same chirality. It was stated that these amino acids were prone to form dimers with the molecules connected to each other by hydrogen bonds. Such dimeric molecules, in turn, are mutually interconnected via Van der Waals bonds. This arrangement is exemplified in Figure 11a, which depicts projections of the L-valine crystal structure onto the *ac* plane. It is distinctly seen that the dimer molecules form layers separated by Van der Waals contacts—i.e., the structure as a whole can be regarded as layered.

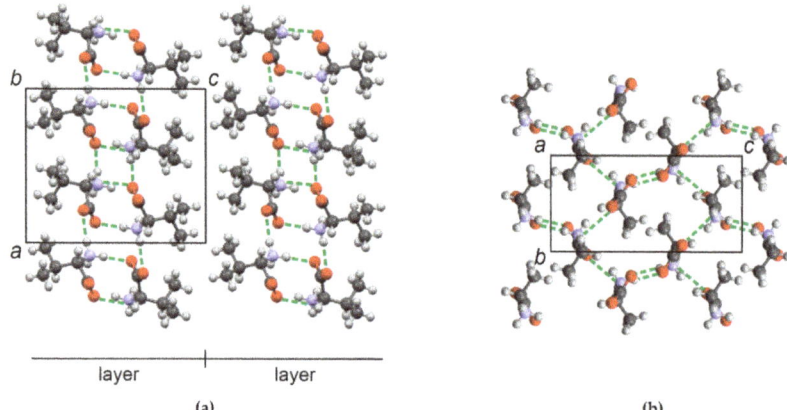

Figure 11. Projection of the L-valine crystal structure onto the *ac* plane of the monoclinic cell [18] (**a**) and projection of the L-alanine crystal structure onto the *bc* plane of the orthorhombic cell (**b**). Dotted lines are hydrogen bonds. The projections of the orthorhombic cell are plotted using the structural data from CSD (identifiers LVALIN01 and LALANINE54 [35]).

Both alanine and valine molecules (Figure 12a,b) contain a methyl (CH_3) end group. However, alanine contains only one methyl on its end, while valine possesses two of them. At the first glance, this difference seems to be insignificant, but nevertheless it results in two amino acids with totally different crystal arrangements. While the crystal structure consisting of the dimeric molecule layers is quite viable for L-val (Figure 11a), this is not possible for L-ala.

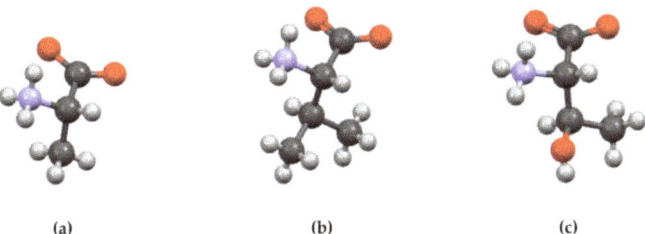

Figure 12. Molecules of alanine (**a**), valine (**b**), and threonine (**c**).

Figure 11b shows the projection of the L-ala crystal structure onto the *bc* plane. It can be seen that there certainly are Van der Waals contacts in the direction of the *b* axis, but each of them is alternated with an N–H . . . O hydrogen bond that results in a much more robust connection. At the same time, in the direction of the *c* axis there are two hydrogen bonds of the above type for every one of the Van der Waals bonds. Therefore, L-ala forms a "network" crystal structure (Figure 11b), similar to those of L-ser (Figure 10) and L-threonine (L-thr) [10], despite the fact that L-ser has a CH_2OH end group (see Figure 2b) and L-thr possesses CH_3 and OH end groups (Figure 12c). Furthermore, the acids L-ala, L-ser, and L-thr have another structural feature in common—that is, they all crystallize in the orthorhombic space group $P2_12_12_1$.

The strong anisotropy of the crystal structure, which also resulted in a negative thermal expansion (contraction in the direction of the *a* axis), was observed as well in the related studies of L-threonine and L-*allo*-threonine [10]. A hinge mechanism can also be used to describe the thermal deformations of these diastereomers. Therefore, each one of L-ala, L-ser, L-thr and L-*allo*-thr has an orthorhombic crystal structure, while the L-valine, L-isoleucine, and L-leucine examined earlier [15–19] crystallize in monoclinic syngony.

Technically, the thermal deformations of the acids of both types ("layered" and "network") are similar to some extent. In both cases, the maximum and minimum (including "negative") thermal expansion is observed in the direction of the weakest and strongest intermolecular contacts, respectively. However, as deducible from changes of the monoclinic angle β at elevated temperatures, the leading role in structural deformations of monoclinic crystal structures belongs to shear deformations.

The observed response of the network orthorhombic structures to rising the temperature allowed to suspect a correlation between the resulting deformation and the range of intermolecular distance variation (chiefly Van der Waals contacts) as a result of deformation of the voids (channels) via a hinged mechanism in the corresponding structures (see, for example, Figures 9 and 10).

The thermal deformations of the crystal structures of L-ala and L-ser discussed in the present article are anisotropic. However, the anisotropy of the L-ala crystals is more evident in comparison with that observed in L-ser which can be caused by the greater concentration of the Van der Waals bonds in the structure of the former amino acid. The same phenomenon can account for significant differences in the changes of the parameters and volume of their orthorhombic cells and hence corresponding thermal expansion coefficients. Here, it is worth noting again that the α_V value for L-ala is almost by a half greater than that of L-ser.

Obviously, the differences in the size and shape of L-ala and L-ser molecules play only a minor role in imposing considerable limitations on the isomorphic miscibility in the L-ala–L-ser system. A much greater part is played by the differences in the nature of the intermolecular contacts in the respective amino acids' structures—namely, by a significantly greater concentration of Van der Waals bonds in the crystal structure of L-ala compared to L-ser.

4. Conclusions

It was established that the L-alanine–L-serine system belongs to systems featuring eutectic behavior with very limited solid solutions (less than 10 mol.%) in the vicinity of the components. Solid solutions shown are the compositions (mol.%): L-ser/L-ala = 93/7 and 7/93. Using the physical mixture L-ser/L-ala = 90/10 as an example, it is demonstrated that the limits of solid solutions expand with increasing temperature. At a temperature of 175 °C, the aforementioned mixture is homogenized.

The heating of the system components L-ser and L-ala revealed a pronounced anisotropy of the thermal deformations of their crystal structures. This is reflected in the values of the parameters of the thermal deformation tensor of L-ala and L-ser and, respectively, in the figures of their thermal expansion coefficients. A multidirectional synchronous change in the CTE values with an intensive increase in the parameter b and a decrease in the parameter a (up to negative values) indicates a hinge mechanism of thermal deformations.

The results obtained add new insights in the structure-property relationships of amino acid systems as compounds of biological, geochemical, and industrial importance.

Author Contributions: Conceptualization, E.K. and H.L.; methodology, E.K., H.L., L.K. and R.S.; formal analysis, R.S. and L.K.; investigation, R.S., L.K. and E.K.; writing—original draft preparation, E.K.; writing—review and editing, E.K., H.L. and R.S.; visualization, E.K., R.S., L.K.; supervision, E.K. and H.L. All authors have read and agreed to the published version of the manuscript.

Funding: This research was supported by the President of Russian Federation grant to leading scientific schools (NSh-2526.2020.5).

Acknowledgments: The investigations were performed using equipment of the Resource Centre "Centre for X-ray Diffraction Studies" of St. Petersburg State University and support of the MPI Magdeburg PCF lab. The authors thank M. G. Krzhizhanovskaya, A. I. Isakov and A. A. Zolotarev Jr. for collaboration.

Conflicts of Interest: The authors declare no conflict of interest.

References

1. Giron, D. Polymorphism in the pharmaceutical industry. *Therm. Anal. Calorim.* **2001**, *64*, 37–60. [CrossRef]
2. Murakami, H. From Racemates to Single Enantiomers—Chiral Synthetic Drugs over the last 20 year. *Top. Curr. Chem.* **2006**, *269*, 273–299.
3. Bredikhin, A.A.; Bredikhina, Z.A.; Zakharychev, D. Crystallization of chiral compounds: Thermodynamical, structural and practical aspects. *Mendeleev Commun.* **2012**, *22*, 171–180. [CrossRef]
4. Robins, J.; Jones, M.; Matisoo-Smith, E. *Amino Acid Racemization Dating in New Zealand: An Overview and Bibliography*; Auckland Univ.: Auckland, New Zealand, 2010.
5. Killops, S. Introduction to Organic Geochemistry, 2nd edn (paperback). *Geofluids* **2005**, *5*, 236–237. [CrossRef]
6. Torres, T.; Ortiz, J.; Arribas, I.; Delgado, A.; Julià, R.; Rubí, J.A.M. Geochemistry of Persististrombus latus Gmelin from the Pleistocene Iberian Mediterranean realm. *Lethaia* **2009**, *43*, 149–163. [CrossRef]
7. Saha, B.K. Thermal Expansion in Organic Crystals. *J. Indian Inst. Sci.* **2017**, *97*, 177–191. [CrossRef]
8. Kotelnikova, E.N.; Isakov, A.; Lorenz, H. Thermal deformations of crystal structures formed in the systems of malic acid enantiomers and l-valine–l-isoleucine enantiomers. *CrystEngComm* **2018**, *20*, 2562–2572. [CrossRef]
9. Nicolaï, B.; Barrio, M.; Tamarit, J.-L.; Céolin, R.; Rietveld, I.B. Thermal expansion of L-ascorbic acid. *Eur. Phys. J. Spéc. Top.* **2017**, *226*, 905–912. [CrossRef]
10. Taratin, N.; Lorenz, H.; Binev, D.; Seidel-Morgenstern, A.; Kotelnikova, E. Solubility Equilibria and Crystallographic Characterization of the l-Threonine/l-allo-Threonine System, Part 2: Crystallographic Characterization of Solid Solutions in the Threonine Diastereomeric System. *Cryst. Growth Des.* **2015**, *15*, 137–144. [CrossRef]
11. Isakov, A.; Kotelnikova, E.N.; Kryuchkova, L.Y.; Lorenz, H. Effect of crystallization conditions on polymorphic diversity of malic acid RS—Racemate. *Trans. Tianjin Univ.* **2013**, *19*, 86–91. [CrossRef]
12. Isakov, A.; Kotelnikova, E.; Bocharov, S.; Zolotarev, A., Jr.; Lorenz, H. Thermal deformations of the crystal structures of L-valine, L-isoleucine and discrete compound V2I. In Proceedings of the 23rd International Workshop on Industrial Crystallization (BIWIC-2016), Magdeburg, Germany, 6–8 September 2016; pp. 7–12.

13. Koolman, H.C.; Rousseau, R.W. Effects of isomorphic compounds on the purity and morphology of L-isoleucine crystals. *Aiche J.* **1996**, *42*, 147–153. [CrossRef]
14. Kurosawa, I.; Teja, A.; Rousseau, R. Solubility measurements in the l-Isoleucine + l-Valine + Water System at 298 K. *Eng. Chem. Res.* **2005**, *44*, 3284–3288. [CrossRef]
15. Isakov, A.; Kotelnikova, E.; Muenzberg, S.; Bocharov, S.; Lorenz, H. Solid Phases in the System L-Valine—L-Isoleucine. *Cryst. Growth Des.* **2016**, *16*, 2653–2661. [CrossRef]
16. Kotelnikova, E.; Isakov, A.; Lorenz, H. Non-equimolar discrete compounds in binary chiral systems of organic substances (Highlight). *Cryst. Eng. Comm.* **2017**, *19*, 1851–1869. [CrossRef]
17. Kryuchkova, L.; Kotelnikova, E.; Zolotarev, A., Jr.; Lorenz, H. Limits of solid solutions in the L-leucine—isoleucine system according to PXRD and SCXRD data. In Proceedings of the 25rd International Workshop on Industrial Crystallization (BIWIC-2018), Rouen, France, 5–7 September 2018; pp. 225–226.
18. Isakov, A.I.; Lorenz, H.; Zolotarev, A.A.; Kotelnikova, E.N. Heteromolecular compounds in binary systems of amino acids with opposite and same chiralities. *CrystEngComm* **2020**, *22*, 986–997. [CrossRef]
19. Kotelnikova, E.; Isakov, A.; Kryuchkova, L.; Zolotarev, A., Jr.; Bocharov, S.; Lorenz, H. Acids with Chiral Molecules as Essential Organic Compounds of Biogenic–Abiogenic Systems. In *Processes and Phenomena on the Boundary between Biogenic and Abiogenic Nature*; Frank-Kamenetskaya, O.V., Vlasov, D., Panova, E.G., Lessovaia, S.N., Eds.; Springer Nature Switzerland: Berlin, Germany, 2020; Volume 37, pp. 695–719.
20. Dalhus, B.; Görbitz, C.H. Molecular aggregation in crystalline 1:1 complexes of hydrophobic D- and L-amino acids. *Acta Crystallogr. Sect. B Struct. Sci.* **1999**, *55*, 424–431. [CrossRef]
21. Dalhus, B.; Görbitz, C. Molecular aggregation in selected crystalline 1:1 complexes of hydrophobic D- and L-amino acids. II. The D-norleucine series. *Acta Crystallogr.* **1999**, *C55*, 1105–1112.
22. Dalhus, B.; Görbitz, C.H. Molecular aggregation in selected crystalline 1:1 complexes of hydrophobic D- and L-amino acids. III. The L-leucine and L-valine series. *Acta Crystallogr.* **1999**, *C55*, 1547–1555.
23. Görbitz, C.H. Crystal structures of amino acids: From bond lengths in glycine to metal complexes and high-pressure polymorphs. *Crystallogr. Rev.* **2015**, *21*, 1–53. [CrossRef]
24. Simpson, H.J.; Marsh, R.E. The crystal structure of L -alanine. *Acta Crystallogr.* **1966**, *20*, 550–555. [CrossRef]
25. Kistenmacher, T.J.; Rand, G.A.; Marsh, R.E. Refinements of the crystal structures of DL-serine and anhydrous L-serine. *Acta Crystallogr. Sect. B Struct. Crystallogr. Cryst. Chem.* **1974**, *30*, 2573–2578. [CrossRef]
26. Moggach, S.A.; Allan, D.R.; Morrison, C.A.; Parsons, S.; Sawyer, L. Effect of pressure on the crystal structure of L-serine-I and the crystal structure of L-serine-II at 5.4 GPa. *Acta Crystallogr. Sect. B Struct. Sci.* **2005**, *61*, 58–68. [CrossRef] [PubMed]
27. Zakharov, B.; Kolesov, B.A.; Boldyreva, E.V. Effect of pressure on crystalline L- and DL-serine: Revisited by a combined single-crystal X-ray diffraction at a laboratory source and polarized Raman spectroscopy study. *Acta Crystallogr. Sect. B Struct. Sci.* **2012**, *68*, 275–286. [CrossRef] [PubMed]
28. Kimura, F.; Oshima, W.; Matsumoto, H.; Uekusa, H.; Aburaya, K.; Maeyama, M.; Kimura, F. Single crystal structure analysis via magnetically oriented microcrystal arrays. *CrystEngComm* **2014**, *16*, 6630–6634. [CrossRef]
29. Costa, S.N.; Sales, F.A.M.; Freire, V.N.; Maia, F.F.; Caetano, E.; Ladeira, L.O.; Albuquerque, E.L.; Fulco, U.L. l-Serine Anhydrous Crystals: Structural, Electronic, and Optical Properties by First-Principles Calculations, and Optical Absorption Measurement. *Cryst. Growth Des.* **2013**, *13*, 2793–2802. [CrossRef]
30. Rajesh, K.; Kumar, P.P. Structural, Linear, and Nonlinear Optical and Mechanical Properties of New Organic L-Serine Crystal. *J. Mater.* **2014**, *2014*, 1–5. [CrossRef]
31. Podder, J.; Akhtar, F. A Study on Growth, Structural, Optical and Electrical Characterization of L-alanine Single Crystal for Optoelectronic Devices. *Res. J. Phys.* **2012**, *6*, 31–40. [CrossRef]
32. Vijayan, N.; Rajasekaran, S.; Bhagavannarayana, G.; Babu, R.R.; Gopalakrishnan, R.; Palanichamy, M.; Ramasamy, P. Growth and Characterization of Nonlinear Optical Amino Acid Single Crystal: L-Alanine. *Cryst. Growth Des.* **2006**, *6*, 2441–2445. [CrossRef]
33. Sapoundjiev, D.; Lorenz, H.; Seidel-Morgenstern, A. Solubility of Chiral Threonine Species in Water/Ethanol Mixtures. *J. Chem. Eng. Data* **2006**, *51*, 1562–1566. [CrossRef]
34. Taratin, N.; Binev, D.; Lorenz, H.; Seidel-Morgenstern, A.; Kotelnikova, E. Characterization and limits of solid solutions in binary systems of threonine diastereomers according to X-ray diffraction data. In Proceedings of the III International Conference Crystallogenesis and Mineralogy, Novosibirsk, Russia, 27 September–1 October 2013; pp. 341-I–342-II.

35. CSD files (identifiers): LALANINE54 (L-alanine), LSERIN41 (L-serine) and LVALIN01 (L-valine). Available online: https://www.ccdc.cam.ac.uk (accessed on 16 July 2020).
36. Filatov, S.; Bubnova, R. The nature of special points on unit cell parameters temperature dependences for crystal substances. *Z. Kristallogr.* **2007**, *26*, 447–452. [CrossRef]
37. Filatov, S.; Krivovichev, S.; Bubnova, R. *General Crystal Chemistry*; Publishing House of St. Petersburg University: St. Petersburg, Russia, 2018; pp. 222–224.

© 2020 by the authors. Licensee MDPI, Basel, Switzerland. This article is an open access article distributed under the terms and conditions of the Creative Commons Attribution (CC BY) license (http://creativecommons.org/licenses/by/4.0/).

Article

Stereoselective Crystallization of Chiral 3,4-Dimethylphenyl Glycerol Ether Complicated by Plurality of Crystalline Modifications

Alexander A. Bredikhin *, Dmitry V. Zakharychev, Zemfira A. Bredikhina, Alexey V. Kurenkov, Aida I. Samigullina and Aidar T. Gubaidullin

Arbuzov Institute of Organic and Physical Chemistry, FRC Kazan Scientific Center of RAS, Arbuzov St., 8, 420088 Kazan, Russia; dmzakhar@gmail.com (D.V.Z.); zemfira@iopc.ru (Z.A.B.); alexeykurenkov84@gmail.com (A.V.K.); a_samigullina@iopc.ru (A.I.S.); aidar@iopc.ru (A.T.G.)
* Correspondence: baa@iopc.ru

Received: 28 February 2020; Accepted: 12 March 2020; Published: 14 March 2020

Abstract: Spontaneous resolution of Pasteur's salt was historically the first way to obtain pure enantiomers from the racemate. The current increase in interest in the direct racemates resolution during crystallization is largely due to the opened prospects for the industrial application of this approach. The chiral 3-(3,4-dimethylphenoxy) propane-1,2-diol 1 is a synthetic precursor of practically useful amino alcohols, the enantiomers of which exhibit different biological effects. In this work, it was first discovered that racemic diol 1 is prone to spontaneous resolution. However, the crystallization process is complicated by the existence, along with the conglomerate, of two other crystalline forms. Using the differential scanning calorimetry (DSC) approach, methods have been developed to obtain individual metastable phases, and all identified modifications ((R)-1, (R+S)-1, α-rac-1, β-rac-1) were ranked by energy. The IR spectroscopy and powder X-ray diffraction (PXRD) methods demonstrated the identity of the first two forms and their proximity to the third, while β-rac-1 is significantly different from the rest. The crystal structure of the forms (R)-1 and α-rac-1 was established by the single crystal X-ray diffraction (SC-XRD) method. Preliminary information on the structure of β-rac-1 phase was obtained by the PXRD approach. Based on the information received, the experimental conditions for a successful direct resolution of racemic 1 into individual enantiomers by a preferential crystallization procedure were selected.

Keywords: chirality; deracemization; preferential crystallization; racemic conglomerate; phase behavior; polymorphism; aryl glycerol ethers

1. Introduction

Chirality (the ability of an object to exist in the form of non-superimposable mirror copies) is a fundamental property that has numerous manifestations, including various biological effects of enantiomers on a living organism [1]. For this reason, since the beginning of the century, among the new active pharmaceutical ingredients (APIs), chiral substances, represented by a single enantiomer, have dominated [2]. This trend continues to the present. For instance, 45 new drugs have been approved in the USA in 2015, 33 of which were monomeric chiral compounds and, with only one exception, were pure enantiomers [3]. According to the information provided in the review [4], 30 new chiral APIs of a monomeric nature were registered in USA in 2018, of which only two were approved as racemates.

Crystallization, employed in batch or continuous format, is used almost universally for the purification and isolation of solid crystalline APIs [5,6]. Crystallization is widely used for the deracemization of chiral APIs through the separation of their diastereomeric derivatives [7–9].

Compared with the classical methods for the preparation of non-racemic substances [10], direct methods of racemates resolution based on the preferential crystallization of one of the enantiomers from racemic solutions (less often melts) have a number of significant advantages [7,8,11]. The direct methods of resolution of the racemic APIs themselves or their racemic synthetic precursors have been used for quite some time [12]. In the early stages, this approach was applied mainly on an empirical basis. Recently, there has been a steady trend towards turning it into modern technology. This is expressed in the mathematization of the description of the process itself [13–15], in the increasingly sophisticated use of the phase diagram technique [16,17], in the designing of specific reactors for implementing one or another modification of the deracemization process [18,19], finally, in enlarging one-time downloads of racemic raw materials [20]. But with all the modern improvements, an indispensable condition for the implementation of a particular type of direct resolution is the crystallization of the racemic starting material in the form of a conglomerate (i.e., a mixture of enantiopure crystals). Additionally, the search for new conglomerates, in particular, structurally related to bioactive substances, still does not lend itself to strict forecasts.

Expectorant guaifenesin, 3-(2-methoxyphenoxy) propane-1,2-diol, the object of deracemization in [20], as well as in our earlier work [21], refers to chiral glycerol aromatic ethers $ArOCH_2CH(OH)CH_2OH$. This series is notable, on the one hand, in that among its representatives there are registered APIs (for example, guaifenesin, mephenesin, chlorphenesin [22]), as well as drug precursors with different activities [12,23–25]. On the other hand, in this series the phenomenon of spontaneous resolution of enantiomers during crystallization is much more common than average [26]. Thus, among the 2,6-, 2,3- and 3,5-dimethylphenyl ethers of glycerol, which serve as the precursors of APIs mexiletine [27], xibenolol [28,29], and metaxalone [30] (Scheme 1), the first two diols crystallize as conglomerates and were obtained by us in an enantiopure form by preferential crystallization.

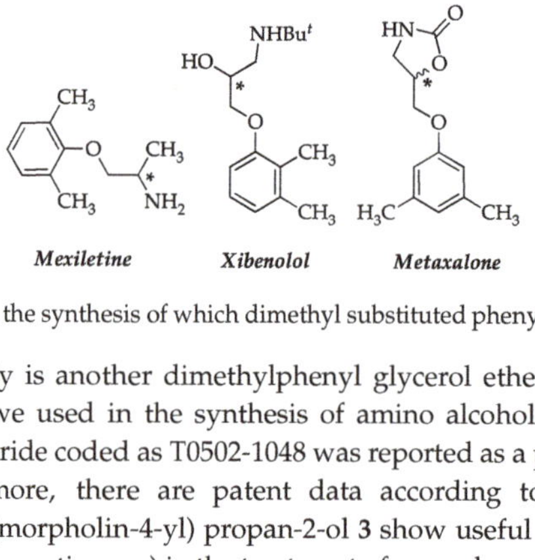

Scheme 1. Chiral drugs, in the synthesis of which dimethyl substituted phenyl glycerol ethers are used.

The object of this study is another dimethylphenyl glycerol ether, 3-(3,4-dimethylphenoxy) propane-1,2-diol **1**, which we used in the synthesis of amino alcohols **2** and **3** (Scheme 2) [31]. Aminopropanol **2** hydrochloride coded as T0502-1048 was reported as a promising β_2-adrenoceptor antagonist [32]. Furthermore, there are patent data according to which stereoisomers of 1-(3,4-dimethylphenoxy)-3-(morpholin-4-yl) propan-2-ol **3** show useful activities (but different for the racemate and individual enantiomers) in the treatment of neurodegenerative and neuromuscular disorders, as well as of Friedreich's ataxia [33].

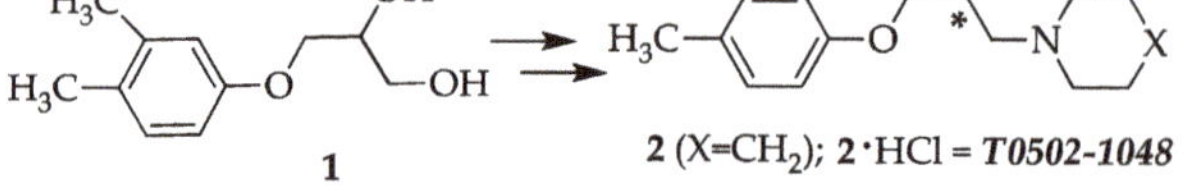

Scheme 2. 3,4-Dimethylphenyl ether of glycerol **1** and related bioactive aminopropanols **2** and **3**.

Previously we have obtained diol **1** by Sharpless asymmetric dihydroxylation of 3,4-dimethylphenyl allyl ether [31]. We have noticed that the melting point of diol (*R*)-**1** (96–98 °C) was noticeably higher than that of rac-**1** (75–77 °C). This situation is characteristic of organic compounds prone to spontaneous resolution. For this reason, we decided to study the possibility of direct resolution of its racemate and find out the features of phase behavior for this substance.

2. Materials and Methods

2.1. Instrumentation

The IR spectra of the polycrystalline samples of rac- and (R)-diols **1** under investigations in KBr pellets were recorded on a Bruker Tensor 27 spectrometer (Bruker Optic GmbH, Cermany). Optical rotations were measured on a Perkin–Elmer model 341 polarimeter (PerkinElmer, USA). The melting curves were measured on a NETZSCH 204 F1 Phoenix DSC differential scanning calorimeter (NETZSCH-Gerätebau GmbH, Germany) in sealing aluminum pans with the rate of heating of 5 °C·min^{-1}. The mass of the samples amounted to approximately ~ 1 mg in determining the enthalpies and temperatures of the phase transitions and ~ 7 mg when measuring the specific heats of the samples and was controlled with Sartorius CPA 2P balance (Sartorius AG, Goettingen, Germany). The heat capacity was measured by a continuous method, subtracting the previously measured heat capacity of the empty cell. Temperature scale and heat flux were calibrated against the data for indium and naphthalene. When measuring the heat capacities, the heat capacity of the corundum sample was used for calibration. HPLC analyses were performed on a Shimadzu LC-20AD system controller (SHIMADZU CORPORATION, Kyoto, Japan), UV monitor 275 nm was used as detector. The column used, from Daicel Inc., was Chiralcel OD (Daicel Chemical Industries, LTD - Chiral Technologies Inc., West Chester, PA, USA), (0.46 × 25 cm); eluent—hexane/2-propanol (7:3), flow rate—1 mL·min^{-1}, column temperature—22 °C.

2.2. Starting Materials

Racemic 3-chloropropane-1,2-diol and 3,4-dimethylhenol (Acros Organics) as well as (R)- and (S)-3-chloropropane-1,2-diol (Alfa Aesar) were commercially available.

2.3. Synthesis and Samples Preparation

Racemic and enantiopure 3-(3,4-dimethylphenoxy)propane-1,2-diols, rac-**1** and (R)- or (S)-**1**, used as seed, were prepared from rac- and (R)- or (S)-3-chloropropane-1,2-diol by analogy with published procedure [21]. The crude diols were purified by recrystallization from hexane/EtOAc (3:1). Yield 67%–69%. Characteristics of the obtained diols are shown below:

rac-3-(3,4-Dimethylphenoxy)propane-1,2-diol, rac-**1**: mp 75–78 °C. (Lit. [31] mp 75–77 °C).
(R)-3-(3,4-Dimethylphenoxy)propane-1,2-diol, (R)-**1**: mp 96–98 °C, $[\alpha]_D^{20}$ −7.7 (c 1.0, EtOH), 99.9% ee (chiral HPLC analysis, t$_R$ 7.7 min). [Lit. [31] mp 96–98 °C, $[\alpha]_D^{20}$ −7.7 (c 1.0, EtOH), 99.7% ee].
(S)-3-(3,4-Dimethylphenoxy)propane-1,2-diol, (S)-**1**: mp 96–97 °C, $[\alpha]_D^{20}$ +7.3 (c 1.0, EtOH), 99.2% ee (chiral HPLC analysis, t$_R$ 10.7 min). [Lit. [31] mp 96–97.5 °C, $[\alpha]_D^{20}$ +6.7 (c 1.0, EtOH), 99.5% ee].

2.4. Single Crystal X-ray Analysis

The crystals of (R)-**1** for single crystal X-ray diffraction (SC-XRD) analysis were prepared by slow evaporation of the saturated solution of the corresponding sample in mixture of ethyl acetate/hexane. The single crystal of α-rac-**1** form was randomly selected from a racemic polycrystalline sample prepared by rapid crystallization of a saturated solution in the same solvent.

The X-ray diffraction data for these crystals were collected on a Bruker Smart Apex II CCD diffractometer (Bruker AXS GmbH, Karlsruhe, Germany) in the ω-scan and φ-scan modes using graphite monochromated Mo Kα (λ 0.71073 Å) radiation at 296(2) K. The crystal data, data collection, and the refinement parameters are given in Table 1.

Table 1. Crystallographic data for (R)-1 and α-rac-1 modifications of diol 1.

Compound	(R)-1	α-rac-1
Formula	$C_{11}H_{16}O_3$	$C_{11}H_{16}O_3$
Formula weight (g/mol)	196.24	196.24
Temperature, K	296(2)	296(2)
Crystal class	Orthorhombic	Monoclinic
Space group	$P2_12_12_1$	$P2_1/n$
Crystal size	0.32 × 0.21 × 0.14 mm^3	0.24 × 0.13 × 0.04 mm^3
Z, Z'	4, 1	4, 1
Cell parameters	a = 4.9382(4) Å, b = 7.2807(6) Å, c = 28.728(2) Å	a = 4.9563(10) Å, b = 7.2972(18) Å, c = 28.491(7) Å, $β$ = 93.559(15)°
Volume, Å^3	1032.88(14)	1028.4(4)
F(000)	424	424
Calculated density, g/cm^3	1.262	1.267
μ, cm^{-1}	0.91	0.91
Theta range for data collection, deg	3.985–31.291	2.149–27.868
Reflections measured	14762	12777
Independent reflections/R(int)	3206/0.0374	2290/0.1433
Number of parameters/restraints	191/0	139/1
Reflections [I > 2σ(I)]	2669	1592
Final R indices, R_1/wR_2 [I > 2σ(I)]	0.0446/0.0965	0.1493/0.3564
Final R indices, R_1/wR_2 (all reflections)	0.0595/0.1026	0.1803/0.3791
Goodness-of-fit on F^2	1.042	1.372
Largest diff. peak and hole, $ρ_{max}/ρ_{min}$ (eÅ^{-3})	0.236/−0.185	0.782/−0.535

Data were corrected for the absorption effect using SADABS program [34]. The structures were solved by direct method and refined by the full matrix least-squares using SHELX [35] and WinGX [36] programs. All non-hydrogen atoms were refined anisotropically. All hydrogen atoms in (R)-1 were located from difference maps and refined isotropically. In α-rac-1 hydrogen atoms were inserted at calculated positions and refined as riding atoms except the hydrogens of OH groups which were located from difference maps and refined isotropically. All figures were made using Mercury program [37]. Molecular structures and conformations were analyzed by PLATON [38].

Crystallographic data for the structure of (R)-1 and α-rac-1 reported in this paper were deposited with the Cambridge Crystallographic Data Centre as supplementary publication numbers CCDC **1984093** and **1985618**, respectively. Copies of the data can be obtained, free of charge, on application to CCDC, 12 Union Road, Cambridge CB2 1EZ, UK, (fax: +44-(0)1223-336033 or e-mail: deposit@ccdc.cam.ac.uk).

2.5. Powder X-ray Diffraction Investigations

Powder X-ray diffraction (PXRD) data were collected on a Bruker D8 Advance X-ray diffractometer (Bruker AXS GmbH, Karlsruhe, Germany) equipped with a Vario attachment and Vantec linear PSD, using Cu radiation (40 kV, 40 mA) monochromated by a curved Johansson monochromator (λ Cu $K_{α1}$ 1.5406 Å). Room-temperature data were collected in the reflection mode with a flat-plate sample. Samples were applied on the surface of a standard zero diffraction silicon plate. The samples were kept spinning (15 rpm) throughout the data collection. Patterns were recorded in the 2θ range between 3° and 90°, in 0.008° steps, with a step time of 0.1–4.0 s. Several diffraction patterns in various experimental modes were collected for the samples. Processing of the data obtained was performed using EVA [39], indexing of powder data and crystal structure solving of β-rac-1 were carried out with TOPAS [40], and EXPO2014 [41] software packages.

3. Results and Discussion

3.1. Solubility Test and Preliminary Entrainment Experiment

The tests showed that at room temperature *rac*-1 has fairly good solubility in ethyl acetate, chloroform, acetonitrile and poor solubility in water and hexane. Diol *rac*-1 is moderately soluble in CCl_4, toluene and methyl *tert*-butyl ether (MTBE). Carbon tetrachloride turned out to be inconvenient for further experiments, since during the crystallization process, the formed crystals float to the surface of the solution and hardly form a distributed suspension. When cooling heated saturated solutions of diol 1 in toluene, an emulsion instead of suspension is initially obtained, which crystallizes only after a lapse of time. Of these three solvents, the reverse to the dissolution process, that is, effective spontaneous crystallization, runs smoothly only in MTBE. Therefore, a more detailed study of the dissolution and crystallization of both racemic and enantiopure 1 was carried out in this solvent.

Figure 1 shows the temperature dependence of the solubility of diol 1 stereoisomers in this solvent. The solid red circles identify the end of dissolution, and the red hollow circles—the starting points of crystallization for the racemic samples. Similarly, the behavior of enantiopure diol in the "dissolution–crystallization" cycle is characterized in blue. Concentration values for racemate are given per individual stereoisomer (i.e., the real values of racemate concentration are halved).

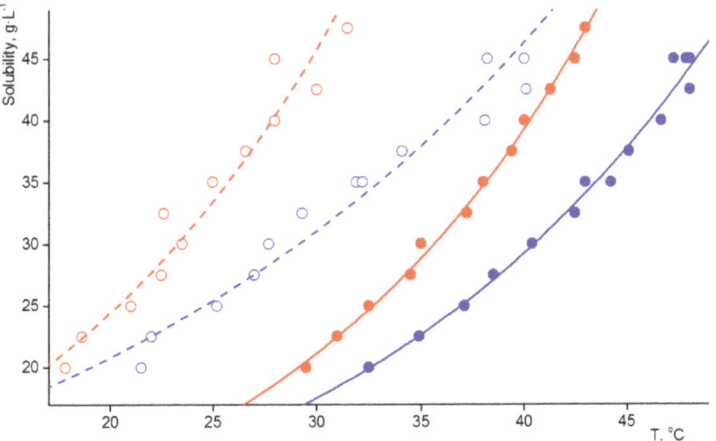

Figure 1. Temperature dependence of solubility (per single enantiomer, solid lines and circles) and for onset of crystallization (dashed lines, hollow circles) of *rac*-1 (red) and (*S*)-1 (blue) in methyl *tert*-butyl ether.

From the data obtained, it follows that the width of metastable zone for both *rac*-1 and (*S*)-1 varies slightly in the studied temperature range and amounts to 15 ± 2 °C. The chart analysis shows, that the experimental points, corresponding to the process of dissolution of enantiopure and racemic samples, do not lie on the common curve, and with increasing temperature the difference in solubility increases. This means that, for the system under study, Meyerhoffer's rule, according to which the racemate solubility twice as high the solubility of single enantiomer [42,43], is not performed. In turn, this means that the dissolution process for diol 1 is complicated by some, for the moment, unobvious effects. Meyerhoffer's coefficient high values are associated with a decrease of the metastable zone width and, consequently, with reduced spontaneous resolution efficiency [7]. Since in our case the width of the metastable zone is not very dependent on temperature, in the test experiment we carried out the crystallization stage during racemic diol 1 resolution by entrainment process at room temperature. The initial concentration of the individual stereoisomers was about 25 $g \cdot L^{-1}$, and the enantiomeric composition of the mother liquor was monitored by HPLC. The results for four cycles (eight runs) of resolution are shown in Table 2.

Table 2. Resolution by entrainment of *rac*-3-(3,4-dimethylphenoxy)propane-1,2-diol, *rac*-1 in methyl *tert*-butyl ether (20 mL, 5 mg of crystal seeds on every run; crystallization temperature 23.5 ± 0.5 °C).

Run	Added Amount of *rac*-1, mg	Operation Amount of Enantiomers, mg		Resolution Time, min	(R)-1 and (S)-1 Obtained			
		(R)-1	(S)-1		Yield, mg	ee [1], %	YE [2]	
							mg	%
1	1000	500	500	90	(R) 23	72.7	12	2.4
2	18	495	505	138	(S) 50	69.0	30	5.9
3	45	513	487	143	(R) 46	82.2	33	6.4
4	41	488	512	303	(S) 51	79.0	35	6.8
5	46	516	484	462	(R) 55	85.6	42	8.1
6	50	487	513	157	(S) 69	76.8	48	9.4
7	64	517	483	198	(R) 94	69.4	60	11.6
8	89	481	519	236	(S) 101	64.8	60	11.6

[1] ee: enantiomeric excess (HPLC). [2] YE: Yield of enantiomer; YE(mg) = [Yield (mg) × ee (%)]/100 − 5 (seed weight); YE(%) = [YE(mg) × 100]/Operation amount of (R)- or (S)-1(mg).

The results presented in the Table 2 allow us to state with certainty that spontaneous resolution during crystallization of *rac*-1 does occur. At the same time, the realized procedure cannot be considered as a satisfactory one. The individual runs are, firstly, too long, and secondly, irreproducible. The latter applies to the enantiomeric composition of the crystalline crop as well. All this, together with the non-compliance of Meyerhoffer's rule, prompted us to study in more detail the phase behavior of diol **1**.

3.2. Thermochemical Investigations

Figure 2 shows differential scanning calorimetry (DSC) thermograms of enantiopure (curve 1) and racemic (curves 2-4) samples of diol **1**. The thermogram of an enantiopure sample (R)-**1** is represented by a single narrow peak, which indicates its phase uniformity. The thermochemical parameters for this sample are presented in Table 3. The same parameters for the sample (R)-**1** obtained by crystallization from the melt practically do not differ from those for the sample crystallized from solution.

On the contrary, a thermogram of a freshly-prepared chemically-pure racemate obtained by crystallization from a hexane/ethyl acetate mixture demonstrates the complex contour of the melting process curve 2. The main peak, located in the temperature range of ~ 75 °C, has a leading edge with a pronounced kink, which is preceded by a minor endothermic peak, observed at a significantly lower temperature (~ 68 °C). This behavior is characteristic of phase-inhomogeneous systems represented by a mixture of several crystalline modifications. Intensive mixing of a suspension of such a sample in hexane for several hours at room temperature results in a thermodynamically equilibrium phase-homogeneous racemic sample, the thermogram of which is a narrow peak with a regular shape (curve 3). The demonstration of spontaneous resolution of *rac*-**1** (Section 3.1) suggests that the main crystalline form of the racemate is a conglomerate, which we denote by the symbol (R+S). The melting point of the brought to equilibrium *rac*-**1** sample, calculated as the intersection of the free energy curves of the crystalline enantiopure phase and melted (R+S)-**1** (Figure 4) was 76.2 °C, which practically coincides with the actually observed for conglomerate (76.3 °C) (Table 3).

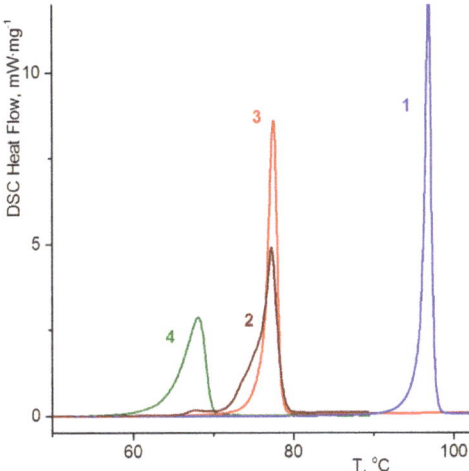

Figure 2. Differential scanning calorimetry (DSC)traces of an enantiopure sample (*R*)-**1** (blue curve 1), a freshly recrystallized racemic sample *rac*-**1** (wine curve 2), the same sample after prolonged stirring of a suspension in hexane (red curve 3), and after crystallization of the racemate from the melt (olive curve 4). Sample weight ~ 1 mg.

Table 3. Thermochemical characteristics of the identified crystalline forms of 3-(3,4-dimethylphenoxy) propane-1,2-diol **1**.

Forms	Fusion Temperature (T_f), °C	Enthalpy of Fusion (ΔH_f), J mole^{-1}
α-*rac*	74.4	33.3
β-*rac*	66.1	27.8
(*R*)	96.5	38.2
(*R*+*S*)	76.3(76.2 [1])	34.9(35.2 [1])

[1] In parentheses are the values calculated on the basis of the DSC data for the dependence of the thermodynamic potentials of the phases on temperature (Section 3.3, Figure 3).

If a racemic sample of diol **1** of any prehistory is melted and quickly cooled to room temperature, then the supercooled melt in the range of ~10 min undergoes spontaneous crystallization. The melting of the phase thus obtained (β-*rac*-**1**) is described by a peak at ~ 66 °C (curve 4), which on the temperature scale is practically in the same region as the minor endothermic peak in curve 2, Figure 2. This suggests that the β-*rac*-**1** phase is present in the samples, which initially precipitate from solutions, and the minor endothermic peak arises due to this.

The general appearance of melting curve 2 indicates that the phase behavior of the primary *rac*-**1** crystals is even more complex and, in addition to the identified (*R*+*S*)-**1** and β-*rac*-**1**, other crystalline modifications can be present in the system. Indeed, after the β-*rac*-**1** phase is heated to 72 °C and held at this temperature, it melts almost completely, but then crystallizes again. The melting parameters of the resulting α-*rac*-**1** phase (Figure 3, violet) are close, but not identical to those for the equilibrium racemic sample (Figure 3, red; Table 3).

Taken together, the thermochemical data show that for the enantiopure forms of diol **1** the only one crystalline phase is realized, and the racemate can be represented by a thermodynamically equilibrium conglomerate (*R*+*S*)-**1**, two polymorphs α-*rac*-**1** and β-*rac*-**1**, as well as a mixture of these modifications. Table 3 shows the thermochemical parameters of the corresponding systems.

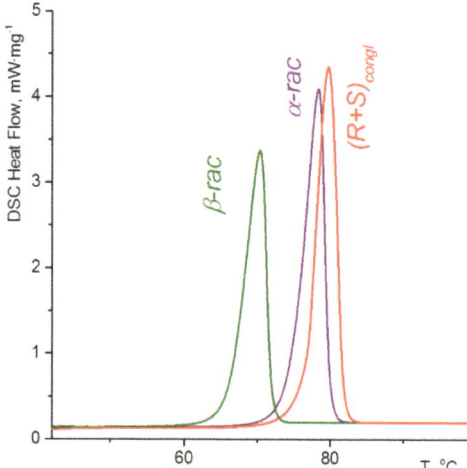

Figure 3. DSC traces of an equilibrium racemic sample identified as a conglomerate (red curve), a racemic sample obtained by crystallization from melt at 72 °C (violet curve), and crystallization from melt upon rapid cooling to 20 °C (olive curve). Sample weight ~ 8 mg.

3.3. Phase Energetics

Taken on their own, the thermochemical characteristics do not say anything about the internal organization of a particular phase, but can be used to assess their thermodynamic stability. For a detailed analysis of the energy ratios between the forms realized in the system, the heat capacities of all the modifications observed in the system were measured by the DSC method in the temperature range from 20 to 110 °C. Changes in the thermodynamic parameters ΔH, ΔS and ΔG of a system with temperature can be described by Equations (1)–(3).

$$\Delta H^{T1/T0} = \left| \begin{array}{l} \int_{T0}^{T1} C_p^{solid}(T)dT, T1 < T^f, \\ \int_{T0}^{T^f} C_p^{solid}(T)dT + \Delta H_f + \int_{T^f}^{T1} C_p^{lq}(T)dT, T1 \geq T^f, \end{array} \right. \quad (1)$$

$$\Delta S^{T1/T0} = \left| \begin{array}{l} \int_{T0}^{T1} \frac{C_p^{solid}(T)}{T}dT, T1 < T^f, \\ \int_{T0}^{T^f} \frac{C_p^{solid}}{T}(T)dT + \frac{\Delta H_f}{T^f} + \int_{T^f}^{T1} \frac{C_p^{lq}(T)}{T}dT, T1 \geq T^f, \end{array} \right. \quad (2)$$

$$\Delta G^{T1/T0} = \Delta H^{T1/T0} - T1 \cdot \Delta S^{T1/T0}. \quad (3)$$

Based on these equations and the totality of the experimentally-obtained thermochemical information for diol **1**, we constructed an energy diagram reflecting the relationship between the free energies of various crystalline modifications at different temperatures (Figure 4). The details of the calculations are given by us earlier [29].

It should be noted that we do not know the absolute values of the standard thermodynamic potentials of the considered forms. For this reason, the enthalpy and entropy of the enantiopure phase (R)-**1** at 20 °C were taken as conventional zero in calculating the thermodynamic potentials. Accordingly, changes in the Gibbs energy were calculated by Equation (4).

$$\Delta G_x^T = (H_x^T - H_R^{20°C}) - T(S_x^T - S_R^{20°C}) = (H_x^T - T \cdot S_x^T) - (H_R^{20°C} - T \cdot S_R^{20°C}). \quad (4)$$

Thus, the Gibbs energy for each phase is calculated relative to a hypothetical system, the enthalpy and entropy of which coincides with the values of these parameters for the enantiopure phase at

20 °C. This approach leaves invariant the relative arrangement of curves and characteristic points on the graph.

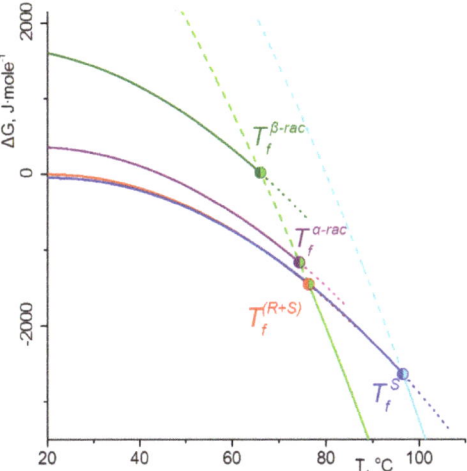

Figure 4. The temperature dependence of the Gibbs free energy change for the enantiopure sample (R)-**1** (blue line), racemic conglomerate (R+S)-**1** (red line), α-rac-**1** (violet line), β-rac-**1** (olive line), racemic melt (green line), and single enantiomeric melt (cyan line). Dashed lines correspond to the metastable supercooled state of melts; dot lines correspond to extrapolations of the free energies of crystalline phases in the temperature range above their melting temperature.

The representation of the free energies of the putative racemic and enantiopure forms of the substance in a single scale is based on the fact that mixtures of enantiomers in the molten state well satisfy the conditions of the ideal solution model (zero enthalpy of mixing, constant (Rln2) entropy of mixing of enantiomers during the formation of racemic melt), which makes it possible to use the energy of the melt level as a reference point for bringing the free energies of all phases realized in the system to a common scale [44].

The complete coincidence of the thermodynamic characteristics of enantiopure crystals (R)-**1** and conglomerate (racemic eutectic of the enantiomers), which finds its expression in the merger of the corresponding curves in Figure 4, means that there are no signs of solid solution formation in the (R+S)-**1** form.

A useful consequence of the ranking of identified crystalline modifications of diol **1** is the ability to evaluate their relative solubility on this basis. Considering that, at least for dilute solutions, the solvation effects for α-rac-**1**, β-rac-**1** and (R+S)-**1** will be the same, their relative solubility will be determined only by differences in the free energies of crystalline modifications. The change in the free energy of the phase x upon dissolution, $\Delta G^T_{x/soln}$, is described by Equation (5), in which $\left[c^{sat}_x\right]$ is the equilibrium concentration of the saturated solution (i.e., solubility) for phase x.

$$\Delta G^T_{x/soln} = -RTln\left[c^{sat}_x\right]. \tag{5}$$

Then the solubility ratio of the two phases $x1$ and $x2$ can be calculated on the basis of the difference of the free energies of these phases as follows:

$$\frac{\left[c^{sat}_{x2}\right]}{\left[c^{sat}_{x1}\right]} = \exp\frac{-(\Delta G^T_{x2} - \Delta G^T_{x1})}{RT}. \tag{6}$$

For α-rac-**1** and (R+S)-**1** forms, the difference calculated from the Gibbs free energies under standard conditions is $\Delta\Delta G^0_{\alpha/(R+S)} \approx 0.36\ kJ \cdot mole^{-1}$. Then the ratio of their solubilities calculated by Equation (6) will be $\frac{[C_\alpha]}{[C_{(R+S)}]} \approx 1.16$. For β-rac-**1** and (R+S)-**1** the calculated values of the corresponding quantities are $\Delta\Delta G^0_{\beta/(R+S)} \approx 1.6\ kJ \cdot mole^{-1}$ and $\frac{[C_\beta]}{[C_{(R+S)}]} \approx 1.9$.

3.4. IR Investigations of All the Detected Phases

Differences in the vibrational spectra of crystalline modifications formed by identical (accurate to configuration) molecules, primarily, although not always explicitly, reflect differences in the internal organization of crystals. Some time ago, we proposed a quantitative and at the same time graphic way of pairwise comparison of such spectra [45,46]. In this case, the correlation coefficient between the spectral curves acts as a quantitative measure of the coincidence of the spectra, and the correlation trajectory, which degenerates into a straight line oriented along the main diagonal of the graph when the spectra are completely identical, serves the purposes of visibility. In Figure 5, the spectrum of the pure racemic conglomerate (R+S)-**1** is alternately compared with the spectra of pure samples (R)-**1**, α-rac-**1** and β-rac-**1**.

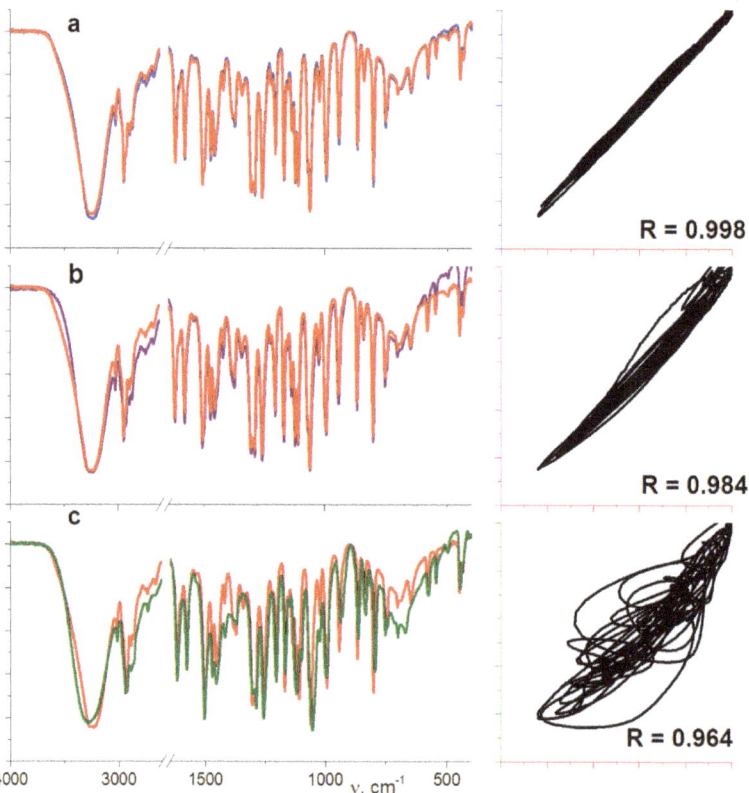

Figure 5. Comparison of the IR spectrum of conglomerate (R+S)-**1** (red curve) with the spectra of samples (R)-**1** (**a**), (blue curve), α-rac-**1** (**b**), (violet curve), and β-rac-**1** (**c**), (olive curve).

It can be seen from the figure that the internal organization of the conglomerate crystals and the enantiopure sample practically coincide, and the forms (R+S)-**1** and α-rac-**1** are structurally close. At the same time, both the correlation coefficient **R** and the appearance of the correlation diagram (the right fragment of the figure) indicate significant differences in the crystalline organization between these modifications and the metastable racemic phase β-*rac*-**1**. The information on similarities and differences revealed in this way is in complete agreement with the above data on the energy of crystalline modifications of diol **1** and with the results of XRD studies below.

3.5. X-ray Diffraction Investigations

As shown in the previous sections, only the β-*rac*-**1** form differs markedly in energy (Section 3.3) and in internal organization (Section 3.4) from the rest studied. This conclusion fully coincides with the data in Figure 6, where the experimental powder diffraction patterns of the crystalline forms α-*rac*-**1**, β-*rac*-**1**, (*R*+*S*)-**1** and (*R*)-**1** are compared. The only curve that differs markedly from the others is the β-*rac*-**1** phase diffractogram.

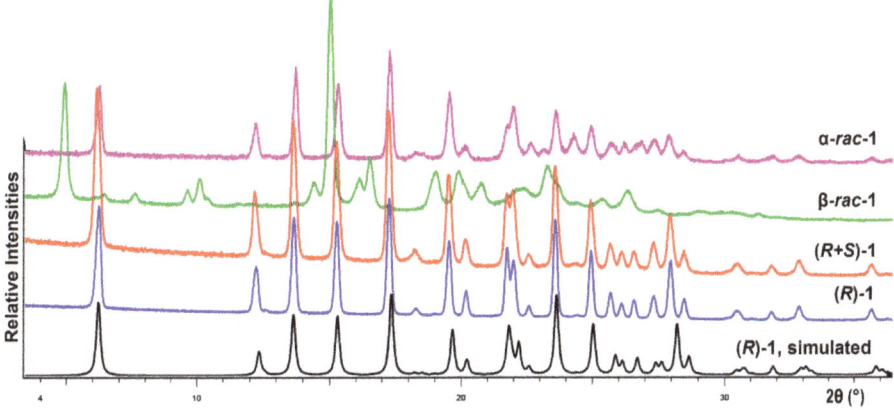

Figure 6. Comparison of experimental powder X-ray diffraction (PXRD) patterns of (*R*)-**1**, (*R*+*S*)-**1**, α-*rac*-**1**, and β-*rac*-**1** forms and simulated PXRD patterns of (*R*)-**1** forms.

X-ray powder diffraction also clearly reveals the metastable nature of this phase. While other diffractograms retain all the main features for a long time, the β-*rac*-**1** phase begins to change already during the experiment. Figure 7 shows the diffraction patterns of this phase which was freshly prepared or stored for two months. While on the diffractogram of the fresh sample there are only traces of the impurity signals (for example, in the region of scattering angles 2θ 6°–7° and 11°–12°), on the diffractogram of the aged sample the peaks belonging to the α-*rac*-**1** phase are clearly visible (if they do not prevail).

As we now know, enantiopure diol **1** exists in a single stable crystalline modification, which allows one to obtain good quality single crystals suitable for X-ray diffraction. The results of an X-ray experiment for (*R*)-**1** crystals are shown in Table 1. Figure 8a shows the only symmetry independent molecule present in the unit cell of these crystals.

In general, the conformation of the glycerol fragment in the molecules of glycerol aromatic ethers can be characterized by torsion angles H1O1C1C2, O1C1C2C3, C1C2C3O3, C2C3O3C4, C3O3C4C5, H2O2C2C3, O2C2C3O3. In the order of listing, for (*R*)-**1** they are 146.4°; 50.9°; 53.2°; 175.4°; −177.6°; 155.8°; and 175.9°, which corresponds respectively to *ac*, *sc*, *sc*, *ap*, *ap*, *ac*, and *ap* conformation. According to our previous experience, such a conformation is inherent in compounds that form a homochiral guaifenesin-like supramolecular motif in their crystals [47]. The main supramolecular synthon in such crystals is the sequence of intermolecular hydrogen bonds {O1−H1···O2′, O2′−H2′···O1′, O1′−H1′···O2″}, that is, the $C_2^2(4)$ chain formed around a screw axis 2_1 parallel to the *b* axis. Donor (O2−H2 and O1′−H1′) and acceptor (O2 and O1′) fragments that are not involved in the construction of this chain form another chain $C_2^2(4)$ around the adjacent axis 2_1. Together, the guaifenesin-like motif represents a bilayer parallel to *0ba* plane. It is this motif that is realized in (*R*)-**1** crystals.

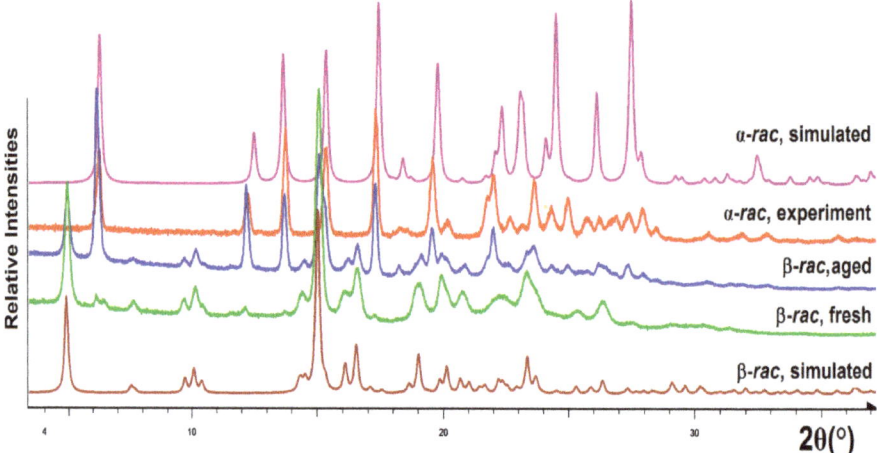

Figure 7. Comparison of experimental and simulated PXRD patterns for α-*rac*-**1** and β-*rac*-**1** polymorphs.

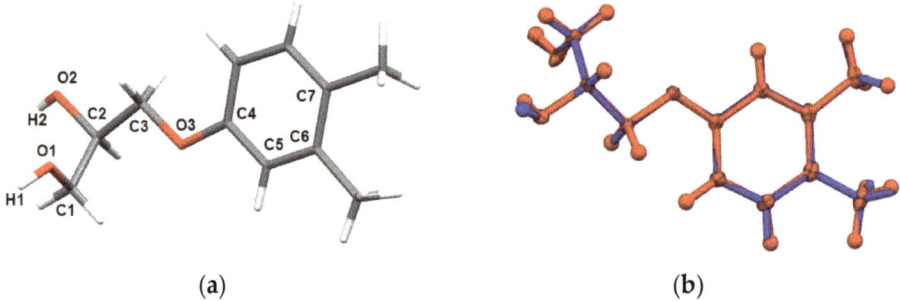

Figure 8. (**a**) Geometry of the molecules in (*R*)-**1** crystals. (**b**) Conditional superposition of *R*-enantiomers in (*R*)-**1** (blue) and α-*rac*-**1** crystals (red).

The investigated crystal of the racemic sample α-*rac*-**1** was of lower quality and with noticeable twinning, which is not surprising for the metastable phase. The structure was solved in monoclinic syngony with cell parameters close to the parameters of the enantiopure orthorhombic crystal (Table 1), except that the angle β = 93.559° differs significantly from 90°. The structure of α-*rac*-**1** was refined in the space group $P2_1/n$ with the only symmetry independent molecule. The experimental powder diffraction patterns of the α-*rac*-**1** form are generally consistent with the calculated one (Figure 7). Some differences in peak intensities are associated with twinning of crystals and the presence of insignificant texturing of the sample.

The geometry of the symmetrically-independent *R*-enantiomer molecule in α-*rac*-**1** crystals turned out to be almost identical to that just described for the independent molecule in (*R*)-**1** crystals. A visual evidence of such an identity is a conditional superposition of the *R*-enantiomers present in crystals of both forms (Figure 8b). For systems with close molecular geometry and close cell parameters, it is natural to expect a similar supramolecular organization. Indeed, the same guaifenesin-like motif is realized in α-*rac*-**1** crystals as in (*R*)-**1** crystals. The only, but important, difference between the internal organization of this pair is that in (*R*)-**1** crystals all homochiral bilayers are formed by the same enantiomers, and in crystals of α-*rac*-**1** each individual bilayer is homochiral, but adjacent bilayers are formed by opposite enantiomers (Figure 9). A similar situation was described by us in detail on the example of 3-(4-*n*-buthylphenoxy)propane-1,2-diol [47].

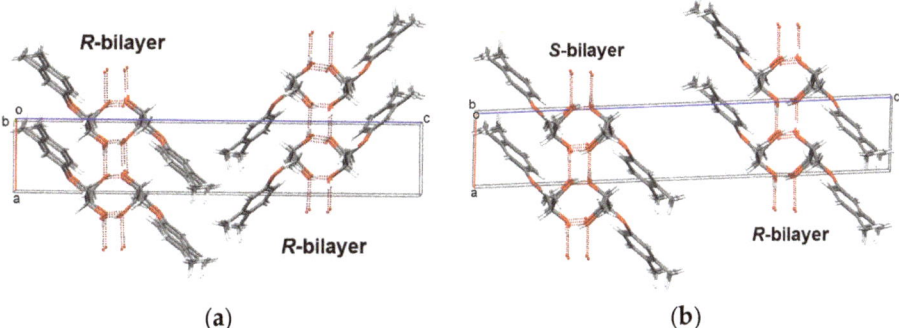

Figure 9. Stacking of H-bonded homochiral bilayers in (*R*)-**1** (**a**) and α-*rac*-**1** (**b**) crystals.

The process of the formation of crystals of homochiral and racemic samples can be represented as layer-by-layer stacking of homochiral 2D bilayer structures along the direction 0c (Figure 9). In the first case, the bilayers are connected by screw axes 2_1, and in the second, by inversion centers. Although the calculated packing coefficients in crystals of enantiopure and racemic forms (69.9% and 70.0%, respectively) practically coincide, the second packing method, apparently, required a certain shift of the 2D bilayers relative to each other, which has resulted in a deviation of the monoclinic angle from 90°. Perhaps the same effect also explains the fact of a noticeable twinning of the crystals of the racemic sample in comparison with the enantiopure ones.

From the entire preceding text, it is obvious that the structure of the crystalline β-*rac*-**1** polymorph should be noticeably different from the structure of other identified modifications. Its metastable nature does not allow to obtain stable crystals of the required quality for the study of their internal structure by SC-XRD method. However, despite weak scattering (Figure 7), we tried to index diffractogram and solve the structure of β-*rac*-**1** form from powder diffraction data. It should be added that this form is difficult to obtain as a pure one-component system and it always contains other phases in impurity quantities. Over time, the content of these phases grows, as can be seen from a comparison of the diffraction patterns of β-*rac*-**1** samples freshly prepared and stored for some time (Figure 7). However, knowledge of the position of the peaks for known phases allows us to ignore these reflections at the stage of indexing and the structure solving.

Indexing of the powder diffraction pattern of the β-*rac*-**1** form by several independent software packages made it possible to index it in a triclinic cell whose parameters (a = 19.27(1)Å, b = 12.38(7)Å, c = 5.54(7)Å, α = 93.22(4), β = 94.14(7), γ = 72.87(3)°, V = 1258(1) Å^3) differ markedly from those for the enantiopure and α-*rac*-**1** forms. According to preliminary results obtained using the EXPO 2014 software package [41], the β-*rac*-**1** structure was solved in the space group *P*-1 with two independent molecules in an asymmetric unit. A good coincidence of the experimental powder diffraction pattern of β-*rac*-**1** polycrystalline sample and the diffraction pattern calculated from the atom coordinates of the molecules in this cell testify to the correct choice of the cell and the determined geometry of molecular fragments (Figure 7). According to preliminary data, two independent A and B molecules of diol **1**, noticeably different in their geometry, are present in β-*rac*-**1** crystal (Figure 10).

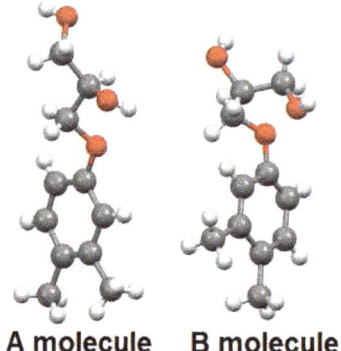

Figure 10. Probable geometry of two symmetrically-independent *R*-molecules in β-*rac*-**1** crystals.

The principal supramolecular motif in β-*rac*-**1** crystals is illustrated in Figure 11. As can be seen from the Figure, each independent molecule due to the classical hydrogen bonds O–H⋯O′ is bonded with its enantiomer into a separate centrosymmetric dimer, and already A-A and B-B dimers act as subunits in the formation of the 1D construct oriented along the crystallographic direction *b*.

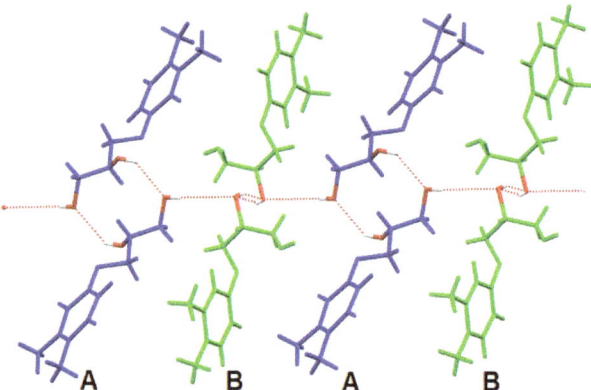

Figure 11. The principal supramolecular motif in β-*rac*-**1** crystals.

As recommended by Bernstein et al. [48], such a hydrogen bonding pattern may be called a "chain of rings". Extending somewhat the system of symbols proposed in the review [48], such a motif with two different rings can be designated as $C_2^2(11)[R_2^2(4)R_2^2(10)]$. Such a sophisticated packing based on a one-dimensional motif can hardly be dense. Indeed, preliminary calculations point that the Kitaygorodsky packing index is below 60% (KPI = 56.1%).

3.6. Direct Resolution of Rac-1 by Entrainment Procedure

In general, our study of the phase behavior of 3-(3,4-dimethylphenoxy)propane-1,2-diol **1** showed that the conglomerate (*R*+*S*)-**1** is the most stable crystalline modification of the racemate, which does not contain signs of a solid solution and is not prone to phase transformations. Other detected racemic forms are metastable and more soluble than conglomerate. Therefore, an increase in the crystallization temperature of the solution should help to increase the efficiency of the process of direct resolution of racemic **1** by reducing the supersaturation for undesirable forms and approximating the crystallization conditions to equilibrium ones. Further, the results of a pilot experiment (Section 3.1) showed that crystallization of pure racemic **1** is accompanied by a significant and irreproducible induction period duration. We believed that some supersaturation of the initial solution with the target component should contribute to a decrease in the influence of this factor on the crystallization stage of a specific enantiomer. Finally, an increase in the relative amount of introduced crystal seeds should be an important factor that favorably affects the kinetics of the process. With all that in mind, we planned and implemented an experiment the details of which are shown in Table 4.

Table 4. Resolution by entrainment of *rac*-3-(3.4-dimethylphenoxy)propane-1,2-diol, *rac*-1 in methyl *tert*-butyl ether (60 mL, 75 mg of crystal seeds on every run; crystallization temperature 27 °C).

Run	Added Amount of *rac*-1, g	Operation Amount of Enantiomers, g		Resolution Time, min	(R)-1 and (S)-1 Obtained			
		(R)-1	(S)-1		Yield, g	ee, %	YE [2]	
							g	%
1	3.00 [1]	1.42	1.58	110	(S) 0.50	73.8	0.29	18.5
2	0.42	1.57	1.43	65	(R) 0.40	78.0	0.24	15.2
3	0.33	1.44	1.56	80	(S) 0.40	78.0	0.24	15.3
4	0.33	1.58	1.42	115	(R) 0.43	75.0	0.25	15.7
5	0.35	1.46	1.57	95	(S) 0.43	74.8	0.25	15.6
6	0.35	1.59	1.41	85	(R) 0.49	71.6	0.28	17.3

[1] Sample slightly enriched with (S)-enantiomer (5.4% *ee*). [2] YE: Yield of enantiomer; YE(g) = [Yield × *ee*]/100 − 0.075; YE(%) = [YE(g) ×100]/Operation amount of (R)- or (S)-1.

In Figure 12, the same process of separation of racemic *rac*-1 is clearly illustrated by changes in the enantiomeric excess values of its mother liquor. Solid circles indicate *ee* values, upon reaching which the process was interrupted and the precipitate formed was filtered off. Then, compensating amounts of *rac*-1 and solvent were added to the heated mother liquor, after which the process was repeated.

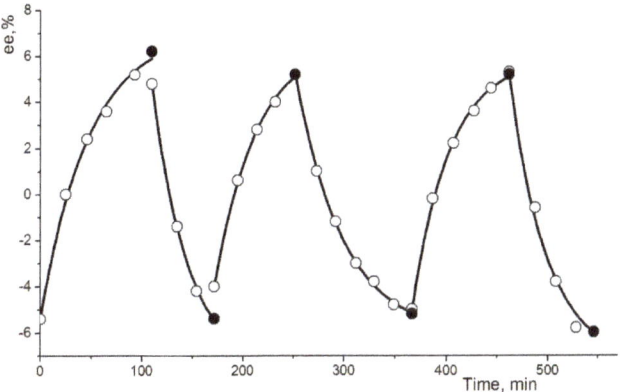

Figure 12. Mother liquor enantiomeric excess vs time of preferential crystallization of slightly enantiomerically enriched diol **1**. Closed circles represent the values of *ee*, on reaching which the process was interrupted.

A comparison of the data of Tables 2 and 4 shows that the yield of the pure enantiomer increases from 11% to 17%–18%, the process proceeds more reproducibly, in a constant temperature range and with a smaller scatter in the *ee* values (72%–78%) of its filtered precipitates. In principle, the proposed resolution procedure can be scaled and repeated as many times as necessary. A high degree of enantiomeric purity of collected (R)- and (S)-diols can be achieved by crop recrystallization from mixture of EtOAc:hexane (1:2).

4. Conclusions

An experimental study of the phase behavior of chiral 3-(3,4-dimethylphenoxy) propane-1,2-diol **1** showed that enantiopure samples exist in a single stable crystalline form. In contrast, racemic **1** during crystallization under different conditions can exist in at least three modifications, namely stable racemic conglomerate (R+S)-**1** and two metastable racemic compounds α-*rac*-**1** and β-*rac*-**1**. With rapid crystallization from racemic solutions, all three phases can crystallize simultaneously.

Using the DSC approach, methods have been developed to obtain individual metastable phases, and all identified homochiral and racemic crystal modifications were ranked by energy. The IR spectroscopy and PXRD methods demonstrated the identity or similarity of all forms except β-*rac*-**1**, which is significantly different from the rest. The crystal structure of the forms (R)-**1** and α-*rac*-**1** was established by the SC-XRD method. It was found that supramolecular crystal-forming motifs in both forms are fundamentally the same and only symmetries differ. Preliminary information on the structure of β-*rac*-**1** phase was obtained by the PXRD approach.

Based on the information received, the experimental conditions for a successful direct resolution of racemic **1** into individual enantiomers by a preferential crystallization procedure were selected.

Author Contributions: Conceptualization, writing—review and editing, A.A.B.; methodology, DSC investigation, thermodynamics calculations, D.V.Z.; methodology, proofread the manuscript, resolution experiments, Z.A.B.; samples preparation and resolution, A.V.K.; XRD investigation, A.I.S.; XRD investigation, proofread the manuscript, A.T.G. All authors have read and agreed to the published version of the manuscript.

Funding: This research received no external funding.

Acknowledgments: The authors thank I.I. Vandyukova for valuable help with IR spectra analysis. The authors are grateful to the Assigned Spectral-Analytical Center of FRC Kazan Scientific Center of RAS for technical assistance in research.

Conflicts of Interest: The authors declare no conflicts of interest.

References

1. Wagniere, G.H. *On Chirality and the Universal Asymmetry: Reflections on Image and Mirror Image*; Wiley-VCH: Hoboken, NJ, USA, 2007.
2. Murakami, H. From racemates to single enantiomers—Chiral synthetic drugs over the last 20 years. *Top. Curr. Chem.* **2007**, *269*, 273–299. [CrossRef] [PubMed]
3. Calcaterra, A.; D'Acquarica, I. The market of chiral drugs: Chiral switches versus de novo enantiomerically pure compounds. *J. Pharm. Biomed. Anal.* **2018**, *147*, 323–340. [CrossRef] [PubMed]
4. De la Torre, B.G.; Albericio, F. The pharmaceutical industry in 2018. An analysis of FDA drug approvals from the perspective of molecules. *Molecules* **2019**, *24*, 809. [CrossRef] [PubMed]
5. Wood, B.; Girard, K.P.; Polster, C.S.; Croker, D.M. Progress to date in the design and operation of continuous crystallization processes for pharmaceutical applications. *Org. Process Res. Dev.* **2019**, *23*, 122–144. [CrossRef]
6. Gao, Z.G.; Rohani, S.; Gong, J.B.; Wang, J.K. Recent developments in the crystallization process: Toward the pharmaceutical industry. *Engineering* **2017**, *3*, 343–353. [CrossRef]
7. Lorenz, H.; Seidel-Morgenstern, A. Processes to separate enantiomers. *Angew. Chem. Int. Ed.* **2014**, *53*, 1218–1250. [CrossRef]
8. Coquerel, G. Chiral discrimination in the solid state: Applications to resolution and deracemization. In *Advances in Organic Crystal Chemistry*; Tamura, R., Miyata, M., Eds.; Springer: Berlin/Heidelberg, Germany, 2015; pp. 393–420. [CrossRef]
9. Wang, Y.; Chen, A. Crystallization-Based Separation of Enantiomers. In *Stereoselective Synthesis of Drugs and Natural Products*; Andrushko, V., Andrushko, N., Eds.; John Wiley and Sons, Inc.: Hoboken, NJ, USA, 2013. [CrossRef]
10. Faigl, F.; Fogassy, E.; Nogradi, M.; Palovics, E.; Schindler, J. Strategies in optical resolution: A practical guide. *Tetrahedron Asymmetry* **2008**, *19*, 519–536. [CrossRef]
11. Palmans, A.R.A. Deracemisations under kinetic and thermodynamic control. *Mol. Syst. Des. Eng.* **2017**, *2*, 34–46. [CrossRef]

12. Bredikhin, A.A.; Bredikhina, Z.A. Stereoselective crystallization as a basis for single-enantiomer drug production. *Chem. Eng. Technol.* **2017**, *40*, 1211–1220. [CrossRef]
13. Levilain, G.; Eicke, M.J.; Seidel-Morgenstern, A. Efficient resolution of enantiomers by coupling preferential crystallization and dissolution. Part 1: Experimental proof of principle. *Cryst. Growth Des.* **2012**, *12*, 5396–5401. [CrossRef]
14. Eicke, M.J.; Levilain, G.; Seidel-Morgenstern, A. Efficient resolution of enantiomers by coupling preferential crystallization and dissolution. Part 1: A parametric simulation study to identify suitable process conditions. *Cryst. Growth Des.* **2013**, *13*, 1638–1648. [CrossRef]
15. Eicke, M.J.; Levilain, G.; Seidel-Morgenstern, A. Efficient resolution of enantiomers by coupling preferential crystallization and dissolution. Part 2: A parametric simulation study to identify suitable process conditions. *Cryst. Growth Des.* **2014**, *14*, 4872. [CrossRef]
16. Cascella, F.; Seidel-Morgenstern, A.; Lorenz, H. Exploiting ternary solubility phase diagrams for resolution of enantiomers: An instructive example. *Chem. Eng. Technol.* **2020**, *43*, 329–336. [CrossRef]
17. Oketani, R.; Marin, F.; Tinnemans, P.; Hoquante, M.; Laurent, A.; Brandel, C.; Cardinal, P.; Meekes, H.; Vlieg, E.; Geerts, Y.; et al. Deracemization in a complex quaternary system with a second-order asymmetric transformation by using phase diagram studies. *Chem. Eur. J.* **2019**, *25*, 13890–13898. [CrossRef]
18. Binev, D.; Seidel-Morgenstern, A.; Lorenz, H. Continuous separation of isomers in fluidized bed crystallizers. *Cryst. Growth Des.* **2016**, *16*, 1409–1419. [CrossRef]
19. Galan, K.; Eicke, M.J.; Elsner, M.P.; Lorenz, H.; Seidel-Morgenstern, A. Continuous preferential crystallization of chiral molecules in single and coupled mixed-suspension mixed-product-removal crystallizers. *Cryst. Growth Des.* **2015**, *15*, 1808–1818. [CrossRef]
20. Cascella, F.; Temmel, E.; Seidel-Morgenstern, A.; Lorenz, H. Efficient resolution of racemic guaifenesin via batch-preferential crystallization processes. *Org. Process Res. Dev.* **2020**, *24*, 50–58. [CrossRef]
21. Bredikhina, Z.A.; Novikova, V.G.; Zakharychev, D.V.; Bredikhin, A.A. Solid state properties and effective resolution procedure for guaifenesin, 3-(2-methoxyphenoxy)-1,2-propanediol. *Tetrahedron Asymmetry* **2006**, *17*, 3015–3020. [CrossRef]
22. O'Neil, M.J. (Ed.) *The Merck Index*, 14th ed.; Merck and Co. Inc.: Whitehouse Station, NJ, USA, 2006.
23. Saddique, F.A.; Zahoor, A.F.; Yousaf, M.; Irfan, M.; Ahmad, M.; Mansha, A.; Khan, Z.A.; Naqvi, S.A.R. Synthetic approaches towards the synthesis of beta-blockers (betaxolol, metoprolol, sotalol, and timolol). *Turk. J. Chem.* **2016**, *40*, 193–224. [CrossRef]
24. Agustian, J.; Kamaruddin, A.H.; Bhatia, S. Single enantiomeric beta-blockers—The existing technologies. *Process Biochem.* **2010**, *45*, 587–1604. [CrossRef]
25. Campo, C.; Llama, E.F.; Bermudez, J.L.; Sinisterra, J.V. Methodologies for the stereoselective synthesis of adrenergic beta-blockers: An overview. *Biocatal. Biotransform.* **2001**, *19*, 163–180. [CrossRef]
26. Bredikhin, A.A.; Bredikhina, Z.A.; Zakharychev, D.V. Crystallization of chiral compounds: Thermodynamical, structural and practical aspects. *Mendeleev Commun.* **2012**, *22*, 171–180. [CrossRef]
27. Bredikhina, Z.A.; Kurenkov, A.V.; Krivolapov, D.B.; Bredikhin, A.A. Stereoselective crystallization of 3-(2,6-dimethylphenoxy)propane-1,2-diol: Preparation of the single-enantiomer drug mexiletine. *Tetrahedron Asymmetry* **2015**, *26*, 577–583. [CrossRef]
28. Bredikhin, A.A.; Bredikhina, Z.A.; Kurenkov, A.V.; Gubaidullin, A.T. Synthesis, crystal structure, and absolute configuration of the enantiomers of chiral drug xibenolol hydrochloride. *Tetrahedron Asymmetry* **2017**, *28*, 1359–1366. [CrossRef]
29. Bredikhin, A.A.; Zakharychev, D.V.; Bredikhina, Z.A.; Kurenkov, A.V.; Krivolapov, D.B.; Gubaidullin, A.T. Spontaneous resolution of chiral 3-(2,3-dimethyl phenoxy)propane-1,2-diol under the circumstances of an unusual diversity of racemic crystalline modifications. *Cryst. Growth Des.* **2017**, *17*, 4196–4206. [CrossRef]
30. Bredikhin, A.A.; Zakharychev, D.V.; Gubaidullin, A.T.; Bredikhina, Z.A. Solid phase behavior, polymorphism, and crystal structure features of chiral drug metaxalone. *Cryst. Growth Des.* **2018**, *18*, 6627–6639. [CrossRef]
31. Bredikhina, Z.A.; Kurenkov, A.V.; Bredikhin, A.A. Nonracemic dimethylphenyl glycerol ethers in the synthesis of physiologically active aminopropanols. *Russ. J. Org. Chem.* **2019**, *55*, 837–844. [CrossRef]
32. Hothersall, J.D.; Black, J.; Caddick, S.; Vinter, J.G.; Tinker, A.; Baker, J.R. The design, synthesis and pharmacological characterization of novel β2-adrenoceptor antagonists. *Br. J. Pharmacol.* **2011**, *164*, 317–331. [CrossRef]

33. Araujo, N.; Ferreira da Silva, A.; Qing, Y.; Lifino, M.; Russell, A.J.; Small, B.; Wade-Martins, R.; Wynne, G.M. Therapeutic Compounds. *Int. Patent Appl.* **2015**. No. WO 2015/004485 A1.
34. Sheldrick, G.M. *SADABS, Program for Empirical X-ray Absorption Correction*; Bruker-Nonius: Delft, The Netherlands, 2004.
35. Sheldrick, G.M. A short history of SHELX. *Acta Crystallogr. A* **2008**, *64*, 112–122. [CrossRef]
36. Farrugia, L.J. WinGX suite for small-molecule single-crystal crystallography. *J. Appl. Crystallogr.* **1999**, *32*, 837–838. [CrossRef]
37. Macrae, C.F.; Edgington, P.R.; McCabe, P.; Pidcock, E.; Shields, G.P.; Taylor, R.; Towler, M.; van de Streek, J. Mercury: Visualization and analysis of crystal structures. *J. Appl. Crystallogr.* **2006**, *39*, 453–457. [CrossRef]
38. Spek, A.L. Single-crystal structure validation with the program PLATON. *J. Appl. Crystallogr.* **2003**, *36*, 7–13. [CrossRef]
39. Bruker AXS. DIFFRAC Plus Evaluation package EVA, Version 11. In *User's Manual*; Bruker AXS: Karlsruhe, Germany, 2005; p. 258.
40. Bruker AXS. TOPAS v3: General profile and structure analysis software for powder diffraction data. In *Technical Reference*; Bruker AXS: Karlsruhe, Germany, 2005; p. 117.
41. Altomare, A.; Cuocci, C.; Giacovazzo, C.; Moliterni, A.; Rizzi, R.; Corriero, N.; Falcicchio, A. EXPO2013: A kit of tools for phasing crystal structures from powder data. *J. Appl. Crystallogr.* **2013**, *46*, 1231–1235. [CrossRef]
42. Meyerhoffer, W. Stereochemische Notizen. *Ber. Dtsch. Chem. Ges.* **1904**, *37*, 2604–2610. [CrossRef]
43. Jaques, J.; Collet, A.; Wilen, S.H. *Enantiomers, Racemates, and Resolutions*; J. Wiley and Sons, Inc.: Hoboken, NJ, USA, 1981; p. 447.
44. Li, Z.J.; Zell, M.T.; Munson, E.J.; Grant, D.J.W. Characterization of racemic species of chiral drugs using thermal analysis, thermodynamic calculation, and structural studies. *J. Pharm. Sci.* **1999**, *88*, 337–346. [CrossRef]
45. Bredikhin, A.A.; Bredikhina, Z.A.; Akhatova, F.S.; Zakharychev, D.V.; Polyakova, E.V. From racemic compounds through metastable to stable racemic conglomerates: Crystallization features of chiral halogen and cyano monosubstituted phenyl glycerol ethers. *Tetrahedron Asymmetry* **2009**, *20*, 2130–2136. [CrossRef]
46. Bredikhin, A.A.; Zakharychev, D.V.; Fayzullin, R.R.; Antonovich, O.A.; Pashagin, A.V.; Bredikhina, Z.A. Chiral para-alkyl phenyl ethers of glycerol: Synthesis and testing of chirality driven crystallization, liquid crystal, and gelating properties. *Tetrahedron Asymmetry* **2013**, *24*, 807–816. [CrossRef]
47. Bredikhin, A.A.; Zakharychev, D.V.; Gubaidullin, A.T.; Fayzullin, R.R.; Samigullina, A.I.; Bredikhina, Z.A. Crystallization of chiral para-n-alkylphenyl glycerol ethers: Phase diversity and impressive predominance of homochiral guaifenesin-like supramolecular motif. *Cryst. Growth Des.* **2018**, *18*, 3980–3987. [CrossRef]
48. Bernstein, J.; Davis, R.E.; Shimoni, L.; Chang, N.L. Patterns in hydrogen bonding—Functionality and graph set analysis in crystals. *Angew. Chem. Int. Ed.* **1995**, *34*, 1555–1573. [CrossRef]

© 2020 by the authors. Licensee MDPI, Basel, Switzerland. This article is an open access article distributed under the terms and conditions of the Creative Commons Attribution (CC BY) license (http://creativecommons.org/licenses/by/4.0/).

Article

Gypsum Crystallization during Reverse Osmosis Desalination of Water with High Sulfate Content in Presence of a Novel Fluorescent-Tagged Polyacrylate

Maxim Oshchepkov [1,2], Vladimir Golovesov [1,3], Anastasia Ryabova [4], Anatoly Redchuk [1], Sergey Tkachenko [1,2], Alexei Pervov [3] and Konstantin Popov [1,*]

1. JSC "Fine Chemicals R&D Centre", Krasnobogatyrskaya, str. 42, b. 1, 107258 Moscow, Russia; maxim.os@mail.ru (M.O); golovesov.vova@mail.ru (V.G.); aredchuk@gmail.com (A.R.); s.tkach.8@gmail.com (S.T.)
2. Mendeleev University of Chemical Technology of Russia, Miusskaya sq. 9, 125047 Moscow, Russia
3. Moscow State University of Civil Engineering, Yaroslavskoe shosse, 26, 129337 Moscow, Russia; ale-pervov@yandex.ru
4. Prokhorov General Physics Institute of the Russian Academy of Sciences, Vavilov str., 38, 119991 Moscow, Russia; nastya.ryabova@gmail.com
* Correspondence: ki-popov49@yandex.ru

Received: 17 March 2020; Accepted: 14 April 2020; Published: 16 April 2020

Abstract: Gypsum scaling in reverse osmosis (RO) desalination process is studied in presence of a novel fluorescent 1,8-naphthalimide-tagged polyacrylate (PAA-F1) by fluorescent microscopy, scanning electron microscopy (SEM), dynamic light scattering (DLS) and a particle counter technique. A comparison of PAA-F1 with a previously reported fluorescent bisphosphonate HEDP-F revealed a better PAA-F1 efficacy, and a similar behavior of polyacrylate and bisphosphonate inhibitors under the same RO experimental conditions. Despite expectations, PAA-F1 does not interact with gypsum. For both reagents, it is found that scaling takes place in the bulk retentate phase via heterogeneous nucleation step. The background "nanodust" plays a key role as a gypsum nucleation center. Contrary to popular belief, an antiscalant interacts with "nanodust" particles, isolating them from calcium and sulfate ions sorption. Therefore, the number of gypsum nucleation centers is reduced, and in turn, the overall scaling rate is diminished. It is also shown that, the scale formation scenario changes from the bulk medium, in the beginning, to the sediment crystals growth on the membrane surface, at the end of the desalination process. It is demonstrated that the fluorescent-tagged antiscalants may become very powerful tools in membrane scaling inhibition studies.

Keywords: reverse osmosis; membrane fouling; gypsum scaling; fluorescent-tagged polyacrylate; fluorescence; scale inhibition mechanisms

1. Introduction

Reverse osmosis (RO) is becoming recently a powerful technology for the purification of sea, brackish and waste water [1–4]. However, one of the major limitations in efficient RO application is the membrane scaling [5–7]. Inorganic scaling, occurs when the solubility limits are exceeded. The most common scales are represented by calcium carbonate, calcium sulfate and silica [6]. As a result of inorganic fouling, the operation cost of an RO plant increases due to higher consumption of energy and expenses of membrane cleaning. The most common method in mitigating scaling in RO facilities is an application of antiscalants. Among these, polycarboxylates (polyacrylates, polyaspartates, etc.) and phosphonates are found to be highly efficient [6–11].

However, in spite of numerous relevant studies, some controversy regarding both the dominant scaling mechanism in particular situations and the mechanism of antiscalant activity still exists [12–18].

Recent reviews on scale formation control in RO technologies [6,19] mention two main hypothetic mechanisms of inhibition: (i) antiscalant molecules adsorb on the active growth sites at the crystal surface of sparingly soluble inorganic salt and retard nucleation and crystal growth by distorting its crystal structure; (ii) antiscalant molecules provide similar electrostatic charge, and thus, repulsion between particles prevents them from agglomeration.

Nevertheless, our recent static [20,21] and RO [22] experiments operating gypsum as a model scale in presence of a novel fluorescent-tagged bisphosphonate antiscalant 1-hydroxy-7-(6-methoxy-1,3-dioxo-1H-benzo[de]isoquinolin-2(3H)-yl)heptane-1,1-diyl-bis(phosphonic acid), HEDP-F (H_4hedp-F) revealed a paradoxical effect: an antiscalant does not interact with gypsum at all, but provides nevertheless retardation of corresponding deposit formation. According to the classical crystallization theory [23], this is possible only in the case, when gypsum passes bulk heterogeneous nucleation, and exactly the "nanodust" plays the role of the solid phase template. Indeed, it is demonstrated that HEDP-F molecules being immersed into the stock solution (undersaturated against gypsum) occupy a significant part of "nanodust" crystallization centers and form there their own solid phase Ca_2hedp-F·nH_2O. However, polyacrylates are much less sensitive to calcium environment than phosphonates [20,21]. In this way, it was reasonable to study the traceability of phosphorus-free fluorescent polymeric antiscalants in RO process.

The present study is focused on the scale inhibitor visualization during RO treatment of model water sample, with high sulfate content, in the presence of a fluorescent antiscalant- 1,8-naphthalimide-tagged polyacrylate, PAA-F1, Figure 1.

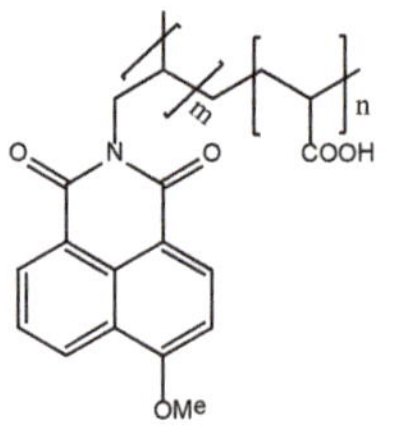

Figure 1. 1,8-Naphthalimide-tagged polyacrylate molecular structure.

The gypsum scale was taken as a model of a sparingly soluble salt due to: (i) its importance for the RO and other water treatment technologies [10–16]; (ii) its poor dependence on pH; (iii) its easily detectable crystal shapes; and (iv) the nucleation of gypsum has been investigated extensively in the past [10,11,14,16–18,24–33]. On the other hand PAA-F1 is expected to be a better antiscalant for $CaSO_4·2H_2O$ deposits relative to HEDP-F. This was demonstrated for the non-fluorescent prototypes 1-hydroxyethane-1,1-bis(phosphonic acid) (HEDP) and polyacrylate (PAA) [34].The fluorescent-tagged polyacrylates have gained increasing interest as the reagents for on line antiscalant concentration monitoring in water treatment applications recently [35]. However, they have not been applied, so far, for scale formation mechanisms studies. As far as we know, this is the first communication on polymer antiscalant visualization in a RO experiment with gypsum scaling.

2. Materials and Methods

2.1. Reagents, Membrane Material and Model Solutions

Antiscalant PAA-F1 (Figure 1) was synthesized by our group as described elsewhere [36] along with its scale inhibition efficiency against gypsum scaling and its fluorescent properties. It has the mean molecular mass 4000 Da with c.a. 1% mass of 1,8-naphthalimide moiety. This corresponds randomly to c.a. 0.2 fluorescent fragments per one molecule of polyacrylate.

For model scaling solutions, the reagent grade $CaCl_2·2H_2O$ and Na_2SO_4 were used in crystalline form and were separately dissolved in distilled water (conductivity 2 μS/cm) to prepare stock solutions of 0.04 mol·dm^{-3}. After complete dissolution, stock solutions represented transparent colorless liquids, and were deliberately exposed no filtration for a better imitation of saline or brackish water. For gypsum (CaSO$_4$·2H$_2$O) scaling experiments, each stock solution was combined with distilled water to achieve a total volume of 5 L and a final concentration of 0.015 mol·dm^{-3} [Ca^{2+}] and 0.015 mol·dm^{-3} [SO$_4^{2-}$]. The solvent (distilled water) and all stock solutions were analyzed separately for foreign particles content, Table 1. A particle counter SLS-1100 (Particle Measuring Systems Inc.) reveals a presence of background solid suspended particles ("microdust") in both stock brines, as well as in the distilled water used for the brine preparation, Table 1.

Table 1. Initial stock solutions characterization by particle counter.

Solvent/Solution	Concentration (mol·dm^{-3})	pH	Cumulative Number of Foreign Particles in 1 mL *			
			≥100 nm	≥200 nm	≥300 nm	≥500 nm
Distilled water for feed solution preparation; 2 µS/cm	55.55	5.5	390,000	97,000	19,600	16,400
CaCl$_2$ stock solution, diluted by distilled water	0.015	7.1	1,800,000	200,000	76,000	39,000
Na$_2$SO$_4$ stock solution, diluted by distilled water	0.015	6.1	1,550,000	185,000	73,000	32,000
PAA-F1 solution	1.7·10^{-6}	6.2	860,000	110,000	38,000	16,000

* The data deviations found for three replicate measurements constituted ±20%.

The values of gypsum solubility in water at 25 °C provided by different research groups, varies from 0.018 to 0.025 mol·dm^{-3} and depend drastically on the background NaCl content [37]. Therefore, the stock calcium and sulfate solutions have been prepared (Table 1) in such concentrations, that being mixed in 1:1 volume ratio they would give 0.015 mol·dm^{-3} gypsum solution, that is a bit below the saturation level. However, the retentate was expected to exceed the saturation level already at saturation coefficient (concentration factor) K = 1.5 and to reach steadily supersaturation S~4 (K = 5) at the end of the experiment (in absence of scaling). Notably, these supersaturation assessments are very approximate ones as NaCl content in retentate is changed in experiment run from 0.03 to c.a. 0.15 mol·dm^{-3}, increasing gypsum solubility.

Here and further saturation coefficient K and saturation ratio S are denoted as:

K = (total initial volume of feeding solution)/(current volume of retentate)
S = (initial gypsum concentration, mol·dm^{-3})/(gypsum solubility at 25 °C, mol·dm^{-3})
Thus for S < 1 the solution is undersaturated, while for S > 1 it is supersaturated.

Notably, an antiscalant, where necessary, was added always initially to the sulfate test solution in amounts that provided its final concentration 7 mg·dm^{-3} (corresponds to c.a. 1.7·10^{-6} mol·dm^{-3} PAA-F1 concentration bearing in mind that the mean molecular mass is 4000 Da) in gypsum scaling experiment, and equilibrated there no less than 30 min.

It is well-known that the heterogeneous nucleation, in the presence of such solid impurities, as clay minerals or other foreign particles, is characterized by a lower free energy barrier than the homogeneous one [23]. Bearing in mind the particle counter data, Table 1, a bulk homogeneous formation of solid gypsum phase is unlikely in our case, while the bulk heterogeneous nucleation is the most likely route. It should be noted that deionization of distilled water leads to a significant reduction of suspended particles concentration. However, this operation fails to remove even "microdust" completely, to say nothing of "nanodust". The latter is likely to be present in any aqueous samples in much higher amounts than "microdust", although the "nanodust" (1 nm < particle sizes < 100 nm) is beyond the detection limit of commercial particle counters. This becomes clear by extrapolating the cumulative number of foreign particles in 1 mL (Table 1) to the 1≤ nm ≤ 100 range. Thus, all the background solid suspended particles are referred to as "nanodust".

A rough estimation indicates that if the number of undetectable by the particle counter "nanodust" particles with a size D < 100 nm is equal to the detected "microdust" ones, than the total heterogeneous impurities concentration constitutes c.a. 3,000,000 units in 1 mL, or $3 \cdot 10^9$ in 1 L. At the same time PAA-F1 concentration corresponds to $3.5 \cdot 10^{15}$ molecules per liter. Thus, there are at least 10^6 molecules of PAA-F1 per one nano/microdust particle in the system.

A detailed analysis of the background solid suspended particles nature in all chemicals used (H_2O; $CaCl_2 \cdot 2H_2O$; Na_2SO_4; PAA-F1) represents a special complicated task, and is hardly possible. It is outside of the frames of the present study. However, some preliminary analyses carried out in [22] indicate that a tentative nature of solid impurities in distilled water might be assigned to Al/Fe hydroxo/oxides and to either SiO_2 or to some silicate solid impurities. At the same time, it should be noted that all the background solid suspended particles, listed in Table 1, correspond to the ppb level, e.g., to the reagent grade purity solutions.

Studies were carried out using commercial spiral wound BLN-type low pressure reverse osmosis membrane produced by CSM Co. (Seoul, South Korea). The membranes are found to have their own non uniform fluorescence. This makes it difficult to observe PAA-F1 location on membrane surface, as the coresponding images are not very clear.

2.2. Reverse Osmosis Membrane System

Gypsum scaling experiments were carried out using an automatically controlled laboratory-scale cross-flow RO spiral wound module RE 1812 tailored with thin film composite BLN membrane (Figure 2). The test unit was operated in circulation mode whereby concentrate after membrane module was returned back to feed water tank. The feed water was added to feed water tank 1. The volume of tank 1 was 5 L. Feed water from tank 1 was supplied by small gear pump 2 to membrane module 3. In all experiments the commercially available spiral-wound filter elements (CSM RE1812-80 GPD) made of polyamide and manufactured by CSM (Seoul, South Korea) were used. These operated at constant feed flow rate of 72.0 ± 0.2 dm^3/h, permeate flow rate of 6.0–6.3 dm^3/hour, constant temperature of 25.0 ± 2 °C; constant pressure 7.0± 0.2 bar in concentration mode.

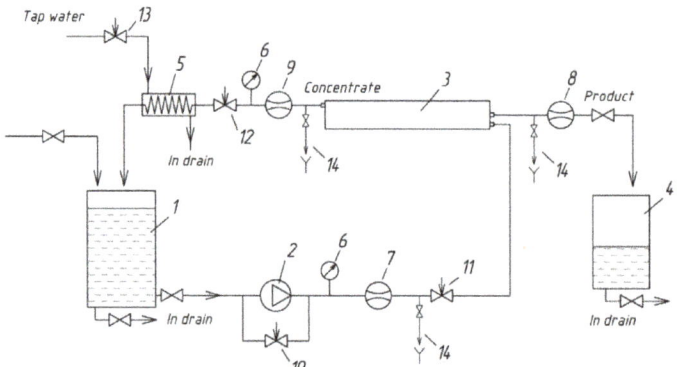

Figure 2. Schematic diagram of laboratory RO unit for membrane scaling tests: 1-feed water tank; 2-pump; 3-spiral wound membrane module; 4-permeate tank; 5-heat exchanger; 6-pressure gauge; 7-feed water flow meter; 8-permeate flow meter; 9-concentrate flow meter; 10-by-pass adjusting valve; 11-feed water adjusting valve; 12-concentrate adjusting valve; 13-cooling water adjusting valve; 14-sampler.

Stabilized salt rejection, at a constant pressure of 7 bar, a solution temperature of 25 °C, and a pH value of 6.5–7.0 is reported to be 96.5–97.0% for a 200 mg·dm^{-3} NaCl solution (manufacturer's data). Each run was performed with a virgin membrane sample.

2.3. Gypsum Scaling Experiments

The gypsum scale formation studies, included three blank experiments (A,B,C) and a gypsum scale inhibition test run in presence of PAA-F1 (GSI), Table 2, Figures 3–9. Each of the blank experiments had its own objective. Blank A experiment was intended to evaluate "free" PAA-F1 sorption by membrane in concentration operating mode. Within the frames of this experiment, the PAA-F1 concentration was monitored directly by fluorescence intensity measurements of aqueous phase in retentate and in permeate tanks, Figure 2. To the extent that no antiscalant was found in permeate, the difference between the calculated total PAA-F1 concentration in retentate + membrane system (Figure 3A, curve a) and its experimentally-measured concentration in retentate (Figure 3A, curve c), indicated an amount of PAA-F1 consumed by membrane (Figure 3A, curve b). For simplicity and clarity, here, and further the membrane-consumed PAA-F1 is expressed in units of concentration as the loss of the bulk antiscalant concentration relative to the total concentration.

An objective of the Blank B experiment (Figure 4) was to estimate a possible PAA-F1 participation in side reaction with Ca^{2+} ions, followed by its undesirable consumption by calcium due to formation of insoluble $Ca_{0.5x}H_yPAA$-$F1 \cdot nH_2O$ (0.5x + y = 1) salts (these are indicated further as [Ca-PAA-F1]). A PAA-F1 distribution between retentate and membrane (Figure 4) was found in the same way as in a Blank A experiment.

Blank C experiment (Figure 6) was intended to demonstrate a non-inhibited RO membrane gypsum scaling as a reference to the inhibited one. Variations of Ca^{2+} concentration in a Blank C experiment was monitored by an immediate titration of calcium with EDTA in samples taken from retentate and permeate tanks (Figure 6b,c). Then, a total calcium concentration in retentate + membrane was calculated (Figure 6a) as a difference between the calculated total Ca^{2+} concentration in the system and its experimentally found content in permeate (Figure 6c). Then, the calcium content on membrane surface as gypsum (Figure 6d) was found, and expressed in units of calcium concentration decrease in the same way as it was done for PAA-F1 distribution in the Blank A and B experiments.

Individual gypsum scaling experiments were performed with a virgin sample of pre-soaked membrane. Each membrane was initially contacted with distilled water overnight (12–14 h) to allow the membrane permeability to stabilize. The experimental protocol for the scaling tests in cross-flow RO membrane filtration is presented in Table 2. Experiments were run with a single superficial cross-flow velocity and were terminated after reaching K = 5. The cross-flow velocity varied from 3.0 to 3.6 cm/sec, which excluded influence of concentration polarization on gypsum supersaturation at membrane surface. This value adequately fits the range of cross-flow velocities encountered in spiral-wound RO/NF.

At the end of each experimental cycle, scaled membrane samples were carefully extracted from the autopsied membrane element and submerged in an ultrapure water bath for approximately 2 s to prevent further crystallization from evaporation of residual scaling solution. The membrane samples were then air dried for at least 48 h and afterwards cut into ten equally sized pieces of 4 cm × 10 cm. These pieces were stored in a desiccator for at least 24 h. Then the fragments were sent for analysis by scanning electron microscopy (SEM) and by fluorescent microscopy (FM).

Each test was run with a new virgin membrane spiral wound element in two replicates. Pressure and retentate cross-flow rate were monitored through digital sensors. The permeate volume was continuously recorded. The temperature was almost constant (varying by less than 2 °C) during each experiment at a level of 25 °C. The liquid phase was periodically sampled and also examined by laser confocal microscopy, dynamic light scattering (DLS), and the current Calcium content was measured by titration with EDTA. During the blank experiment with PAA-F1, content of antiscalant was monitored by fluorescence intensity measurements (Shimadzu RF-6000).

The concentration of 0.015 mol·dm^{-3} of the starting gypsum solution was chosen to be used in the experiments, which corresponds to an undersaturated state. In all runs with antiscalant, the PAA-F1 solution was initially added to the sulfate brine, equilibrated, there for half an hour, and only after that the calcium brine was added to obtain the total 5 L volume of a feeding solution. Totally four cross-flow tests have been run it two replicates each, Table 2.

Table 2. Experimental protocol for the scaling tests in cross-flow RO membrane filtration.

Membrane Preparation and Treatment					
Step	Task	Feed	Pressure	Duration	Comments
1	Membrane preparation	Distilled water	-	20–30 minutes	Washing with distilled water for preservatives removal.
2	Membrane conditioning	Distilled water	-	8–12 hours	Covered by distilled water layer for preservatives removal.
3	RO membrane desalination in concentration mode continuous monitoring	Feed solution circulation mode; well-controlled feed solution: composition, pH, T, etc.	88 to 110 psi depending on desirable flux	5–6 hours duration of the entire experiment for all runs	Feed, permeate and retentate samples collected and analyzed for Ca content, for PAA-F1 concentration (by fluorimeter) and by fluorescent microscope.
4	Post experiment treatment	-	-	24	RO membrane is removed, gently rinsed out with distilled water; dried at 22–25 °C and segmented. RO membrane segments are characterized by SEM and fluorescent microscopy.
Cross-flow tests					
Test	Initial feed solution		Comments		
Blank A	7 mg·dm^{-3} PAA-F1 in distilled water		Experiment is intended to evaluate "free" PAA-F1 sorption by membrane in concentration operating mode.		
Blank B	0.015 mol·dm^{-3} CaCl$_2$ and 7 mg·dm^{-3} PAA-F1 in distilled water		Experiment is intended to estimate PAA-F1 participation in side reaction with Ca^{2+} ions, followed by its undesirable consumption by calcium due to formation of insoluble Ca$_{0.5x}$H$_y$PAA-F1·nH$_2$O (0.5x + y = 1) salts.		
Blank C	0.015 mol·dm^{-3} CaCl$_2$ and 0.015 mol·dm^{-3} Na$_2$SO$_4$ in distilled water		Experiment has to demonstrate non-inhibited RO membrane gypsum scaling as a reference to the inhibited one.		
Gypsum Scale Inhibition (GSI)	0.015 mol·dm^{-3} CaCl$_2$, 0.015 mol·dm^{-3} Na$_2$SO$_4$ and 7 mg·dm^{-3} PAA-F1 in distilled water		A PAA-F1 inhibited gypsum scaling.		

2.4. Fluorescent Microscopy Measurements

Confocal microscopy measurements have been run with laser scanning confocal microscope LSM-710-NLO (Carl Zeiss MicroImaging GmbH, Jena, Germany), 20× Plan-Apochromat objective

(NA = 0.8). The samples were placed onto the Petri dish with a glass bottom 0.16 mm thick. The fluorescence of the PAA-F1 was recorded in the wavelength range of 500–600 nm, when excited by laser radiation with a wavelength of 488 nm. As a result, overlay of the distribution of PAA-F1 fluorescent image (green pseudo-color in images) and transmitted light image (grey color) was obtained. Notably, all the retentate samples have been taken and analyzed by confocal microscopy within 2–3 min after sampling.

2.5. Fluorescence Intensity Measurements of Aqueousphase

PAA-F1 concentration in aqueous phase was monitored by fluorescence intensity measurements, carried with luminescence spectrometer Shimadzu RF-6000 (Shimadzu Corporation, Kyoto, Japan) operating with a xenon lamp as a light source. All spectral measurements were carried out in a quartz sample cell (path length ℓ = 1 cm) at 20 ± 1 °C in air-saturated solutions. The fluorescence intensity was measured at a wavelength of 462 nm (2 nm slid width).

2.6. SEM Crystal Characterization

The membrane pieces with precipitated solids, after being triply rinsed with deionized water and air dried at 20–25 °C, were characterized by scanning electron microscopy (SEM, TM-3030, Hitachi, Japan). The sample examinations by SEM were done at 15 kV accelerating voltage in a Charge-Up Reduction Mode with crystal phase located on a Conducting Double-Sided Tape and the working distance of 4.1 mm.

2.7. DLS Characterization of Retentate

Liquid phase was monitored by the dynamic light scattering technique. DLS experiments were performed at 25 °C with Malvern Nano ZS instrument (λ = 633 nm, operating power 4 mW) at θ = 173°. Gypsum was always taken as a light scattering material even for the pure PAA-F1 solutions.

3. Results and Discussion

All experiments were monitored in a liquid phase along the retentate saturation (fluorescent microscope, DLS, fluorescence intensity, pH and calcium concentration measurements) and were followed by a final SEM analysis of solid membrane surface after each run.

3.1. Blank A Experiment Results

This blank experiment was intended to evaluate "free" PAA-F1 sorption by membrane in concentration mode run in distilled water, Table 2. The periodic fluorescence intensity measurements of retentate reveal an increasing sorption of PAA-F1 as K is changing from 1 to 5, Figure 3A.

At the same time, no detectable PAA-F1 concentration was found in permeate. Therefore, an antiscalant sorption by membrane was estimated as the difference between the PAA-F1 total content and its real content in liquid phase. Location of PAA-F1 on membrane might provide an isolation of potential gypsum crystallization centers there. At the same time PAA-F1 "free" concentration remains at the level of 7 to 12 mg·dm^{-3}, which is capable to provide scale inhibition in the bulk aqueous phase. Meanwhile DLS reveals no notable content of PAA-F1 globules in aqueous phase.

3.2. Blank B Experiment Results

This experiment is intended to estimate possible PAA-F1 participation in its side reaction with excess of Ca^{2+} ions via formation of soluble Ca_nPAA-F1 complexes. Indeed, Figure 3B demonstrates some changes relative to Figure 3A. It exhibits that calcium ions do interact with PAA-F1 forming colloid solutions. This manifests in some decrease of fluorescence intensity relative to the Blank A experiment already at K = 1, and in arrival of a light scattering band, indicating formation of Ca_nPAA-F1 colloids with a mean size of c.a. 400 nm, Figure 4.

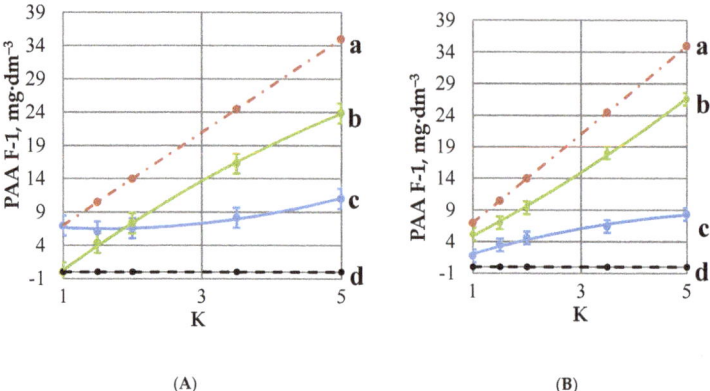

(A)   (B)

Figure 3. Variation of PAA-F1 content monitored by fluorescence intensity within a Blank A (**A**) and Blank B (**B**) experiment: Total content in retentate + membrane (a), its content on membrane, expressed in units of PAA-F1 concentration in retentate (b), in retentate (c), and in permeate (d).

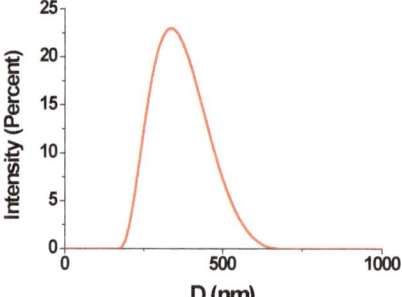

Figure 4. DLS particle size distribution by intensity in retentate in a Blank B experiment for $K = 1$.

Indeed, the individual Ca_nPAA-F1 aggregates are then detected on the membrane surface as bright green spheres with a size ranging from 10 to 20 µm, Figure 5.

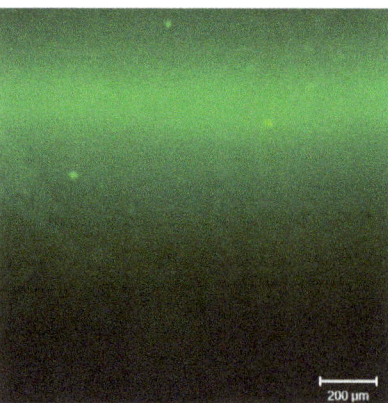

Figure 5. Fluorescent image of membrane surface deposit after Blank B experiment. Scale marker corresponds to 200 µm.

Notably, an excess of calcium ions relative to antiscalant, forces PAA-F1 to concentrate preferably in [Ca-PAA-F1] moieties, located both in a liquid phase and on a membrane surface. Meanwhile PAA-F1 "free" concentration remains at the level of 2 to 7 mg·dm^{-3}, which is still capable to provide scale inhibition in the bulk aqueous phase.

3.3. Blank C Experiment Results

This experiment has to demonstrate non-inhibited RO membrane gypsum scaling as a reference to the inhibited one. Figure 6 reveals a linear increase of Ca^{2+} content from 0.015 up to 0.025 mol·dm^{-3}. When CaSO$_4$·2H$_2$O saturation is achieved (K = 2), the Ca^{2+} concentration reaches the maximum. Then [Ca^{2+}] decreases due to the gypsum crystals deposition, and at the final moment (K = 5) [Ca^{2+}] corresponds almost to its initial level. At the same time 80% of calcium gets deposited as gypsum scale on membrane surface.

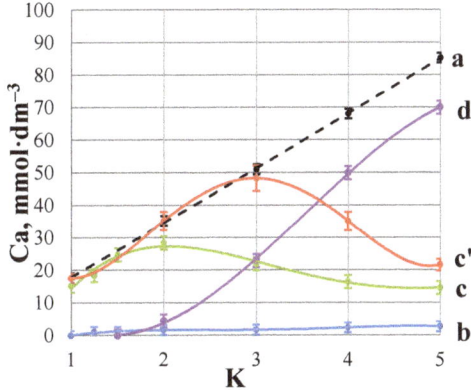

Figure 6. Variation of Ca^{2+} concentration in a Blank C (a,b,c,d) and in GSI experiment (c'): (a) total calcium concentration in retentate + membrane; (b) Ca^{2+} concentration in permeate; (c,c') "free" calcium in retentate; and (d) calcium content on membrane surface as gypsum, expressed in units of calcium concentration.

The corresponding scale on membrane surface fits well typical stick-shaped gypsum crystals morphology [6], Figure 7A,B. These images leave an impression that the scale is formed by crystal precipitation from the bulk, rather than by their initial formation on the membrane surface: several crystals lying on top of each other are clearly visible.

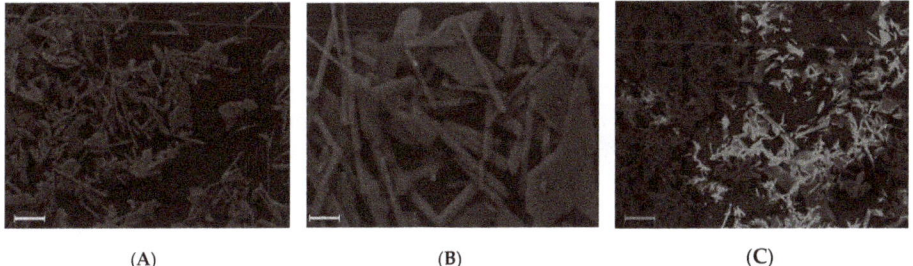

Figure 7. SEM images of gypsum deposit on membrane surface at the end of the Blank C (**A**,**B**) and GSI (**C**) experiment. Scale marker corresponds to 20 (**A**,**C**) and 5 (**B**) μm.

3.4. GSI Experiment Results

Figure 6 demonstrates a PAA-F1 inhibited RO membrane gypsum scaling (Figure 6c') relative to the un-inhibited one (Figure 6c). Like in a blank experiment, an increase of Ca^{2+} content from 15 mmol·dm^{-3} (K = 1) up to 48 mmol·dm^{-3} at K = 3 is observed (Figure 6c,c'). This corresponds to the gypsum saturation SI~2. Then [Ca^{2+}] decreases due to the gypsum crystals deposition, and at the final moment (K = 5) it corresponds almost to its initial level, Figure 6c'. A variation of calcium concentration with K in retentate for a Blank C and GSI experiments is nearly the same for K ≤ 1.5, e.g., before gypsum starts to form, Figure 6c,c'. When K > 1.5 the crystals of $CaSO_4 \cdot 2H_2O$ start to form in the Blank C experiment, while in presence of PAA-F1 this process starts at K>3, and the gypsum formation goes slower relative to the blank run. A significant shift of curve "c' " (GSI experiment) relative to curve "c" (blank experiment) clearly indicates that an effective inhibition takes place. For K = 3 PAA-F1 reveals c.a. 90% inhibition, and for K= 4 – c.a. 60%.

Fluorescent images of retentate (Figure 8) correspond well to the calcium content data, Figure 6c'. Indeed, there are no any crystals in the stock solution (K = 1) and at K = 2 saturation level, Figure 8A,B. At K = 3 gypsum deposition starts. The corresponding image (Figure 8C) indicates the landslide formation of numerous gypsum stick-like crystals with a mean size c.a. 10 to 20 μm. These are much smaller than those found later on the membrane surface after GSI experiment is finished, Figure 7C. Images (Figure 8C) leave no doubt that the major location of gypsum crystals formation is the bulk retentate solution, but not the membrane surface. Most of them have no any traces of antiscalant presence neither on their surface, nor inside of the crystals, Figure 8C–E. Meanwhile, the big bright green spherical solids with diameter ranging from 10 to 50 μm belong to the solid particles of pure [Ca-PAA-F1] complexes, which do not have any gypsum inclusions, Figure 8C. Bearing in mind that there are 360 g of gypsum per 1 g of PAA-F1, their size indicates that almost all antiscalant is concentrated in [Ca-PAA-F1] particles.

Indeed, if it is assumed that [Ca-PAA-F1] species form 100 nm size primary spherical particles, then each green sphere presented in Figure 8C corresponds to an aggregate of 10^5–10^7 such particles. Thus most of PAA-F1 and gypsum seem to form solids by itself with no interaction with each other. This observation is very similar to that one found by us previously for HEDP-F/gypsum system [22]. For K = 4 most of gypsum and of [Ca-PAA-F1] complexes are deposited on membrane surface. Therefore much less gypsum crystals remain in the bulk solution, Figure 8D. At K = 5 only a few gypsum crystals remain in the bulk retentate, while the rest are completely deposited on membrane, Figures 7C and 8E.

Notably, the size and shapes of gypsum crystals deposited in presence of PAA-F1 (Figure 7C) are similar to those, observed in a Blank C experiment, Figure 7A. This indicates that there is very little interaction of antiscalant with gypsum if any during its growth stage. Meanwhile, the size of $CaSO_4 \cdot 2H_2O$ crystals at the end of GSI experiment (Figure 7C) is at least twice bigger than of those formed in the bulk solution at K = 3 (Figure 8C). Thus, it is likely that, after fast formation in the bulk medium at K=3, the gypsum crystals pass sedimentation and proceed to grow already on membrane surface.

Although, there is no bulk crystal formation detected by fluorescent microscopy for K < 3 (Figure 8A,B), the DLS experiment, run in a parallel way, reveals an intensive formation and aggregation of colloids already at K = 1 and K = 2, Figure 9a,b.

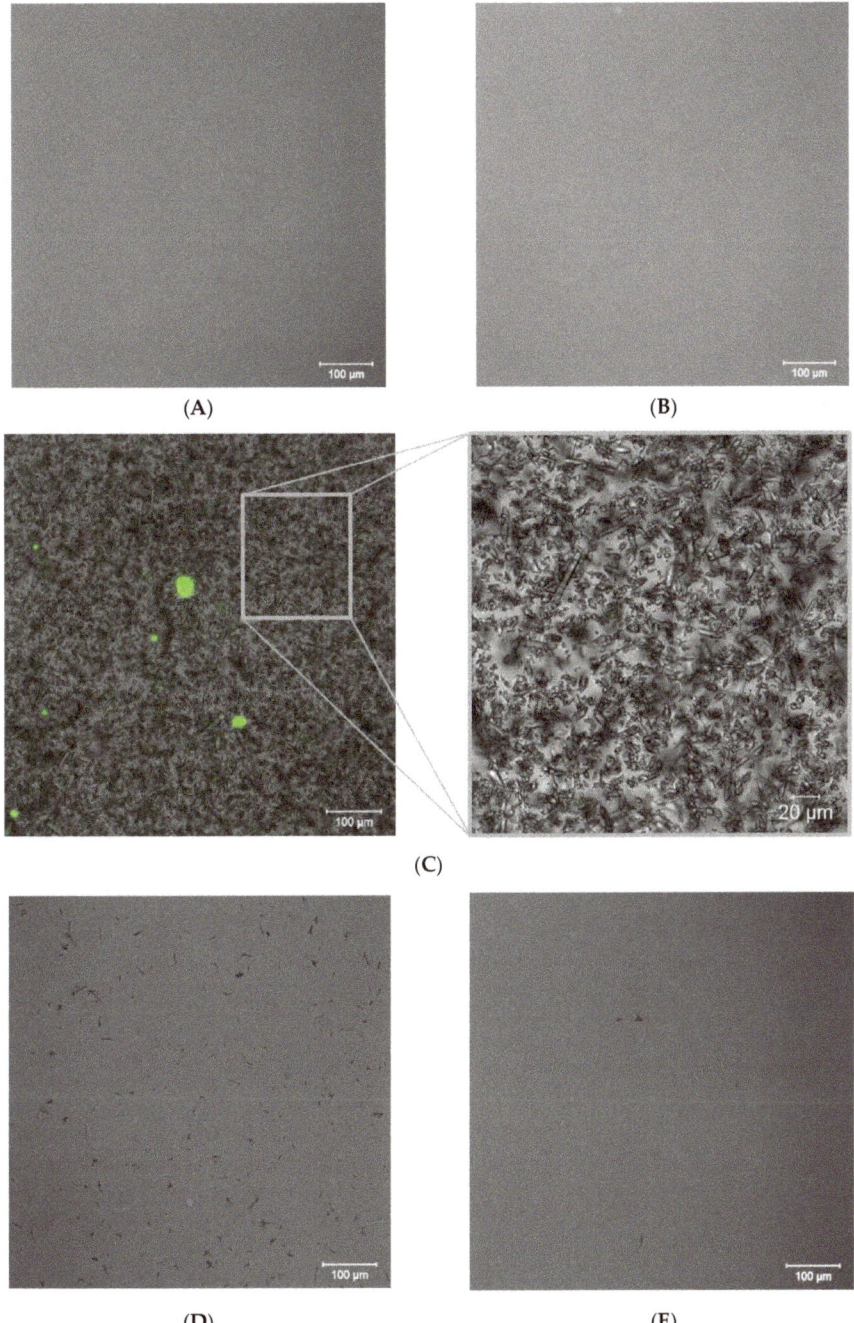

Figure 8. Fluorescent images of initial undersaturated gypsum solution droplets (**A**), K = 1, and of retentate at K = 2 (**B**), K = 3 (**C**), K = 4 (**D**), K = 5 (**E**) within the GSI experiment.

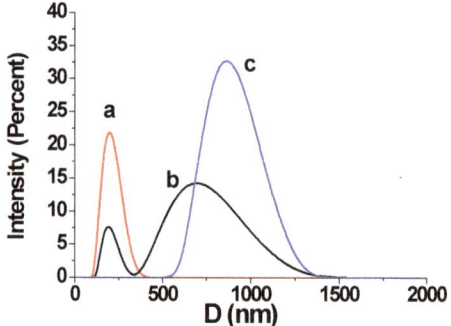

Figure 9. DLS particle size distribution by intensity in retentate of the GSI experiment for K = 1 (a), K = 2 (b) and K = 3 (c).

These data are unable to distinguish gypsum and [Ca-PAA-F1] particles, but they provide some additional information on what happens in a transparent retentate before visible gypsum crystals appear. However, DLS gives some independent approval of heterogeneous mechanism of gypsum particles in the bulk: it clearly indicates that $CaSO_4 \cdot 2H_2O$ and/or [Ca-PAA-F1] aggregates appear in retentate almost immediately after saturation starts. Indeed, according to the classical crystallization theory [23], this is possible only in the case, when gypsum passes bulk heterogeneous nucleation, and exactly the "nanodust" plays the role of the solid phase template.

It should be noted that PAA-F1 is more efficient than HEDP-F in a gypsum scale formation inhibition, reported in [22]. In a similar experiment, run under the same conditions, HEDP-F provides supersaturated gypsum solution stabilization only for $1 < K \leq 2$ [22], while PAA-F1 is effective for $1 < K \leq 3$. This result is in a good agreement with a sequence found earlier in the batch experiments for the non-fluorescent analogues HEDP and PAA: PAA>>HEDP [34].

3.5. Tentative Mechanism of Gypsum Membrane Fouling Inhibition by PAA-F1 in RO Process

PAA-F1 has definitely proved itself as an effective antiscalant in gypsum brine RO desalination, Figure 6. This was also confirmed earlier by the static experiment tests [36]. However, the visualization of PAA-F1 molecules indicates clearly that there is no definite interaction between antiscalant and gypsum along the brine RO treatment. The same result was obtained earlier for HEDP-F/gypsum RO desalination process [22] as well as for batch static experiments with gypsum [20] and barite [21] in presence of HEDP-F. A tentative mechanism of gypsum inhibition in RO membrane fouling is proposed [22] and our present data for PAA-F1 give a further approval to this hypothesis.

This mechanism involves interaction of foreign solid impurities ("nanodust"), which are always present in RO brines (Table 1), with antiscalant. In the absence of scale inhibitor the gypsum nucleation has a heterogeneous origin with solid foreign particles ("nanodust") serving as nucleation centers in the bulk retentate solution. Antiscalant molecules block these nucleation centers partly or completely, via sorption on their surface before retentate gets supersaturated relative to gypsum. Thus, when gypsum solution gets supersaturated, the potential sorption centers on the surface of "nanodust" particles become much less available for gypsum layers formation. This hampers and retards the process of scale formation.

Indeed, as it was mentioned earlier, there are at least 10^6 molecules of PAA-F1 (with a mean number of 50 monomer units, e.g., 4000 Da) per one nano/microdust particle in the system studied. A simple calculation indicates, that one PAA-F1 molecule is capable to cover 4 nm^2 of a particle surface, being completely stretched, Assuming that all nano/microdust particles have an equal size of 100 nm and an ideally spherical form, there are only $7.9 \cdot 10^3$ molecules of PAA-F1 needed to cover the whole single particle surface by a monolayer. An option to occupy only some active centers diminishes this number, while globular conformation of polymer molecule increases it. Evidently PAA-F1 is capable to cover all potential nucleation centers several times. Anyhow a supposition that it blocks a sufficient part of them is a quite realistic one.

On the other hand, although a high excess of PAA-F1 over nano/microdust particles surface area slows down gypsum scale formation, it does not stop this process. Therefore, both $CaSO_4 \cdot 2H_2O$ and [Ca-PAA-F1] phases are formed in a parallel way braking each other, as they compete for one and the same set of natural nucleation centers (colloid impurities) present in retentate. Notably, our conclusions derived from antiscalant visualization are perfectly supported by the independent DLS studies [38,39].

At the same time our results conflict somehow with the conclusions of numerous reports on static gypsum crystals formation in supersaturated aqueous solutions [24–33]. All these studies are built on the grounds of homogeneous nucleation scenario, excluding the possibility of "nanodust" presence. Meanwhile, the "nanodust" was surely present in these experiments, that all use Sigma-Aldrich high purity reagents (>98–99%), a double-deionised boiled water, and (in some cases) stock solutions filtration operating 200 nm filter. However, none of these solutions was then examined for the residual solid nanoparticles content. In order to make the situation clear, we have done a blank test, operating model Sigma-Aldrich KCl salt (ACS reagent, 99.0–100.5% CAS 7447-40-7) and a particle counter. Then KCl was dissolved in deionized water (340 particles bigger than 100 nm in 1 mL) to make 0.1 mol·dm^{-3} solution. This KCl solution revealed 268000 particles bigger than 100 nm in 1 mL. This solution was filtered with 220 nm filter and a "purified" solution demonstrated still 1540 particles bigger than 100 nm in 1 mL. A homogeneous scenario was unlikely to take place in [24–33] as an energy barrier for crystals nuclei formation is much lower for heterogeneous scenario, than for homogeneous one [23].

It should be noted that in an excellent study by Nicoleau, Van Driessche and Kellermeier [30] on static gypsum crystallization, in the presence of polyacrylate and of some other polymers, run on the other grounds, a conclusion was partly similar and partly alternative to ours. It was indicated that the polymers do not change the nature of the nucleating primary species, but rather modulate their subsequent growth and/or aggregation increasing the viscosity of the solution. However, the authors of this study [30] did not control "nanodust" content and could not monitor the polymer location. On the other hand, we did not control the viscosity. Our data could be a valuable supplement to the studies [24–33].

Our recent results [20–22], and the data of a present study, indicate the importance of natural background particles for scale inhibitors application strategies. As ppb impurities, these are always present in any ultrapure reagent or solvent, specially prepared for microelectronics, to say nothing of technical grade purity reagents and brackish water commonly used in RO technologies. The particular chemical nature of this "nanodust" is a challenge for researchers, as far as it is hardly possible to isolate them completely from a liquid phase. In our opinion, these solid impurities are chemically non-uniform, and their different ingredients have different affinity towards scale material and antiscalants. Their composition may vary between water samples. At present, only a rough and incomplete estimation of its chemical composition and particle size distribution is feasible. However, even the treatment of "nanodust" as a "blackbox" may become very fruitful.

At the same time, the fluorescent antiscalants may become a promising tool in scale formation studies. This method has very high sensitivity and is widely used in analytical applications [35] as well as a powerful traceability approach in medicine [40]. Normally, the method sensitivity corresponds to the ppb level for both solid and liquid samples. For example, the detection limit of Rhodamine is in the range of 0.01 ppb in distilled water (25 mm cuvette) [35], and fluorescence quantum yield of our fluorescent inhibitor is quite close to Rhodamine [35,36]. Thus, localization of fluorescent inhibitor upon crystals/particles of scale is a valid approach. On the other hand, an absence of fluorescence is a clear indication that the fluorescent inhibitor is not present in the solution but either forms self-aggregates or participates in the crystal formation.

4. Conclusions

Visualization of fluorescent-tagged antiscalants provides a deeper insight of antiscaling mechanisms in reverse osmosis facility desalination. A case study of gypsum scale formation revealed a nonconventional mechanism of antiscalant efficacy. Scaling supposedly takes place in the bulk retentate phase via heterogeneous nucleation step. The "nanodust" particles play a key role as gypsum nucleation centers. It is demonstrated that contrary to popular belief an antiscalant interacts not with gypsum nuclei, but with "nanodust" particles, isolating them from calcium and sulfate ions sorption. Therefore, the number of gypsum nucleation centers are reduced, and in turn, the overall scaling rate is diminished.

At the same time, significant amounts of work is still necessary, in order to understand how much the case of gypsum is a universal one. Therefore, a study of fluorescent-tagged polyacrylates and phosphonates in RO desalination of carbonate brines is underway now.

Author Contributions: Conceptualization, K.P.; methodology; writing—original draft, supervision, V.G.; Investigation (all RO membrane operations); Methodology; Writing—Original Draft (Figures preparation); M.O., Investigation (SEM; calcium concentration measurements, water ICP analysis); writing—review and editing; resources (PAA-F1 synthesis); S.T., Investigation (fluorescence intensity measurements), writing—reviewing and Editing; A.R. (Anatoly Redchuk), Investigation (particle counter measurements; polymer conformation modeling); A.R. (Anastasia Ryabova): Investigation (confocal fluorescent microscope measurements); A.P.: Conceptualization, Supervision, Writing—Review & Editing. All authors have read and agreed to the published version of the manuscript.

Funding: The authors would like to thank Russian Science Foundation, Project No. 19-79-10220 for the financial support of the present study.

Acknowledgments: This work was carried out with the using equipment of the Center for collective use No. 74834 "Technological and diagnostic center for the production, research and certification of micro and nanostructures" in GPI RAS.

Conflicts of Interest: The authors declare no conflict of interest.

References

1. Anis, S.F.; Hashaikeh, R.; Hilal, N. Reverse osmosis pretreatment technologies and future trends: A comprehensive review. *Desalination* **2019**, *452*, 159–195. [CrossRef]
2. Badruzzaman, M.; Voutchkov, N.; Weinrich, L.; Jacangelo, J.G. Selection of pretreatment technologies for seawater reverse osmosis plants: A review. *Desalination* **2019**, *449*, 78–91. [CrossRef]
3. Zhang, P.; Hu, J.; Li, W.; Qi, H. Research Progress of brackish water desalination by reverse osmosis. *J. Water Resour. Prot.* **2013**, *5*, 304–309. [CrossRef]
4. Li, X.; Hasson, D.; Semiat, R.; Shemer, H. Intermediate concentrate demineralization techniques for enhanced brackish water reverse osmosis water recovery—A review. *Desalination* **2019**, *466*, 24–35. [CrossRef]
5. Goh, P.S.; Lau, W.J.; Othman, M.H.D.; Ismail, A.F. Membrane fouling in desalination and its mitigation strategies. *Desalination* **2018**, *425*, 130–155. [CrossRef]
6. Matin, A.; Rahman, F.; Shafi, H.Z.; Zubair, S.M. Scaling of reverse osmosis membranes used in water desalination: Phenomena, impact, and control; future directions. *Desalination* **2019**, *455*, 135–157. [CrossRef]
7. Demadis, K.D.; Neofotistou, E.; Mavredaki, E.; Tsiknakis, M.; Sarigiannidou, E.-M.; Katarachia, S.D. Inorganic foulants in membrane systems: Chemical control strategies and the contribution of green chemistry. *Desalination* **2005**, *179*, 281–295. [CrossRef]
8. Rabizadeh, T.; Morgan, D.J.; Peacock, C.L.; Benning, L.G. Effectiveness of Green Additives vs. Poly(acrylic acid) in Inhibiting Calcium Sulfate Dihydrate Crystallization. *Ind. Eng. Chem. Res.* **2019**, *58*, 1561–1569. [CrossRef]
9. Pramanik, B.K.; Gao, Y.; Fan, L.; Roddick, F.A.; Liu, Z. Antiscaling effect of polyaspartic acid and its derivative for RO membranes used for saline wastewater and brackish water desalination. *Desalination* **2017**, *404*, 224–229. [CrossRef]
10. Ali, S.A.; Kazi, I.W.; Rahman, F. Synthesis and evaluation of phosphate-free antiscalants to control $CaSO_4 \cdot 2H_2O$ scale formation in reverse osmosis desalination plants. *Desalination* **2015**, *357*, 36–44. [CrossRef]

11. Rahman, F. Calcium sulfate precipitation studies with scale inhibitors for reverse osmosis desalination. *Desalination* **2013**, *319*, 79–84. [CrossRef]
12. Abdel-Aal, E.A.; Abdel-Ghafar, H.M.; El Anadouli, B.E. New Findings about Nucleation and Crystal Growth of Reverse Osmosis Desalination Scales with and without Inhibitor. *Cryst. Growth Des.* **2015**, *15*, 5133–5137. [CrossRef]
13. Ying, W.; Siebdrath, N.; Uhl, W.; Gitis, V.; Herzberg, M. New insights on early stages of RO membranes fouling during tertiary wastewater desalination. *J. Membr. Sci.* **2014**, *466*, 26–35. [CrossRef]
14. Shmulevsky, M.; Li, X.; Shemer, H.; Hasson, D.; Semiat, R. Analysis of the onset of calcium sulfate scaling on RO membranes. *J. Membr. Sci.* **2017**, *524*, 299–304. [CrossRef]
15. Benecke, J.; Haas, M.; Baur, F.; Ernst, M. Investigating the development and reproducibility of heterogeneous gypsum scaling on reverse osmosis membranes using real-time membrane surface imaging. *Desalination* **2018**, *428*, 161–171. [CrossRef]
16. Cai, Y.; Schwartz, D.K. Single-nanoparticle tracking reveals mechanisms of membrane fouling. *J. Membr. Sci.* **2018**, *563*, 888–895. [CrossRef]
17. Kim, H.; Park, S.; Choi, Y.; Lee, S.; Choi, J. Fouling due to $CaSO_4$ scale formation in forward osmosis (FO), reverse osmosis (RO), and pressure assisted forward osmosis (PAFO). *Desalin. Water Treat.* **2018**, *104*, 45–50. [CrossRef]
18. Benecke, J.; Rozova, J.; Ernst, M. Anti-scale effects of select organic macromolecules on gypsum bulk and surface crystallization during reverse osmosis desalination. *Sep. Purif. Technol.* **2018**, *198*, 68–78. [CrossRef]
19. Hoang, T.A. Mechanisms of Scale Formation and Inhibition. In *Mineral Scales and Deposits, Scientific and Technological Approaches*, 1st ed.; Amjad, Z., Demadis, K., Eds.; Elsevier: Amsterdam, The Netherlands, 2015; pp. 47–83. [CrossRef]
20. Oshchepkov, M.; Kamagurov, S.; Tkachenko, S.; Ryabova, A.; Popov, K. An Insight into the Mechanisms of the Scale Inhibition. A Case Study of a Novel Task-specific Fluorescent-tagged Scale Inhibitor Location on Gypsum Crystals. *ChemNanoMat* **2019**, *5*, 586–592. [CrossRef]
21. Oshchepkov, M.; Popov, K.; Ryabova, A.; Redchuk, A.; Tkachenko, S.; Dikareva, J.; Koltinova, E. Barite Crystallization in Presence of Novel Fluorescent-tagged Antiscalants. *Int. J. Corros. Scale Inhib.* **2019**, *8*, 998–1021. [CrossRef]
22. Oshchepkov, M.; Golovesov, V.; Ryabova, A.; Tkachenko, S.; Redchuk, A.; Rudakova, G.; Pervov, A.; Rönkkömäki, H.; Popov, K. Visualization of a novel fluorescent-tagged bisphosphonate behavior during reverse osmosis desalination of water with high sulfate content. *Sep. Purif. Technol.* **2020**, in press.
23. Sosso, G.C.; Chen, J.; Cox, S.J.; Fitzner, M.; Pedevilla, P.; Zen, A.; Michaelides, A. Crystal nucleation in liquids: Open questions and future challenges in molecular dynamics simulations. *Chem. Rev.* **2016**, *116*, 7078–7116. [CrossRef] [PubMed]
24. He, S.; Oddo, J.E.; Tomson, M.B. The nucleation kinetics of calcium sulfate dehydrate in NaCl solution up to 6 m and 90 °C. *J. Coll. Interface Sci.* **1994**, *162*, 297–303. [CrossRef]
25. Klepetsanis, P.G.; Dalas, E.; Koutsoukos, P.G. Role of temperature in the spontaneous precipitation of calcium sulfate dehydrate. *Langmuir* **1999**, *15*, 1534–1540. [CrossRef]
26. Lancia, A.; Musmarra, D.; Prisciandaro, M. Measuring induction period for calcium sulfate dehydrate precipitation. *AICHE J.* **1999**, *45*, 390–397. [CrossRef]
27. Prisciandaro, M.; Lancia, A.; Musmarra, D. Gypsum nucleation into sodium chloride solutions. *AICHE J.* **2001**, *47*, 929–934. [CrossRef]
28. Alimi, F.; Elfil, H.; Gadri, A. Kinetics of the precipitation of calcium sulfate dihydrate in a desalination unit. *Desalination* **2003**, *57*, 9–16. [CrossRef]
29. Van Driessche, A.E.S.; Stawski, T.M.; Kellermeier, M. Calcium sulfate precipitation pathways in natural and engineered environments. *Chem. Geol.* **2019**, *530*, 119274. [CrossRef]
30. Nicoleau, L.; Van Driessche, A.E.; Kellermeier, M. Kinetic analysis of the role of polymers in mineral nucleation. The example of gypsum. *Cem. Concr. Res.* **2019**, *124*, 105837. [CrossRef]
31. Van Driessche, A.E.; García-Ruiz, J.M.; Delgado-López, J.M.; Sazaki, G. In Situ Observation of Step Dynamics on Gypsum Crystals. *Cryst. Growth Des.* **2010**, *10*, 3909–3916. [CrossRef]
32. Stawski, T.M.; Van Driessche, A.E.; Ossorio, M.; Rodriguez-Blanco, J.D.; Besselink, R.; Benning, L.G. Formation of calcium sulfate through the aggregation of sub-3 nanometre primary species. *Nat. Commun.* **2016**, *7*, 11177. [CrossRef] [PubMed]

33. Stawski, T.M.; Van Driessche, A.E.; Besselink, R.; Byrne, E.H.; Raiteri, P.; Gale, J.D.; Benning, L.G. The Structure of CaSO$_4$ Nanorods: The Precursor of Gypsum. *J. Phys. Chemi. C* **2019**, *123*, 23151–23158. [CrossRef]
34. Popov, K.; Rudakova, G.; Larchenko, V.; Tusheva, M.; Kamagurov, S.; Dikareva, J.; Kovaleva, N.A. Comparative Performance Evaluation of Some Novel "Green" and Traditional Antiscalants in Calcium Sulfate Scaling. *Adv. Mat. Sci. Eng.* **2016**, *2016*, 7635329. [CrossRef]
35. Oshchepkov, M.; Tkachenko, S.; Popov, K. Synthesis and applications of fluorescent-tagged scale inhibitors in water treatment. A review. *Int. J. Corros. Scale Inhib.* **2019**, *8*, 480–511. [CrossRef]
36. Popov, K.; Oshchepkov, M.; Kamagurov, S.; Tkachenko, S.; Dikareva, Y.; Rudakova, G. Synthesis and properties of novel fluorescent-tagged polyacrylate-based scale inhibitors. *J. Appl. Polym. Sci.* **2017**, *134*, 45017. [CrossRef]
37. Raju, K.; Atkinson, G.J. The thermodynamics of «scale» mineral solubilities. 3. Calcium sulfate in aqueous sodium chloride. *Chem. Eng. Data* **1990**, *35*, 361–367. [CrossRef]
38. Popov, K.I.; Oshchepkov, M.S.; Shabanova, N.A.; Dikareva, Y.M.; Larchenko, V.E.; Koltinova, E.Y. DLS study of a phosphonate induced gypsum scale inhibition mechanism using indifferent nanodispersions as the standards of a light scattering intensity comparison. *Int. J. Corros. Scale Inhib.* **2018**, *7*, 9–24. [CrossRef]
39. Popov, K.; Oshchepkov, M.; Afanas'eva, E.; Koltinova, E.; Dikareva, Y.; Rönkkömäki, H. A new insight into the mechanism of the scale inhibition: DLS study of gypsum nucleation in presence of phosphonates using nanosilver dispersion as an internal light scattering intensity reference. *Colloids Surf. A* **2019**, *560*, 122–129. [CrossRef]
40. Oshchepkov, A.; Oshchepkov, M.; Pavlova, G.; Ryabova, A.; Kamagurov, S.; Tkachenko, S.; Frolova, S.; Redchuk, A.; Popov, K.; Kataev, E. Naphthalimide-functionalized Bisphosphonates for Fluorescence Detection of Calcification in Soft Tissues. *Sens. Actuators B Chem.* **2020**, 128047, in press. [CrossRef]

© 2020 by the authors. Licensee MDPI, Basel, Switzerland. This article is an open access article distributed under the terms and conditions of the Creative Commons Attribution (CC BY) license (http://creativecommons.org/licenses/by/4.0/).

Article

Purification of Curcumin from Ternary Extract-Similar Mixtures of Curcuminoids in a Single Crystallization Step

Elena Horosanskaia [1], Lina Yuan [1], Andreas Seidel-Morgenstern [1,2] and Heike Lorenz [1,*]

[1] Max Planck Institute for Dynamics of Complex Technical Systems, Sandtorstrasse 1, 39106 Magdeburg, Germany; horosanskaia@mpi-magdeburg.mpg.de (E.H.); lina.yuan@novartis.com (L.Y.); anseidel@ovgu.de (A.S.-M.)

[2] Otto von Guericke University, Institute of Process Engineering, Universitätsplatz 2, 39106 Magdeburg, Germany

* Correspondence: lorenz@mpi-magdeburg.mpg.de

Received: 21 February 2020; Accepted: 14 March 2020; Published: 16 March 2020

Abstract: Crystallization-based separation of curcumin from ternary mixtures of curcuminoids having compositions comparable to commercial extracts was studied experimentally. Based on solubility and supersolubility data of both, pure curcumin and curcumin in presence of the two major impurities demethoxycurcumin (DMC) and bis(demethoxy)curcumin (BDMC), seeded cooling crystallization procedures were derived using acetone, acetonitrile and 50/50 (wt/wt) mixtures of acetone/2-propanol and acetone/acetonitrile as solvents. Starting from initial curcumin contents of 67–75% in the curcuminoid mixtures single step crystallization processes provided crystalline curcumin free of BDMC at residual DMC contents of 0.6–9.9%. Curcumin at highest purity of 99.4% was obtained from a 50/50 (wt/wt) acetone/2-propanol solution in a single crystallization step. It is demonstrated that the total product yield can be significantly enhanced via addition of water, 2-propanol and acetonitrile as anti-solvents at the end of a cooling crystallization process.

Keywords: curcumin; purification; ternary mixture of curcuminoids; crystallization

1. Introduction

Curcumin (abbreviated CUR), known as diferuloyl methane, is an intense orange-yellow solid and a natural ingredient of the plant rhizome of *Curcuma Longa* L. Two derivatives of CUR, demethoxycurcumin (abbreviated DMC) and bis(demethoxy)curcumin (abbreviated BDMC), can be found in the plant as well. Altogether they are known as curcuminoids (abbreviated CURD). Depending on the soil condition, the total content of CURDs in the plant rhizome varies between 2 and 9%. With approximately 70% of the total CURD content CUR represents the major component in turmeric [1–3]. As highlighted in Figure 1, the presence or absence of a methoxy functional group on o-position to a phenolic group represents the only difference in the chemical structure of the three CURDs. The molecular structure of CUR comprising two equally substituted aromatic rings linked together by a diketo group, which exhibits keto-enol tautomerism, plays a crucial role in the reactivity of CUR [4,5].

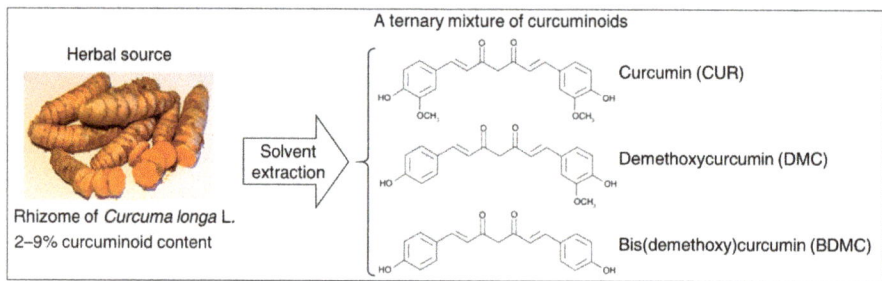

Figure 1. Curcuminoids extracted from the rhizome of Turmeric (*Curcuma longa* L.) and molecular structures of the three major constituents (curcumin (CUR), demethoxycurcumin (DMC) and bis(demethoxy)curcumin (BDMC)).

Studies show that CUR can be potentially used to treat over 25 diseases due to its anti-oxidative, immunosuppressive, wound-healing, anti-inflammatory and phototoxic effects [6–8]. These include, in particular, neurodegenerative diseases, such as Alzheimer's and Parkinson's diseases, diabetes, heart sickness, bacterial, viral and fungal diseases, AIDS and over 20 different cancers [9–12]. In addition to CUR, also the potential use of DMC and BDMC in the prevention of cancer was emphasized [13–15]. It was reported that DMC has the stronger effect on the inhibition of human breast tumor cells, followed by CUR and BDMC [16]. Ruby et al. described the higher bioavailability and cytotoxic activity of BDMC in animal cells [17].

Due to the higher reactivity of CUR associated with the stronger pharmacological activity on the human body comparable to the two other derivatives, CUR currently remains the targeted turmeric compound [18]. Despite the diverse pharmacological effects, the practical insolubility of CUR in water results in a very low bioavailability of the molecule and therewith leads to a limited usage as a drug [19]. To improve the bioavailability, formulation of curcumin nanoparticles or metal complexes were successfully implemented [20,21]. In addition, the application of CUR together with artemisinin in a CUR-artemisinin combination therapy against malaria was reported to decrease the drug resistance [22]. Moreover, the formulation of a CUR-artemisinin co-amorphous solid showed a higher therapeutic effect in the treatment of cancer than the single drug formulation [23]. For each of the application, CUR has to be available in chemically pure form and in sufficient amount.

H.J.J. Pabon described the preparation of synthetic CUR and related compounds [24]. Kim et al. recently published a process for production of CURDs in engineered Escherichia coli [25]. Nevertheless, the separation of CUR by means of solvent extraction from the plant rhizome still represents the most economical way of CUR production. In addition to plant proteins, oils and fats, the final extract contains 80% of the ternary CURD mixture [26]. In this mixture CUR is the major component with approximately 64% share of the total CURD content, together with 21% DMC and 15% BDMC [27]. Commercially available mixture usually contains 77% CUR, 17% DMC and 6% BDMC [28]. Consequently, CUR has to be purified from the ternary mixture.

There are two methods for separation of CUR from the mixture of CURDs described in the literature: by means of column or thin layer chromatography and by crystallization from solution.

For the chromatographic separation of CUR, silica gel (untreated or impregnated with sodium hydrogen phosphate) is commonly used as a stationary phase and various binary solvent mixtures of dichloromethane, chloroform, methanol, acetic acid, ethyl acetate and hexane as the mobile phase [29]. At the end of the process, three chromatographic fractions are enriched with the three CURDs, respectively [30,31]. Usually crystallization is applied as the final formulation step providing the solid product with desired specifications.

In the last decade, crystallization as a single separation technique was studied to purify CUR from the ternary mixture of curcuminoids [32–34]. Processes were described exploiting anti-solvent addition or system cooling, using methanol, ethanol and 2-propanol as process solvents and water as anti-solvent (Table 1).

Table 1. Overview of the results of published studies on CUR purification via crystallization: References 1–3 relate to [32–34], respectively.

Reference	Raw Mixture Content of		Solvent	Crystallization Method	No. of Crystallization Steps	Product Content of		Total Yield %
	CUR %	DMC %				CUR %	DMC %	
1	/[1]	/[1]	Methanol	Anti-solvent addition, water	3	92.2	7.8	40
2	82.0	16.0	Ethanol	Cooling, 70 °C to 5 °C	2	96.0	4.0	/[1]
3	78.6	17.7	2-Propanol	Cooling, 60 °C to 20 °C	3	>98/ 99.1[2]	<2/ 0.9[2]	50[2]

[1] not specified; [2] optimized crystallization conditions.

As summarized in Table 1, from initial CURD mixtures crystalline CUR with purities of 92.2%, 96.0% and 99.1% at overall yields between 40 and 50% were obtained. The used separation methods were implemented as multi-step processes consisting of at least two successive sub-steps. It is reported that the main part of BDMC could be depleted after the first separation step, full removal was achieved after the second crystallization step [33,34]. DMC was always present in the final product. Ukrainczyk et al. observed an exponential decrease of the removal efficiency of DMC with increasing number of successive crystallization steps [34].

In order to reach the desired product purity and also to improve the overall process yield, a combination of the two separation techniques, chromatography and crystallization, was recently studied. Horvath et al. successfully implemented this integrated process for recovery of 99.1% pure artemisinin from an effluent of a photocatalytic reactor with 61.5% yield [35]. Heffernan et al. demonstrated the purification of single CURDs from the crude curcumin extract. There, the firstly performed crystallization process comprised three crystallization cycles, which provided 99.1% pure CUR in the final crystalline product. In the second process step, the remaining mother liquor was processed by column chromatography to isolate DMC and BDMC with purities of 98.3% and 98.6% and yields of 79.7% and 68.8%, respectively [36].

As has been demonstrated for other natural product mixtures, crystallization is a powerful technique to isolate a target compound from a multicomponent mixture within a single crystallization step [37,38]. Due to the fact that a 98% minimum purity of CUR is already sufficient for further drug application in pharmaceutical preparations [22], this study is directed to develop a separation process for isolation of pure crystalline CUR from the ternary mixture of CURDs within a single crystallization step.

To separate a target compound from a multi-component mixture, seeded cooling crystallization is preferably applied. Anti-solvent is usually added either at the beginning of the cooling step to generate the supersaturation in the solution or at the end of the process to increase the overall crystallization yield [39].

To purify CUR from the crude CURD mixture, seeded cooling crystallization processes were designed on the basis of solubility and nucleation measurements of pure CUR and CUR in presence of the CURDs mixture components in acetone, acetonitrile, ethanol, methanol, 2-propanol and selected binary mixtures thereof. Finally, with respect to the solubility results, 2-propanol, acetonitrile and water were considered as anti-solvents to improve the overall process yield.

2. Materials and Methods

2.1. Materials

Solid standards of curcumin, demethoxycurcumin (both >98%, TCI Chemicals) and bis(demethoxy)curcumin (>99%, ChemFaces China) were used as standards for HPLC and X-ray powder diffraction (XRPD) analysis. The solid standard of curcumin was also used to determine the solubility and nucleation behaviors. During the study, four crude solid mixtures of CURDs were purchased from Sigma Aldrich and Acros. The content of CUR, DMC and BDMC in the solids, determined by means of HPLC, is summarized in Table 2.

Table 2. Comparison of the crude solids purchased from Sigma Aldrich (crude solids No. 1–3) and Acros (crude solid no. 4), each representing a ternary mixture of the three CURDs.

Crude Solid No.	CUR Content wt%	DMC Content wt%	BDMC Content wt%
1	67.2	25.5	7.3
2	70.8	23.5	5.7
3	75.0	19.2	5.8
4	80.7	16.5	2.8

The highest CUR content of 80.7% was found in the crude solid obtained from Acros. The CUR content in the crude solids from Sigma Aldrich varies between 67.2% and 75.0% depending on the purchased charge, but is most similar to that of plant extract [28]. Accordingly, the solids from Sigma Aldrich were used as crude mixture for crystallization experiments. It should be emphasized that the analyzed significant differences of the CUR content in the three solid charges made the implementation of the designed crystallization process more challenging.

Acetone, acetonitrile, ethanol, methanol and 2-propanol (>99.8%, HiPerSolv CHROMANORM, VWR Chemicals, Germany) were used for solubility studies and for the crystallization experiments.

2.2. Analytical Methods

An analytical HPLC unit (Agilent 1200 Series, Agilent Technologies Germany GmbH) was used to characterize the solid standards, to quantify the CUR, DMC and BDMC contents in the crude mixtures as well as in the final crystallization products. The reversed phase method reported by Jadhav et al. [40] was adjusted as follows: the mobile phase composition was fixed to 50/50 (vol/vol) acetonitrile/0.1% acetic acid in water. Before usage water was purified via Milli-Q Advantage devices (Merck Millipore). The eluent flow-rate was set to 1 mL/min. Solid samples preliminarily dissolved in acetonitrile were injected (injection volume 1 µL) in the column (LUNA C18, 250 × 4.6 mm, 10 µm, Phenomenex GmbH, Germany, column temperature 25 °C) and analyzed at a wavelength of 254 nm. Figure 2 shows chromatograms of the solid standards of BDMC, DMC and CUR compared to a ternary mixture of CURDs (exemplarily crude solid No. 3).

X-ray powder diffraction (XRPD) was applied to characterize the purchased solid standards, solid fractions obtained during the solubility studies and the crystallization products. For the measurements, solid samples were ground in a mortar and prepared on background-free Si single crystal sample holders. Data were collected on an X'Pert Pro diffractometer (PANalytical GmbH, Germany) using Cu-Kα radiation. Samples were scanned in a 2Theta range of 4 to 30° with a step size of 0.017° and a counting time of 50 s per step.

2.3. Solubility and Metastable Zone Width Measurements

Solubility investigations of pure CUR in acetone, acetonitrile, ethanol, methanol and 2-propanol were carried out via the classical isothermal method [41]. To evaluate the impact of the main impurities (DMC and BDMC) on the solubility behavior of CUR, the crude mixture of CURDs no. 2 was used in selected process solvents. Suspensions containing excess of solid CUR and 5 mL solvent were introduced in glass vials. To guarantee efficient mixing of the prepared suspensions, vials were equipped with a magnetic stirrer and sealed. Samples were placed in a thermostatic bath and allowed to equilibrate at constant temperatures between 5 and 70 °C for at least 48 h under stirring. Afterwards, samples of equilibrated slurries were withdrawn with a syringe and filtered through a 0.45 μm PTFE filter. Obtained liquid phases were analyzed for solute content by HPLC. To preserve equilibrium conditions for low temperature samples, syringes and filters were precooled before usage. The corresponding wet solid fractions were characterized by XRPD.

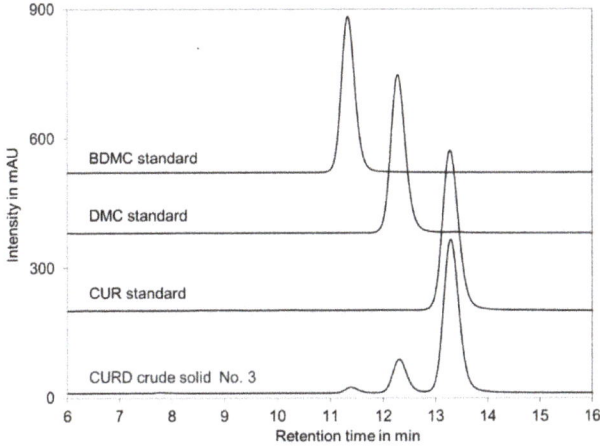

Figure 2. Analytical HPLC chromatograms of solid standards of BDMC, DMC, CUR and of the crude mixture No. 3 (see Table 1).

Metastable zone width data of pure CUR in selected process solvents were acquired by means of the multiple reactor system Crystal16™ (Avantium Technologies BV, Amsterdam). Suspensions containing known excess amount of solid in solvent were prepared in standard HPLC glass vials, equipped with magnetic stirrers and subjected to a heating step from 5 to 60 °C and a subsequent cooling step from 60 to −15 °C, both at a moderate rate of 0.1 °C/min. Temperatures of a "clear" and "cloud" point representing the respective saturation and nucleation temperatures were obtained via turbidity measurement.

Batch crystallization experiments were conducted in a jacketed 200 mL glass vessel equipped with a Pt-100 resistance thermometer (resolution 0.01 °C) connected to a thermostat (RP845, Lauda Proline, Germany) to control the system temperature. A magnetic stirrer was used for agitation.

With respect to the determined solubility behavior of CUR, four process solvents were selected. Consequently, four cooling crystallization processes were derived and conducted. Table 3 gives an overview of the chosen process solvents and the CURD mixtures to be separated. Exact solution composition data (Table 5) and the applied crystallization procedures are presented and discussed in connection with crystallization process design in Sections 3.2 and 3.3.

Table 3. Overview of the selected process solvents and the corresponding crude solids.

Process No.	Process Solvent	Crude Solid No.
1	Acetone	3
2	50/50 (wt/wt) acetone/2-propanol	2
3	50/50 (wt/wt) acetone/acetonitrile	1
4	Acetonitrile	1

3. Results and Discussion

3.1. Selection of Solvents for Crystallization

To design a crystallization-based purification process, the selection of an appropriate solvent is crucial. The operation parameters for the crystallization process are established based on the specific solubility and nucleation behavior of the target compound in the corresponding solvent.

3.1.1. CUR Solubility in Acetone, Acetonitrile, Methanol, Ethanol and 2-propanol

Acetone, acetonitrile, methanol, ethanol and 2-propanol were selected as possible process solvents because of their low toxicity. CUR solubilities determined in these solvents are shown in Figure 3. As seen CUR solubilities increase with increasing temperature in all solvents. Compared to acetone, CUR is significantly less soluble in the other solvents (less than 1 wt%, except in acetonitrile at 40 °C). Hence, acetone was chosen as a suitable solvent for seeded cooling crystallization and acetonitrile, methanol, ethanol and 2-propanol were considered as potential anti-solvents. According to the published very poor solubility of CUR in water (approx. 1.3×10^{-7} wt% at 25 °C) water was also taken into account as anti-solvent without extra solubility studies [20].

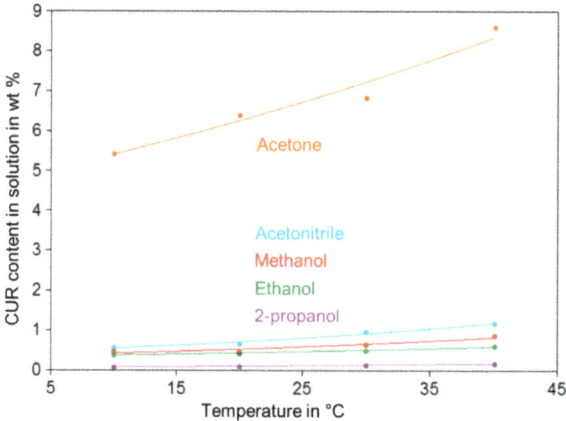

Figure 3. Solubility behavior of CUR in acetone, acetonitrile, methanol, ethanol and 2-propanol. Symbols represent experimental data, fitted curves just serve as guide to the eyes.

In Figure 4, CUR solid phase XRPD patterns are shown obtained from isothermal equilibration of CUR suspensions (Figure 4a), and by (polythermal) cooling of saturated CUR/solvent mixtures (Figure 4b–f). Measured patterns are compared with references for the three CUR polymorphs derived from single crystal data given in the Cambridge Structural Database (CSD) [42].

Commercial solid standard of CUR, which represents the initial solid for isothermal solubility studies, and all CUR solid phases obtained in equilibrium with saturated solutions in the solvents studied (Figure 4a) perfectly match the pattern of the known CUR polymorph I. XRPD patterns obtained for CUR recrystallized polythermally from acetone and acetonitrile solutions (Figure 4b,c) can be assigned to CUR I as well. CUR phases obtained by cooling of saturated methanol, ethanol and 2-propanol solutions (Figure 4d–f) do not match any shown reference phase, but (except a small missing reflex at 6.8° in the ethanol pattern) are identical to each other. Aside from that, their XRPD patterns differ from the CUR I phase only by some additional reflexes in the 2Theta range of 6°–8°. One hypothesis explaining this behavior might be incorporation of small amounts of respective alcohol molecules in the crystal structure without changing the structure type. Further, according to the known complex solid phase behavior of CUR [42–48] and BDMC [49,50], also the formation of a new metastable form of CUR in the three alcohols or a solvate phase from ethanol are possible explanations. Since elucidation of the CUR phase behavior was not the main focus of the present study, this issue has to be verified in future investigations.

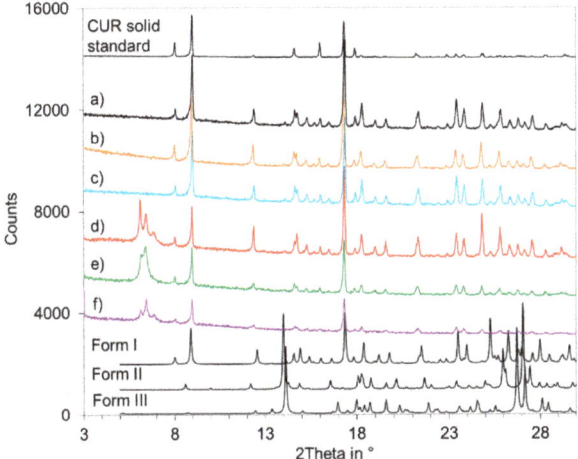

Figure 4. X-ray powder diffraction (XRPD) patterns of CUR crystalline phases (**a**) obtained from isothermal equilibration of CUR suspensions, and (**b–f**) recrystallized by cooling of saturated CUR solutions in acetone (**b**), acetonitrile (**c**), methanol (**d**), ethanol (**e**), 2-propanol (**f**). The topmost diffractogram refers to the CUR solid standard. The three lowermost diffractograms specify the reference crystal structures of CUR polymorphs I-III simulated from CSD single crystal data [42].

With the aim to selectively crystallize pure CUR (form I) from the crude CURD solution and to suppress spontaneous nucleation of undesired DMC and BDMC components, seeding with CUR solid standard (form I) was applied in cooling crystallization experiments.

To evaluate the anti-solvents effect on the CUR solubility in acetone, saturation concentrations of CUR (solid standard) were measured at 30 °C in the 50/50 (wt/wt) acetone/anti-solvent mixtures, exemplarily. Figure 5 shows that the obtained solubility data of CUR in the four binary solvent mixtures deviate from the ideal linear behaviors. Moreover, it is seen that the addition of methanol, ethanol and 2-propanol induces a dilution effect rather than the expected supersaturation of the solution. Since the relative dilution effect of ethanol and methanol is larger than that of 2-propanol, they are not considered further for crystallization process design. In contrast, the addition of acetonitrile increases the supersaturation of CUR in acetone. Therefore, a high product yield can be expected. Consequently, the following four process solvents were selected to conduct the seeded cooling crystallization of CUR: pure acetone and acetonitrile as well as 50/50 (wt/wt) mixtures of acetone/2-propanol and acetone/acetonitrile.

3.1.2. Effect of DMC and BDMC on CUR Solubility in the Selected Process Solvents

Before designing seeded cooling crystallizations, the solubility behavior of CUR was evaluated in presence of the main impurities DMC and BDMC in acetone, 50/50 acetone/2-propanol, 50/50 acetone/acetonitrile and acetonitrile. In Figure 6, the resulting solubility data are compared with the solubility values of pure CUR in the respective solvents.

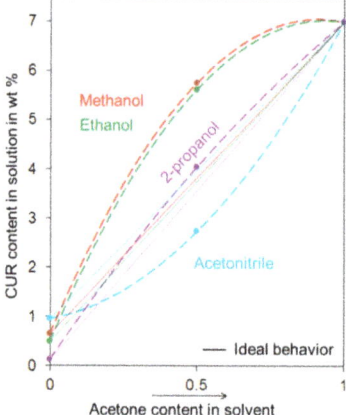

Figure 5. Effect of anti-solvents methanol, ethanol, 2-propanol and acetonitrile on the solubility of CUR in acetone at 30 °C. Symbols represent experimental data, curves are just guide to the eyes. Dashed curves originate from experimentally determined solubility values in the respective 50/50 (wt/wt) acetone/anti-solvent mixtures. Thin solid lines represent ideal linear solubility behaviors.

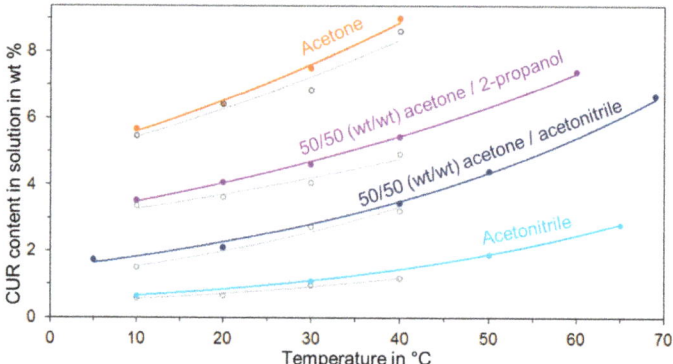

Figure 6. Comparison of solubility curves determined for pure CUR (empty circles, grey curves), and CUR in presence of the main impurities (solid circles, colored curves). Symbols represent experimental data, fitted curves serve as guide to the eyes.

As seen the solubilities of CUR in presence of DMC and BDMC slightly exceed those of pure CUR in the four solvents. Moreover, comparison of the CUR solubility in 50/50 acetone/2-propanol and 50/50 acetone/acetonitrile shows that with the use of acetonitrile as anti-solvent, a higher supersaturation of CUR in the solution can be obtained resulting in a higher product yield. This observation confirms the behavior of pure CUR in the binary solvents discussed in Figure 5.

3.2. Design of the Seeded Cooling Crystallization for Separation of CUR

Based on the solubility curves of CUR in presence of the main impurities and the observed nucleation behavior of pure CUR in the respective solvents, four seeded cooling crystallization processes were derived to separate CUR from the CURD mixtures as illustrated in Figure 7.

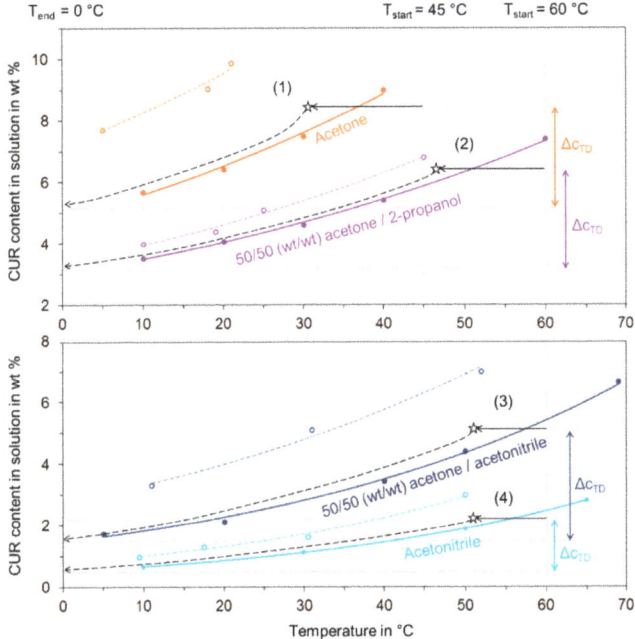

Figure 7. Design of the seeded cooling crystallization processes of CUR based on the solubility curves of CUR in presence of the main impurities (solid lines) and the nucleation border of pure CUR (dashed lines) in acetone (1), 50/50 acetone/2-propanol (2), 50/50 acetone/acetonitrile (3) and acetonitrile (4). Black solid/dashed lines with arrows are imaginary curves representing the variation of the CUR concentration during the crystallization process. (T_{start}/T_{end}: start/end temperature of the cooling step; stars: temperature of seed addition; Δc_{TD}: maximal depletion of CUR from solution).

The starting temperatures of the crystallization processes in the 50/50 mixtures of acetone/2-propanol and acetone/acetonitrile and in acetonitrile were set at 60 °C. To avoid uncontrolled evaporation of acetone, 45 °C was chosen as the starting temperature in this solvent.

The temperatures at which seeds of pure CUR (form I) were introduced into the acetone, 50/50 acetone/acetonitrile and acetonitrile solutions were chosen to be at least 5 K below the saturation temperature of CUR (approximately in the first third of the metastable region). However, the metastable region of CUR in 50/50 acetone/2-propanol (Figure 7, purple lines) is significantly closer than for the other three solvent systems. Therefore, the seeds were added at approximately half of the metastable region. An overview of the selected process parameters for the four seeded cooling crystallization processes is given in Table 4.

Table 4. Overview of the selected crystallization process parameters.

Process No.	Process Solvent	T_{start} °C	T_{sat} °C	T_{seeds} °C	T_{end} °C	Cooling Rate K/h
1	Acetone	45	37	30	0	10
2	50/50 acetone/2-propanol	60	51	46	0	10
3	50/50 acetone/acetonitrile	60	57	51	0	10
4	Acetonitrile	60	56	51	0	10

T_{start}/T_{end}: start/end temperature of cooling step; T_{sat}: CUR saturation temperature; T_{seeds}: seeding temperature.

The initial concentrations of CUR in the crude CURD mixtures were selected in accordance with the set starting temperatures to guarantee undersaturation of CUR in the starting solutions. The amounts of the crude solids, the process solvents and the calculated initial CUR content in the four starting solutions are listed in Table 5.

Table 5. Amount of initial substances used in the four crystallization processes.

Process No.	m (CURD) g	m (Solvent) g	c_{start} (CUR) wt%	$c_{end} = c_{sat}$ (CUR) wt%	Δc_{TD} (CUR) wt%	m_{max} (CUR) g
1	19.0	150	8.4	5.1	3.3	5.7
2	14.0	150	6.4	3.2	3.2	5.1
3	11.5	140	5.1	1.5	3.6	5.5
4	5.2	150	2.2	0.5	1.7	2.7

m(CURD), m(Solvent): amounts of CURD mixture and solvent used for the starting solution; c_{start}(CUR): calculated concentration of CUR in the starting solution; $c_{end} = c_{sat}$: concentration of CUR at the end of the cooling process, equal to the respective saturation concentration, from solubility study; Δc_{TD}(CUR): max. possible change of CUR concentration at the end of the cooling process, calculated based on the thermodynamic values; m_{max}: maximal achievable mass of CUR, calculated based on the thermodynamic values.

3.3. Implementation of the Purification Process

In the first step, the four initial crude solutions were prepared using the corresponding amount of the crude solid mixture in the respective solvent (Table 5). The seeded cooling crystallization of CUR was conducted in a second step following the four process trends shown in Figure 7. Starting at set temperatures, the unsaturated clear solutions were cooled down to 0 °C at a linear rate of 10 K/h. After exceeding the corresponding saturation temperature, seed crystals of pure CUR form I (ca. 50 mg) were introduced into the supersaturated solution at T_{seeds} (Table 4). At the end of the cooling process at 0 °C, the obtained product suspensions were stirred for further 0.5 h. Subsequently, solid-liquid phase separation was carried out on suction filters (pore size of filter paper 0.6 µm). To remove adhering mother liquor from the filter cake, the collected crystals were washed with about 100 g of cold acetone (< 0 °C, in processes 1-3) or with acetonitrile (< 0 °C in process 4). Then, dried at 40 °C, the purity of CUR and the yield were analyzed. During the washing process with acetone a visible dissolution of the filter cake was observed, caused by the high solubility of CUR in acetone (about 5 wt% at 0 °C). Accordingly, lower product yields could be assumed in the processes of using acetone as washing solvent (in processes 1-3).

The results of the four conducted seeded cooling crystallizations are summarized in Table 6. The maximum thermodynamically possible yield of CUR η_{TD} was calculated according to Equation (1), the total product yield of CUR η according to Equation (2).

$$\eta_{TD}(CUR) = \frac{m_{product} \cdot CUR\ product\ content}{m_{max}} \quad (1)$$

$$\eta(CUR) = \frac{m_{product} \cdot CUR\ product\ content}{m_{start}(CUR)} \quad (2)$$

Table 6 shows that in the 50/50 acetone/2-propanol mixture (process 2), the highest purity of CUR (99.4%) in the crystalline product was achieved. However, only 13% of the initial CUR content in the crude mixture was recovered. Crystalline CUR with decreasing purity of 95.7%, 92.3% and 90.1% but increasing total product yields of 31%, 55% and 62% was obtained from acetone, acetonitrile and 50/50 acetone/acetonitrile, respectively. The lower total yields from acetone and acetone/2-propanol solutions are partly associated with the enhanced CUR solubility at the final process temperature compared to the acetonitrile-containing solutions (see Figure 7), which, however, does not explain the extremely low yield achieved in the latter case.

The obtained purity results further verify that BDMC could completely be removed from the crystalline products within a single separation step, while the content of DMC was noticeably reduced. The presence of DMC as impurity in the products can be probably attributed to the most similar molecular structure of DMC and CUR (Figure 1). It can be postulated that DMC molecules compete with CUR in the solution upon forming the main crystal lattices. To ascertain, whether DMC is present near CUR in the crystalline form or as amorphous phase, the four crystallization products were analyzed by XRPD. In Figure 8 the corresponding patterns are compared with the commercial solid standards of DMC and CUR.

Table 6. Results of the four seeded cooling crystallization processes. Table columns containing the products purity and yield are highlighted in grey.

	Ternary Mixture of Curcuminoids						Crystalline Product					
Process No.	Process Solvent	m_{start} (CUR)	CUR Content	DMC Content	BDMC Content	m_{max} (CUR)	m (Product)	CUR Content	DMC Content	BDMC Content	η_{TD} (CUR)	η (CUR)
		g	%	%	%	g	g	%	%	%	%	%
1	Acetone	14.3	75.0	19.2	5.8	5.7	4.6	95.7	4.3	0	77	31
2	50/50 acetone/ 2-propanol	9.9	70.8	23.5	5.7	5.1	1.3	99.4	0.6	0	25	13
3	50/50 acetone/ acetonitrile	7.7	67.2	25.5	7.3	5.5	5.3	90.1	9.9	0	87	62
4	Acetonitrile	3.5	67.2	25.5	7.3	2.7	2.1	92.3	7.7	0	72	55

m_{start}(CUR): calculated amount of CUR in the mixture of CURDs; m_{max}: max. achievable mass of CUR, based on the thermodynamic values; m(product): mass of the crystalline product gained; η_{TD}(CUR): max. thermodynamically possible yield of CUR; η(CUR): total product yield of CUR.

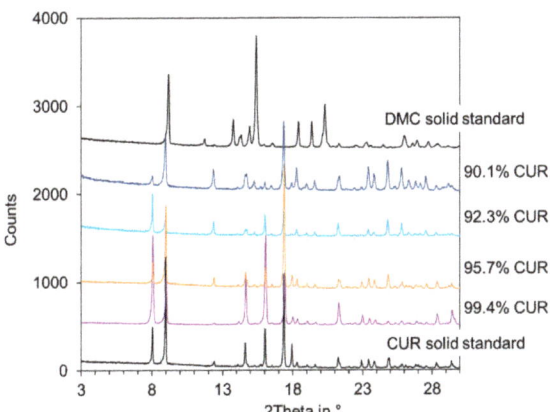

Figure 8. XRPD patterns of four crystallization products with increasing CUR content.

Since all XRPD reflexes in the diffractograms from the crystalline products can be clearly distinguished and are uniformly on the baseline, the presence of an amorphous fraction in the solid products cannot be confirmed. Moreover, all XRPD patterns seem to be identical to the CUR solid standard. Despite the increasing DMC content in the crystalline products (0.6–9.9%), none of the recorded patterns can be clearly assigned to the solid standard of DMC. Only a slight shift of single reflexes of crystalline products is indicated with increased DMC content in the solids.

Due to the strong similarity of CUR and DMC molecules, partial miscibility at the solid state might be a possible explanation here. However, dependent on the instrument and the structural similarity of the compounds used the limit of detection of the XRPD method is known to be 5–7 wt% and 1 wt%

in best cases. Thus, incorporation of DMC molecules is not readily assessable by XRPD at these low contents. Clarifying this issue requires further work which was out of the scope of this paper.

3.4. Improvement of the Total Yield by Means of Anti-Solvent Addition

To increase the product yield of the seeded cooling crystallization, it is suitable to conduct anti-solvent addition to the product suspension at the end of the cooling step [39]. In this work, to improve the overall yields of processes 1, 2 and 4 addition of water, 2-propanol and acetonitrile as anti-solvents of CUR in acetone was investigated. The final solvent/anti-solvent ratio was set to 25/75 (wt/wt). The study was conducted using the crude solid no. 1 with the lowest CUR content of 67.2%.

In the first step three equal starting solutions containing 8.5 wt% CUR in acetone were prepared and three identical seeded cooling crystallization processes were carried out as previously described in process no. 1. When the set end temperature of the cooling profile (0 °C) was reached, cold anti-solvent (< 0 °C) 2-propanol (process 1–2) and acetonitrile (process 1–3) was added to the product suspension, respectively (see Table 7). Afterwards the system was stirred for 3 h at constant 0 °C. The addition of water (process 1–1) was carried out at 26 °C after introducing CUR seeds to the supersaturated acetone solution. Then the suspension was cooled down to the end temperature 0 °C and stirred also for 3 h. Solid-liquid phase separation was performed at the end of each anti-solvent crystallization process. The crystalline products were dried at 40 °C and the CUR purity and yield analyzed. To maintain the total product yield and avoid the previously observed dissolution of the filter cake during washing with cold acetone, the washing step was skipped. The results obtained are summarized in Table 7.

Table 7. Overview of the results to improve total product yield ($\eta(CUR)$) via anti-solvent addition.

Process No.	Solvent	Ternary Mixture of Curcuminoids				Crystalline Product				
		m_{start} (CUR)	CUR content	DMC content	BDMC content	m (product)	CUR content	DMC content	BDMC content	η (CUR)
		g	%	%	%	g	%	%	%	%
1-1	25/75 acetone/water	2.9	67.2	25.5	7.3	2.7	85.5	13.4	1.1	79
1-2	25/75 acetone/2-propanol	2.9	67.2	25.5	7.3	1.1	96.2	3.7	0.1	36
1-3	25/75 acetone/acetonitrile	2.9	67.2	25.5	7.3	2.3	88.3	10.8	0.9	70

In all processes, addition of anti-solvent to the product suspension at the end of the cooling step led to a significant increase of yields. However, the CUR purity in the final crystalline products was noticeably decreased. DMC and in addition low amounts of BDMC (≤1.1%) were present as impurities. In addition to a detrimental effect of abstaining from product washing, an anti-solvent effect on DMC and BDMC cannot be excluded here (even solubility of DMC and BDMC is reported to exceed that of CUR in acetonitrile and isopropanol [34]). However, similar to the cooling crystallization (Table 6), the use of 2-propanol as anti-solvent (process 1–2) provided CUR at highest purity (96.2%) but at significantly lowest yield (36%). Regarding the reduced yield in presence of 2-propanol it can only be presumed at this stage, that, as indicated in polythermal (non-seeded) solubility studies from 2-propanol, an additional metastable (and thus higher soluble) polymorph or solvate phase occurs which causes the respective CUR remaining in the solution phase and thereby reducing the CUR yield in the solid phase.

4. Conclusions

In this work, the solubility behavior of pure CUR and CUR in presence of the two main impurities DMC and BDMC as well as supersolubilities in acetone, acetonitrile, methanol, ethanol, 2-propanol and their binary mixtures were investigated first. Based on the data obtained, seeded cooling crystallizations in four different process solvents (acetone, acetonitrile and 50/50 (wt/wt) mixtures of acetone/2-propanol and acetone/acetonitrile) were designed and implemented. As a result, the purity of CUR could be increased from initial CUR contents of 67–75% in the curcuminoid mixtures up to values of 90.1–99.4% in a single crystallization step. All crystallization processes provided crystalline curcumin (form I) free of BDMC after this single crystallization step. DMC was significantly depleted from initial contents of 19.2–25.5% in the crude mixtures to residual contents of 0.6–9.9%. Total product yields were significantly enhanced to 70–79% via addition of water and acetonitrile as anti-solvents at the end of the cooling crystallization process.

The presence of crystalline or amorphous DMC in the CUR products could not be detected by XRPD analysis. Whether this is caused by experimental detection limits or by potential formation of CUR/DMC mixed crystals has to be clarified in future studies.

Based on the work presented, a seeded cooling crystallization from a 50/50 (wt/wt) acetone/2-propanol solvent mixture is seen as the best purification strategy providing CUR at highest purity of 99.4%, BDMC free in a single crystallization step. However, there is still space for process optimization in particular with respect to yield. This includes application of a reduced cooling rate and a lowered final cooling temperature to increase both crystallization and total yield. Further, to avoid product losses in downstream processing washing the product with an acetone/anti-solvent mixture (for example acetonitrile) is suggested. No information regarding the maximum admissible limit of BDMC and DMC in the crystalline CUR was found in the literature. However, in any case the CUR purification grade obtained within a simple single crystallization step in this study represents a significant improvement compared to alternative process concepts.

Author Contributions: Conceptualization, E.H. and H.L.; investigation, E.H.; supervision, A.S.-M. and H.L.; validation, L.Y., A.S.-M. and H.L.; writing—original draft, E.H. and L.Y.; writing—review and editing, E.H., L.Y. and H.L. All authors have read and agreed to the published version of the manuscript.

Funding: This research received no external funding.

Acknowledgments: We thank Jacqueline Kaufmann and Stefanie Leuchtenberg (Max Planck Institute for Dynamics of Complex Technical Systems, Magdeburg, Germany) for supporting the analytical work. We are also grateful to Minh Tan Nguyen and Dinh Tien Vu (Hanoi University of Science and Technology, Hanoi, Vietnam) for motivating this study and providing an extract sample of *Curcuma Longa* L.

Conflicts of Interest: The authors declare no conflict of interest.

References

1. Abdul, R.; Hatifah, P.L.; Ratna, W.; Muhammad, K. Use of Thin Layer Chromatography and FTIR Spectroscopy Along with Multivariate Calibration for Analysis of Individual Curcuminoid in Turmeric (*Curcuma longa* Linn) Powder. *Int. J. Pharm. Clin. Res.* **2016**, *8*, 419–424.
2. Bagchi, A. Extraction of Curcumin. *IOSR J. Environ. Sci. Toxicol. Food Technol.* **2012**, *1*, 1–16. [CrossRef]
3. Tanaka, K.; Kuba, Y.; Sasaki, T.; Hiwatashi, F.; Komatsu, K. Quantitation of curcuminoids in curcuma rhizome by near-infrared spectroscopic analysis. *J. Agric. Food Chem.* **2008**, *56*, 8787–8792. [CrossRef] [PubMed]
4. Priyadarsini, K.I. Chemical and structural features influencing the biological activity of curcumin. *Curr. Pharm. Des.* **2013**, *19*, 2093–2100. [PubMed]
5. Esatbeyoglu, T.; Hübbe, P.; Ernst, I.M.A.; Chin, D.; Wagner, A.E.; Rimbach, G. Curcumin From Molecule to Biological Function. *Angew. Chem. Int. Ed.* **2012**, *51*, 5308–5332. [CrossRef] [PubMed]
6. Patil, M.B.; Taralkar, S.V.; Sakpal, V.S.; Shewale, S.P.; Sakpal, R.S. Extraction, isolation, and evaluation of anti-inflammatory activity of curcuminoids from Curcuma longa. *Int. J. Chem. Sci. Appl.* **2011**, *2*, 172–174.

7. Esatbeyouglu, T.; Ulbrich, K.; Rehberg, C.; Rohn, S.; Rimbach, G. Thermal stability, antioxidant, and anti-inflammatory activity of curcumin and its degradation product 4-vinyl guaiacol. *Food Funct.* **2015**, *6*, 887–893. [CrossRef]
8. Yanagisawa, D.; Shirai, N.; Amatsubo, T.; Taguchi, H.; Hirao, K.; Urushitani, M.; Morikawa, S.; Inubushi, T.; Kato, M.; Kato, F.; et al. Relationship between the tautomeric structures of curcumin derivatives and their Abeta-binding activities in the context of therapies for Alzheimer's disease. *Biomaterials* **2010**, *31*, 4179–4185. [CrossRef]
9. Aggarwal, B.B.; Gupta, S.C.; Sung, B. Curcumin: An orally bioavailable blocker of TNF and other pro-inflammatory biomarkers. *J. Pharm.* **2013**, *169*, 1672–1692. [CrossRef]
10. Anand, P.; Sundaram, C.; Jhurani, S.; Kunnumakkara, A.B.; Aggarwal, B.B. Curcumin and cancer: An "old-age" disease with an "age-old" solution. *Cancer Lett.* **2008**, *267*, 133–164. [CrossRef]
11. Salem, M.; Rohani, S.; Gillies, E.R. Curcumin, a promising anti-cancer therapeutic: A review of its chemical properties, bioactivity and approaches to cancer cell delivery. *Rsc Adv.* **2014**, *4*, 10815. [CrossRef]
12. Chin, D.; Huebbe, P.; Pallauf, K.; Rimbach, G. Neuroprotective Properties of Curcumin in Alzheimer's Disease–Merits and Limitations. *Curr. Med. Chem.* **2013**, *20*, 3955–3985. [CrossRef]
13. Eckert, G.P.; Schiborr, C.; Hagl, S.; Abdel-Kader, R.; Muller, W.E.; Rimbach, G.; Frank, J. Curcumin prevents mitochondrial dysfunction in the brain of the senescence-accelerated mouse-prone 8. *Neurochem. Int.* **2013**, *62*, 595–602. [CrossRef]
14. Lee, W.H.; Loo, C.Y.; Bebawy, M.; Luk, F.; Mason, R.S.; Rohanizadeh, R. Curcumin and its Derivatives: Their Application in Neuropharmacology and Neuroscience in the 21st Century. *Curr. Neuropharmacol.* **2013**, *11*, 338–378. [CrossRef]
15. Johnson, J.J.; Mukhtar, H. Curcumin for chemoprevention of colon cancer. *Cancer Lett.* **2007**, *255*, 170–181. [CrossRef]
16. Simon, A. Inhibitory effect of curcuminoids on MCF-7 cell proliferation and structure-activity relationship. *Cancer Lett.* **1998**, *129*, 111–116. [CrossRef]
17. Ruby, A.J.; Kuttan, G.; Dinesch Badu, K.; Rajasekharan, K.N.; Kuttan, R. Anti-tumor and antioxidant activity of natural corcuminoids. *Cancer Lett.* **1995**, *94*, 79–83. [CrossRef]
18. Luis, P.B.; Boeglin, W.E.; Schneider, C. Thiol Reactivity of Curcumin and Its Oxidation Products. *Chem. Res. Toxicol.* **2018**, *31*, 269–276. [CrossRef]
19. Modasiya, M.K.; Patel, V.M. Studies on solubility of curcumin. *Int. J. Pharm. Life Sci.* **2012**, *3*, 1490–1497.
20. Carvalho, D.M.; Takeuchi, K.P.; Geraldine, R.M.; Moura, C.J.; Torres, M.C.L. Production, solubility and antioxidant activity of curcumin nanosuspension. *Food Sci. Technol.* **2015**, *35*, 115–119. [CrossRef]
21. Wanninger, S.; Lorenz, V.; Subhan, A.; Edelmann, F.T. Metal complexes of curcumin - synthetic strategies, structures and medicinal applications. *Chem. Soc. Rev.* **2015**, *44*, 4986–5002. [CrossRef] [PubMed]
22. Nandakumar, D.N.; Nagaraj, V.A.; Vathsala, P.G.; Rangarajan, P.; Padmanaban, G. Curcumin-artemisinin combination therapy for malaria. *Antimicrob. Agents Chemother.* **2006**, *50*, 1859–1860. [CrossRef] [PubMed]
23. Chaitanya Mannava, M.K.; Suresh, K.; Bommaka, M.K.; Konga, D.B.; Nangia, A. Curcumin-Artemisinin Coamorphous Solid: Xenograft Model Preclinical Study. *Pharmaceutics* **2018**, *10*, 7. [CrossRef] [PubMed]
24. Pabon, H.J.J. A synthesis of curcumin and related compounds. *Recl. Trav. Chim. Pays-Bas* **1964**, *83*, 379–386. [CrossRef]
25. Kim, E.J.; Cha, M.H.; Kim, B.-G.; Ahn, J.-H. Production of Curcuminoids in Engineered Escherichia coli. *J. Microbiol. Biotechnol.* **2017**, *27*, 975–982. [CrossRef] [PubMed]
26. Grynkiewicz, G.; Ślifirski, P. Curcumin and curcuminoids in quest for medicinal status. *Acta Biochim. Pol.* **2012**, *59*, 201–212. [CrossRef]
27. Pothitirat, W.; Gritsanapan, W. Quantitative analysis of curcumin, demethoxycurcumin and bisdemethoxycurcumin in the crude curcuminoid extract from Curcuma Longa L. in Thailand by TLC-Densitometry. *Mahidol Univ. J. Pharm. Sci.* **2005**, *32*, 23–30.
28. Aggarwal, B.B.; Bhatt, I.D.; Ichikawa, H.; Ahn, K.S.; Sethi, G.; Sandur, S.K.; Sundaram, C.; Seeram, N.; Shishodia, S. Curcumin-Biological and medicinal properties. In *Turmeric the Genus Curcuma*; Ravindran, P.N., Babu, K.N., Sivaraman, K., Eds.; CRC Press: Abingdon, UK, 2007; pp. 297–368.
29. Priyadarsini, K.I. The chemistry of curcumin: From extraction to therapeutic agent. *Molecules* **2014**, *19*, 20091–20112. [CrossRef]

30. Anderson, A.M.; Mitchell, M.S.; Mohan, R.S. Isolation of curcumin from turmeric. *J. Chem. Educ.* **2000**, *77*, 359–360. [CrossRef]
31. Revathy, S.; Elumalai, S.; Benny, M.; Antony, B. Isolation, Purification and Identification of Curcuminoids from Turmeric (*Curcuma longa* L.) by Column Chromatography. *J. Exp. Sci.* **2011**, *2*, 21–25.
32. Péret-Almeida, L.; Cherubino, A.P.F.; Alves, R.J.; Dufossé, L.; Glória, M.B.A. Separation and determination of the physico-chemical characteristics of curcumin, demethoxycurcumin and bisdemethoxycurcumin. *Food Res. Int.* **2005**, *38*, 1039–1044. [CrossRef]
33. Liu, J.; Svard, M.; Hippen, P.; Rasmuson, A.C. Solubility and crystal nucleation in organic solvents of two polymorphs of curcumin. *J. Pharm. Sci.* **2015**, *104*, 2183–2189. [CrossRef] [PubMed]
34. Ukrainczyk, M.; Hodnett, B.K.; Rasmuson, A.C. Process Parameters in the Purification of Curcumin by Cooling Crystallization. *Org. Process Res. Dev.* **2016**, *20*, 1593–1602. [CrossRef]
35. Horvath, Z.; Horosanskaia, E.; Lee, J.W.; Lorenz, H.; Gilmore, K.; Seeberger, P.H.; Seidel-Morgenstern, A. Recovery of Artemisinin from a Complex Reaction Mixture Using Continuous Chromatography and Crystallization. *Org. Process Res. Dev.* **2015**, *19*, 624–634. [CrossRef]
36. Heffernan, C.; Ukrainczyk, M.; Gamidi, R.K.; Hodnett, B.K.; Rasmuson, A.C. Extraction and Purification of Curcuminoids from Crude Curcumin by a Combination of Crystallization and Chromatography. *Org. Process Res. Dev.* **2017**, *21*, 821–826. [CrossRef]
37. Horosanskaia, E.; Triemer, S.; Seidel-Morgenstern, A.; Lorenz, H. Purification of Artemisinin from the Product Solution of a Semisynthetic Reaction within a Single Crystallization Step. *Org. Process Res. Dev.* **2019**, *23*, 2074–2079. [CrossRef]
38. Horosanskaia, E.; Nguyen, T.M.; Vu, T.D.; Seidel-Morgenstern, A.; Lorenz, H. Crystallization-Based Isolation of Pure Rutin from Herbal Extract of Sophora japonica L. *Org. Process Res. Dev.* **2017**, *21*, 1769–1778. [CrossRef]
39. Wiekhusen, D. Development of batch crystallizations. In *Crystallization—Basic Concepts and Industrial Applications*; Beckmann, W., Ed.; Wiley-VCH Verlag GmbH & Co. KGaA: Weinheim, Germany, 2013; pp. 187–202.
40. Jadhav, B.K.; Mahadik, K.R.; Paradkar, A.R. Development and validation of improved reversed phase HPLC method for simultaneous determination of curcumin, demethoxycurcumin and bis-demethoxycurcumin. *Chromatographia* **2007**, *65*, 483–488. [CrossRef]
41. Lorenz, H. Solubility and solution equilibria in crystallization. In *Crystallization – Basic Concepts and Industrial Applications*; Beckmann, W., Ed.; Wiley-VCH Verlag GmbH & Co. KGaA: Weinheim, Germany, 2013; pp. 35–74.
42. Sanphui, P.; Goud, N.R.; Khandavilli, U.B.; Bhanoth, S.; Nangia, A. New polymorphs of curcumin. *Chem. Commun.* **2011**, *47*, 5013–5015. [CrossRef]
43. Gately, S.; Triezenberg, S.J. Solid Forms of Curcumin. U.S. Patent WO2012138907A2, 20 September 2012.
44. Mishra, M.K.; Sanphui, P.; Ramamurty, U.; Desiraju, G.R. Solubility-Hardness Correlation in Molecular Crystals: Curcumin and Sulfathiazole Polymorphs. *Cryst. Growth Des.* **2014**, *14*, 3054–3061. [CrossRef]
45. Thorat, A.A.; Dalvi, S.V. Solid-State Phase Transformations and Storage Stability of Curcumin Polymorphs. *Cryst. Growth Des.* **2015**, *15*, 1757–1770. [CrossRef]
46. Tonnesen, H.H.; Karlsen, J.; Mostad, A. Structural Studies of Curcuminoids. I. The Crystal Structure of Curcumin. *Acta Chem. Scand.* **1982**, *36*, 475–479. [CrossRef]
47. Tonnesen, H.H.; Karlsen, J.; Mostad, A.; Pedersen, U.; Rasmussen, P.B.; Lawesson, S.-O. Structural Studies of Curcuminoids. II. Crystal Structure of 1,7-Bis(4-hydroxyphenyl)-1,6-heptadiene-3,5-dione-Methanol Complex. *Acta Chem. Scand.* **1983**, *37*, 179–185. [CrossRef]
48. Karlsen, J.; Mostad, A.; Tonnesen, H.H. Structural studies of curcuminoids. VI. Crystal structure of 1,7_bis(4-hydroxyphenyl)-1,6-heptadiene-3,5-dione hydrate. *Acta Chem. Scand.* **1988**, *42*, 23–27. [CrossRef]

49. Yuan, L.; Horosanskaia, E.; Engelhardt, F.; Edelmann, F.T.; Couvrat, N.; Sanselme, M.; Cartigny, Y.; Coquerel, G.; Seidel-Morgenstern, A.; Lorenz, H. Solvate formation of bis(demethoxy)curcumin: Crystal structure analyses and stability investigations. *Cryst. Growth Des.* **2019**, *19*, 854–867. [CrossRef]
50. Yuan, L.; Lorenz, H. Solvate formation of bis(demethoxy)curcumin: Screening and characterization. *Crystals* **2018**, *8*, 407. [CrossRef]

© 2020 by the authors. Licensee MDPI, Basel, Switzerland. This article is an open access article distributed under the terms and conditions of the Creative Commons Attribution (CC BY) license (http://creativecommons.org/licenses/by/4.0/).

Article

Freeze Concentration of Aqueous [DBNH][OAc] Ionic Liquid Solution

Nahla Osmanbegovic [1,*], Lina Yuan [2], Heike Lorenz [2] and Marjatta Louhi-Kultanen [1]

1. Department of Chemical and Metallurgical Engineering, School of Chemical Engineering, Aalto University, P.O. Box 16100, FI-00076 Aalto, Finland; marjatta.louhi-kultanen@aalto.fi
2. Max Planck Institute for Dynamics of Complex Technical Systems, Sandtorstrasse 1, 39106 Magdeburg, Germany; lina.yuan@novartis.com (L.Y.); lorenz@mpi-magdeburg.mpg.de (H.L.)
* Correspondence: nahla.osmanbegovic@aalto.fi

Received: 31 January 2020; Accepted: 24 February 2020; Published: 26 February 2020

Abstract: In the present work, freeze crystallization studies, as a novel concentration method for aqueous 1,5-diazabicyclo[4.3.0]non-5-enium acetate ([DBNH][OAc]) ionic liquid solution, were conducted. In order to find the appropriate temperature and composition range for freeze crystallization, the solid–liquid equilibrium of a binary [DBNH][OAc]–water compound system was investigated with differential scanning calorimetry (DSC). Results of this analysis showed that the melting temperature of the pure ionic liquid was 58 °C, whereas the eutectic temperature of the binary compound system was found to be −73 °C. The activity coefficient of water was determined based on the freezing point depression data obtained in this study. In this study, the lowest freezing point was −1.28 °C for the aqueous 6 wt.% [DBNH][OAc] solution. Ice crystal yield and distribution coefficient were obtained for two types of aqueous solutions (3 wt.% and 6 wt.% [DBNH][OAc]), and two freezing times (40 min and 60 min) were used as the main parameters to compare the two melt crystallization methods: static layer freeze and suspension freeze crystallization. Single-step suspension freeze crystallization resulted in higher ice crystal yields and higher ice purities when compared with the single-step static layer freeze crystallization. The distribution coefficient values obtained showed that the impurity ratios in ice and in the initial solution for suspension freeze crystallization were between 0.11 and 0.36, whereas for static layer freeze crystallization these were between 0.28 and 0.46. Consequently, suspension freeze crystallization is a more efficient low-energy separation method than layer freeze crystallization for the aqueous-ionic liquid solutions studied and, therefore, this technique can be applied as a concentration method for aqueous-ionic liquid solutions.

Keywords: melt crystallization; freeze crystallization (FC); recycling; ionic liquid (IL); solid–liquid equilibrium

1. Introduction

Crystallization from melt is one of the separation and concentration techniques used for organic compound solutions [1]. Nevertheless, under certain conditions, several organic compounds and green solvents, such as ionic liquids and deep eutectic solvents, can undergo thermal degradation or hydrolysis [2–6]. Even though it has been reported that low-pressure evaporation and distillation [7–9] can be used as a concentration method in the recycling of ionic liquids from aqueous ionic liquid solutions, the main drawback of such processes is that they are high-energy separation methods due to the high latent heat of evaporation. Consequently, there is a need to find more feasible concentration methods, which also allow application at low-temperature ranges.

Melt crystallization is a low-energy separation method that typically uses a low processing temperature close to room temperature. This is an advantage when working with thermally unstable substances or organic compounds that tend to react and decompose at higher temperatures [10].

Freeze concentration of an aqueous solution is one type of melt-based crystallization method and this is defined as the separation of formed ice from the aqueous solution.

Melt crystallization methods can be generally classified as either layer crystallization or suspension crystallization [11]. In static layer crystallization, the formation and growth of the ice layer occurs at the sub-cooled surface of a crystallizer from stagnant aqueous solution. In contrast, with the suspension crystallization technique, ice crystals form and grow within a sub-cooled solution present inside a crystallizer equipped with a scraper.

Moreover, a recent study [12] has shown that the combination of evaporation and freeze crystallization processes as a new method of recycling an ionic liquid from an aqueous solution is more energy efficient than evaporation-based concentration, primarily as a result of the lower latent heat of freezing when compared with the latent heat of evaporation.

In the present work, an ionic liquid (IL), 1,5-diazabicyclo[4.3.0]non-5-enium acetate ([DBNH][OAc]), was used as a model compound for the investigation of the freeze concentration method. [DBNH][OAc] has been shown to be an efficient solvent for dissolving birch-based cellulose [13] and is considered to be a promising industrial solvent due to its safety, low environmental impact, economic viability, and production of high-quality fiber even from low-refined unbleached pulps [14–17] with good spinnability [18]. Nevertheless, the main challenge to the wider application of IL is that it can be a relatively expensive organic solvent and, therefore, it must be efficiently recycled from aqueous solution as a way to reduce costs [19].

In this work, freeze crystallization techniques as a concentration method for IL recycling were investigated. The solid–liquid equilibria were determined between [DBNH][OAc] and water by the differential scanning calorimetry (DSC) technique.

2. Materials and Methods

This section describes the DSC procedure used for phase diagram construction and outlines the methodologies employed for static layer freeze crystallization, suspension freeze crystallization, and for separation efficiency studies.

2.1. Differential Scanning Calorimetry (DSC)

The solid–liquid equilibrium of binary [DBNH][OAc] and water solution was investigated to obtain phase diagram data that cover both [DBNH][OAc]- and water-enriched solutions. Thermal analyses were conducted by DSC (3, Mettler Toledo (Schwerzenbach, Switzerland)) equipped with an intra-cooler. Individual samples, with a mass of ca. 10 mg, were placed in 25 µL aluminum crucibles that were subsequently sealed before being placed in the experimental chamber with a N_2 flux regulated atmosphere. As low heating and cooling rates offer better information about the thermal behavior of the samples, i.e., whether the ionic liquid is a crystal or a glass formation [20], a heating and cooling rate of 1 K/min was chosen for all experiments.

2.2. Water Activity Coefficient

Water activity coefficients of aqueous [DBNH][OAc] solutions were calculated using Equations (1) and (2). Gmehling et al. [21] derived Equation (1) for the solubility of an organic solute in a solvent starting from the isofugacity condition. It can be used to estimate the activity coefficient of a sub-cooled liquid solvent as a function of the enthalpy of fusion, heat capacity difference between liquid and solid phase, and melting point of the solvent. When the system is close to its melting point, the last two terms of Equation (1) can be neglected and the simplified, Equation (2) is obtained. In the remainder of the article, Equation (1) is referred to as the activity coefficient equation and Equation (2) as the simplified activity coefficient equation.

$$\ln x^L \gamma^L = -\frac{\Delta h_m}{RT}\left(1 - \frac{T}{T_m}\right) + \frac{\Delta c_p}{RT}(T_m - T) - \frac{\Delta c_p}{R}\ln\left(\frac{T_m}{T}\right), \quad (1)$$

$$\ln x^L \gamma^L = -\frac{\Delta h_m}{RT}\left(1 - \frac{T}{T_m}\right), \quad (2)$$

where
- γ^L activity coefficient of water,
- x^L mole fraction of water,
- Δh_m enthalpy of fusion for water at 273.15 K (6009.5 J/mol),
- Δc_p heat capacity difference between water and solid ice (J/molK),
- R   universal gas constant (8.3143 J/molK),
- T_m freezing point of pure water (273.15 K),
- T   freezing point of aqueous ionic liquid solution obtained by DSC (K).

Heat capacity difference, Δc_p, is a function of temperature, and its value changes significantly when the temperature of a system is significantly lower than its melting point. Equation (3), previously reported by Sippola and Taskinen [22], was used to calculate the heat capacity change of water at its freezing point.

$$\Delta c_p = -19656.303 + 98.468097 \cdot (T/[K]) + 234320880 \cdot (T/[K])^{-2} - 0.1386227 \cdot (T/[K])^2, \quad (3)$$
$$237\ K \leq T \leq 273.15\ K.$$

2.3. Layer Freeze Crystallization

Experiments with stagnant 3 wt.% and 6 wt.% [DBNH][OAc] aqueous solutions were conducted in a crystallizer that consisted of a 250 mL jacketed flat-bottom glass vessel equipped with a cylindrical stainless-steel cold finger. The experimental setup is shown in Figure 1.

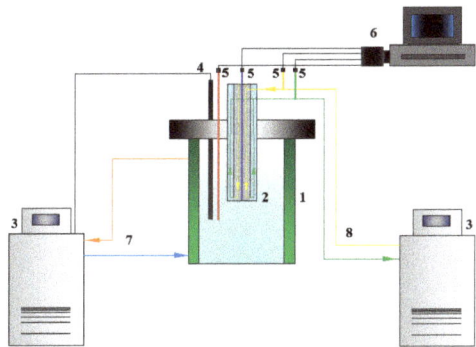

Figure 1. Experimental setup of static layer freeze crystallization: (1) crystallizer, (2) cold finger, (3) thermostats, (4) Pt 100 thermosensor, (5) thermocouples, (6) data processing device, (7) coolant streams circulating through jacketed vessel, (8) coolant streams circulating through cold finger.

Both elements of the crystallizer, the jacketed vessel (1) and the cold finger (2), were connected to a pair of Lauda ECO RE 1050 thermostats (Lauda-Königshofen, Germany) (3). The coolant streams (approx. 50 wt.% aqueous ethylene glycol solution) were circulated at a flow rate of 1.64 L/min through the jacketed vessel and at a flow rate of 0.35 L/min through the cold finger—the flow rates of circulating coolants through the jacketed vessel and cold finger were measured by Kytola EH-5SA and Kytola EH-4AA rotameters (Muurame, Finland), respectively. Thermocouples (5) were used to measure the temperature at four points within the crystallizer: inside the jacketed vessel (representing the temperature of solution), at the inlet of the cold finger coolant line, at the outlet from the cold finger, and inside the cold finger proximal to the tip (representing the temperature of sub-cooling). In addition, the external thermostatic control of the Lauda PT 100 (4) connected to the jacketed vessel was used to measure the solution temperature. Monitoring of temperature and storage of the measured data were performed using LabVIEW (Espoo, Finland) data acquisition software (6).

Studies of the ice layer on the cold finger from [DBNH][OAc] solutions were carried out at five different temperatures of coolant circulating through cold finger (sub-cooling temperatures) with two freezing times of 40 min and 60 min. The coolant temperature in the jacket was kept constant, and each separate temperature of coolant circulating through cold finger was set to be lower than the freezing point of solution, i.e., a sub-cooling temperature. The degree of sub-cooling was varied by altering the temperature of coolant of the cold finger.

For each experiment at a new sub-cooling temperature, the temperature of the thermostat connected to the cold finger was set to the desired value, whereas the temperature of the coolant circulating through the jacketed vessel was kept constant at the freezing point value of the respective solution. In addition, the internal sensor of the thermostat was used to adjust the temperature of coolant circulating through the cold finger at sub-cooling value.

After a constant temperature of solution and temperature of sub-cooling was achieved, freeze crystallization was induced by seeding with an ice crystal. This procedure was performed outside the jacketed vessel by the attachment of a seed ice crystal to the bottom surface of the cold finger, followed by immediate re-immersion in the solution. This immersion time was considered as the starting time in each freezing experiment, and the ice layer was allowed to grow on the cold finger for a pre-selected freezing time. The cold finger surface area where ice layer formation occurred is referred to as the cooling area in the layer freeze crystallization (FC) experiments. The ice seeds were produced by a Scotsman AF 103 Ice Flaker and were transported inside an insulated container.

After the fixed freezing time was complete, the ice layer formed was removed from the cold finger and rinsed with 5 mL of de-ionized water at 0 °C to remove any mother liquor remnants. The dimensions of the ice sample (outer diameter and height) and mass were measured before the sample melted.

2.4. Suspension Freeze Crystallization

The experimental setup of suspension freeze crystallization is shown in Figure 2.

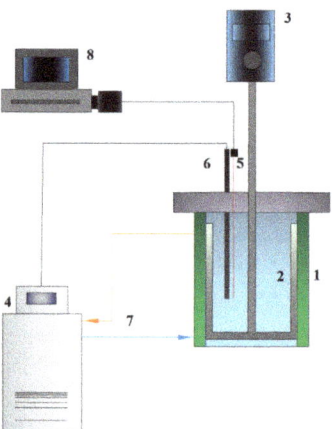

Figure 2. Experimental setup of suspension freeze crystallization: (1) crystallizer, (2) scraper, (3) mixer, (4) thermostat, (5) thermocouple, (6) Pt 100 thermosensor, (7) coolant streams circulating through jacketed vessel, (8) data processing device.

The crystallizer setup consisted of a 250 mL jacketed glass vessel (1) with a rotating scraper (2) that prevents encrustation of ice crystals at the inner surface of the crystallizer. A rotation speed of 18 rpm was set for the scraper (3). The coolant (approx. 50 wt.% aqueous ethylene glycol solution) was circulating at a flow rate of 1.64 L/min through the jacketed vessel connected to a Lauda ECO RE

1050 thermostat (3), and the flow rate of the coolant was measured by a Kytola EH-5SA rotameter. The thermocouple (5) and the Lauda PT 100 thermosensor (6) were used to determine solution temperature. LabVIEW and WinTherm software were used to monitor and store measured temperature data (8).

Suspension crystallization experiments with 3 and 6 wt.% [DBNH][OAc] aqueous solutions were conducted at five different temperatures of sub-cooled solutions for freezing times of 40 and 60 min. In the case of suspension freeze crystallization, the degree of sub-cooling presents the difference between the temperature of the sub-cooled solution and its freezing point. Sub-cooling of the solution depended on the temperature of coolant circulated by the thermostat pump through the jacket of the crystallizer. When the sub-cooling temperature of solution reached the desired value and became stable, ice crystal seeds were placed inside the crystallizer to induce the freeze crystallization. For suspension FC experiments, the cooling area is the vessel inner wall surface area. The ice seeds were produced by a Scotsman AF 103 Ice Flaker and transported inside an insulated container.

As the main challenge was to properly separate ice crystals from the mother liquor, a gravity-based filtration procedure with a perforated plate setup was used (Figure 3), where partial melting of washed ice samples could take place.

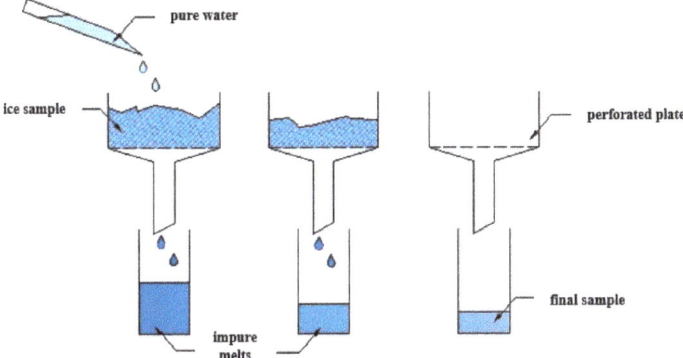

Figure 3. Gravity-based filtration and partial melting of ice crystals.

Approximately 15 g of surface floating ice crystals and mother liquor were taken directly from the reactor with a spoon and pressed on to the perforated plate in order to squeeze the mother liquor from the sample. Additionally, the sample was washed with 7 mL of de-ionized water at 0 °C to remove the remaining mother liquor. Washed samples were then left to partially melt and, with each step, the ice became more purified. The last fraction of ice crystals, with a mass of approx. 5 g, were considered to be pure crystals that had properly separated from the mother liquor, and its melt was used for the ice crystal purity analysis.

2.5. Determining Distribution Coefficient and Crystal Yield

In order to conduct the layer and suspension crystallization experiments, the freezing points of two DBNH[OAc] aqueous solutions were determined in the jacketed crystallizer fitted with a scraper. Upon seeding, ice crystallization commenced and as a result of the heat of crystallization, the solution temperature increased until a constant value was attained, which remained for the duration of crystallization. This temperature was taken as the freezing point and was measured with a thermocouple with a standard uncertainty of 0.01 °C (expanded uncertainty of 0.02 °C).

The ice crystal yield and the distribution coefficient were the main parameters used to assess the efficiency of freeze crystallization as a separation method.

Ice crystal yield was determined based on the mass of a pure ice sample as calculated by Equation (4)

$$Y = \frac{m_{ice}}{m_{water,sol}} \cdot 100, \quad (4)$$

where
- m_{ice} — mass of the pure ice (kg),
- $m_{water,sol}$ — mass of the water in initial solution (kg).

The mass of the ice samples produced by layer freeze crystallization was determined by weighing, whereas the mass of ice samples crystallized in the suspension crystallizer was calculated based on the concentration difference of [DBNH][OAc] in the mother liquor at the end of crystallization and the initial solution.

The distribution coefficient, K, is expressed as the ratio of the impurity in the ice to the initial impurity in the solution:

$$K = \frac{C_{imp}}{C_0}, \quad (5)$$

where
- C_{imp} — concentration of ionic liquid in ice (kg [DBNH][OAc]/kg ice),
- C_0 — initial concentration of ionic liquid in solution (kg [DBNH][OAc]/kg solution).

Purity of the ice samples —which is determined from the concentration of [DBNH][OAc] present in the ice crystals—was analyzed by the measurement of melted ice and mother liquor sample electrical conductivities with a Consort C3050 electrical conductivity meter. To obtain a correlation between electrical conductivity and concentration, the electrical conductivities of six solutions (0, 2, 4, 6, 8, and 10 wt.% [DBNH][OAc]$_{(aq)}$) were measured and are shown in Figure 4.

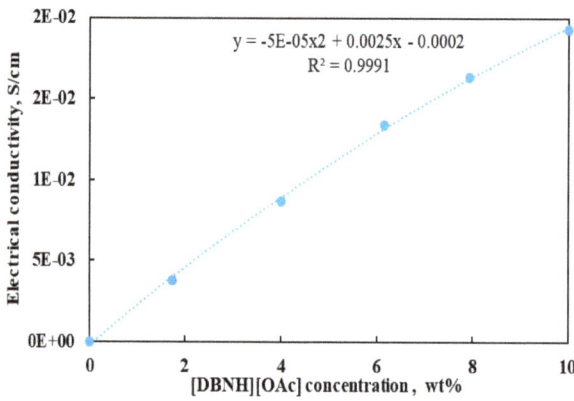

Figure 4. Electrical conductivity of aqueous solutions as a function of ionic liquid concentration.

3. Results

3.1. Phase Equilibria of Binary [DBNH][OAc] and Water System

Mixtures of [DBNH][OAc] and water at different ratios were analyzed by DSC. For each different sample, the corresponding liquidus temperature and/or glass transition (or eutectic temperature) were extracted from DSC curves obtained, and these were then plotted in a phase diagram as shown in Figure S1. The compositions of [DBNH][OAc] and water mixture samples analyzed are provided in Table S1.

The binary phase equilibria between [DBNH][OAc] and water are shown in Figure 5. The phase diagram was constructed based on temperature transition data obtained for the ionic liquid–water mixtures (water content was varied over a range between 0.49 wt.% and 100 wt.%). As can be seen from Figure 5, the black dot highlights glass transition or eutectic temperature, which is at −73 °C and the red dots relate to the liquidus temperatures. For the mixtures of [DBNH][OAc]–H_2O, where water content was less than 54.3 wt.%, only four temperatures were obtained by DSC, due to the ionic liquid glass-formation that occurs with such compositions. Consequently, the ionic liquid liquidus curve was plotted by extrapolation.

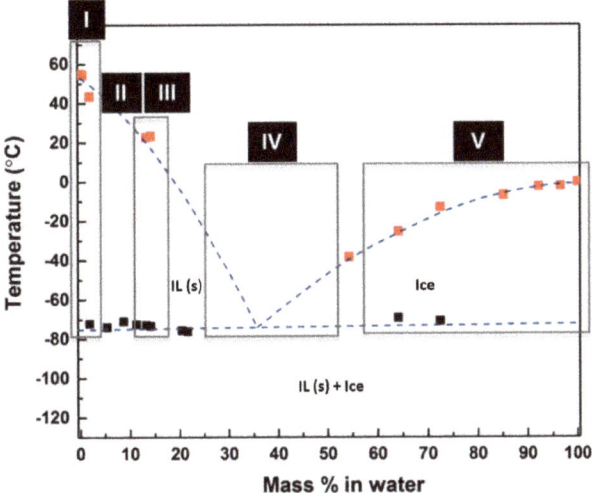

Figure 5. Phase equilibria between [DBNH][OAc] and water with solid forms. Five different crystallization behaviors in the [DBNH][OAc] and water mixtures.

Results from the DSC study show that the [DBNH][OAc]–H_2O mixtures can be divided into five regions based on their different crystallization behaviors. These regions are shown in Figure 5 and relate to the following characteristics:

- Region 1: At lower water content, mixtures were solid at ambient temperature.
- Region 2: No crystallization upon cooling and no recrystallization upon heating were observed, but glass transitions were measured at −73 °C
- Region 3: Crystallization of mixtures occurred upon heating. The mixtures underwent the transitions in following order: glass transition, recrystallization, and finally melting.
- Region 4: Mixtures neither crystallized nor underwent glass transition when they were cooled to −80 °C and heated up to 25 °C during DSC analyses.
- Region 5: For the mixtures with water content was greater than 54.3 wt.%, crystallization occurred upon cooling and melting upon heating. The melting temperatures acquired were used to construct the liquidus line of ice.

The phase equilibrium obtained results show that the appropriate temperature range and composition range for freeze crystallization of aqueous [DBNH][OAc] solution are in Region 5. Aqueous [DBNH][OAc] solutions from this region with freezing points above −10 °C can be feasibly concentrated by freeze crystallization.

Calculated freezing points of an ideal aqueous solution ($\gamma = 1$) based on Equation (2) and experimentally obtained freezing points for an aqueous [DBNH][OAc] solution for the same range of dissolved solute were compared, as shown in Figure 6.

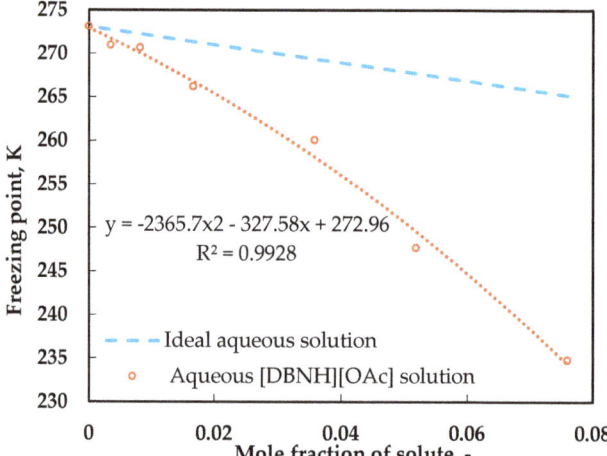

Figure 6. Freezing point depressions of ideal aqueous solutions calculated by simplified activity coefficient equation and aqueous [DBNH][OAc] solutions obtained by differential scanning calorimetry (DSC).

Water activity coefficients for aqueous [DBNH][OAc] solutions from Region 5 of the phase diagram, as calculated by Equations (1) and (2), are presented in Figure 7. It is apparent that aqueous ionic liquid solutions are non-ideal and [DBNH][OAc]–H$_2$O has an attractive interaction, as $\gamma^L < 1$. The freezing point depression data obtained by DSC were used as a basis for the thermodynamic modeling. When the two heat capacity change terms of undercooled water in Equation (2) are considered, this results in a lower level of non-ideality for the studied binary solution in higher concentrations when compared to the model based on Equation (1), where the specific heat capacity change terms are ignored. Furthermore, it is worth noting that Equation (3) was also used to calculate specific heat capacity change at −38.28 °C, even though Equation (3) is only considered to be valid over a temperature range between 0 and −35 °C according to Sippola and Taskinen [22].

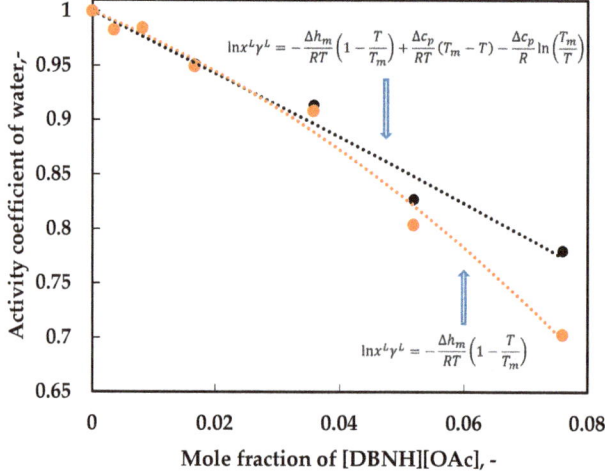

Figure 7. Activity coefficient of water as a function of solute mole fraction.

3.2. Layer Freeze Crystallization Results

In order to make a comparison between crystal yields and distribution coefficients, the experiments for approximately the same set of five sub-cooling degrees for both aqueous [DBNH][OAc] solutions and both freezing times were carried out (n.b., sub-cooling degrees deviated in range between 0.01 °C and 0.09 °C). All experimental and calculated data can be found in the Supplementary Material (Tables S2–S5).

The ice yield and distribution coefficient as a function of sub-cooling degree are presented in Figures 8 and 9. Ice crystal yield as a function of sub-cooling temperature shows a similar linear dependence for both freezing times and for both aqueous [DBNH][OAc] solutions. As expected, the greater the sub-cooling degree, the higher the crystal yield is. The distribution coefficients obtained show the separation efficiency of layer freeze crystallization varied between 0.28 and 0.46. The results indicate that the distribution coefficients for both types of solutions are almost independent of crystallization duration, as there are only negligible differences between the values of distribution coefficients obtained at the same level of sub-cooling for the two different freezing times investigated.

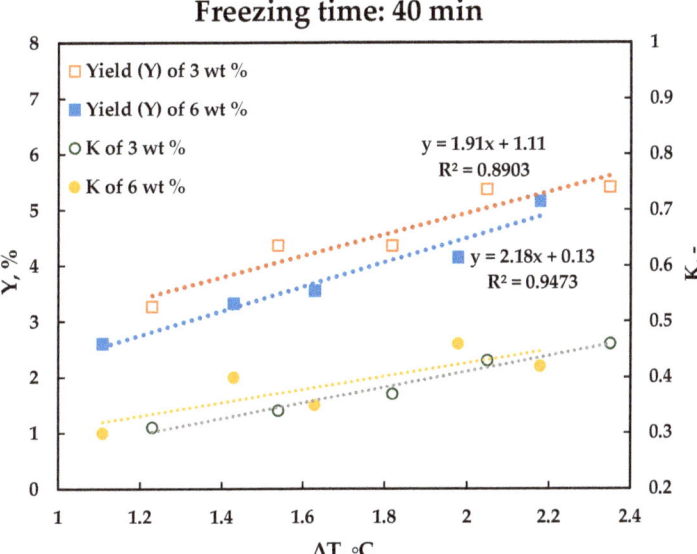

Figure 8. Ice yield and distribution coefficient of static layer freeze crystallization as a function of sub-cooling degree.

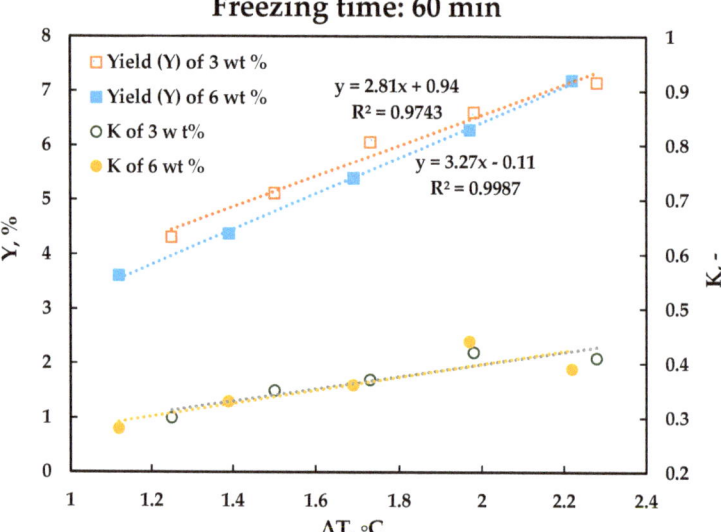

Figure 9. Ice yield and distribution coefficient of static layer freeze crystallization as a function of sub-cooling degree.

3.3. Suspension Freeze Crystallization Results

Experimental and calculated data are presented in the Supplementary Material (Tables S6–S9). Related sub-cooling degrees for both freezing times were found to vary within a range between 0.02 and 0.12.

Figures 10 and 11 show the ice yield and distribution coefficients obtained by the suspension FC experiments as function of sub-cooling degree. The ice crystal yield shows a linear dependence on sub-cooling temperature, and its value increases as the degree of sub-cooling increases. Nevertheless, the distribution coefficient is rather independent of sub-cooling degree and the values vary in a range between 0.11 and 0.36.

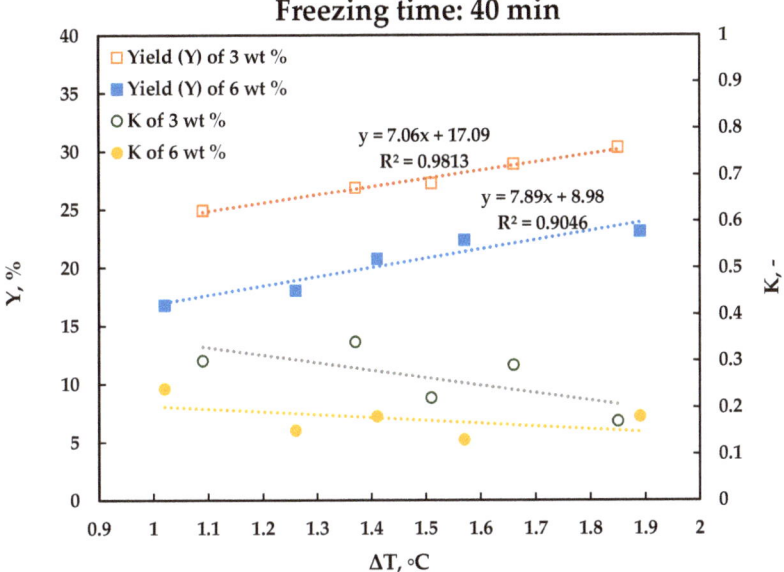

Figure 10. Ice yield and distribution coefficient of suspension freeze crystallization as a function of sub-cooling degree.

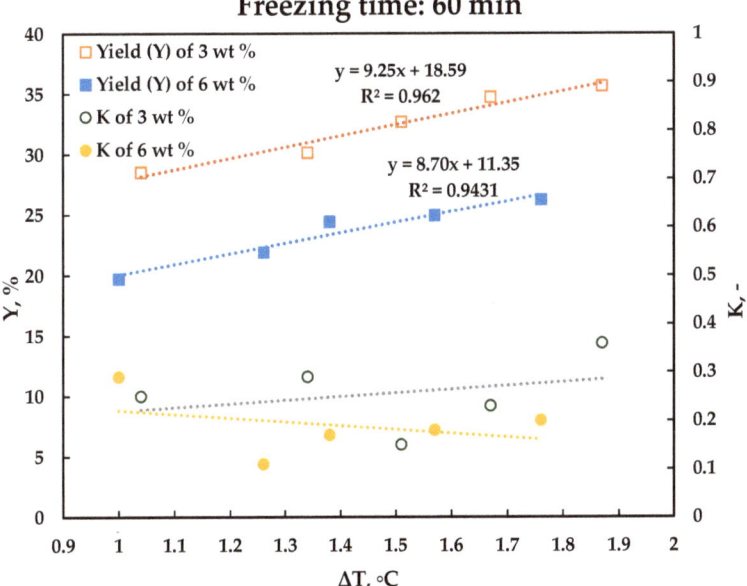

Figure 11. Ice yield and distribution coefficient of suspension freeze crystallization as a function of sub-cooling degree.

4. Discussion

Based on the comparison between static layer FC and suspension FC methods, the following observations are presented.

For the same freezing time and approximately the same sub-cooling temperature, ice crystal yields obtained by suspension freeze crystallization are around four-to-eight times higher than yields obtained by static layer freeze crystallization. This is as a result of the higher consumption of cooling energy and larger cooling surface area needed for the suspension crystallization experiments. In this case, the cooling area for suspension FC experiments was around nine times higher than the cooling area for static layer FC experiments.

From Figures 8–11, it is apparent that the difference between ice crystal yields obtained from the two types of aqueous solutions is greater in the case of suspension crystallization than in the case of layer crystallization. Moreover, the average value of the distribution coefficient is higher for static layer freeze crystallization, which indicates a lower ice purity than in case of suspension crystallization. These observations suggest that the mother liquor remained entrapped within the ice layer formed by the layer FC method. The distribution coefficient of static layer crystallization also shows a tendency to increase with higher undercooling and supersaturation, while for suspension crystallization the case is observed to display the opposite behavior.

For both freeze crystallization methods, the values of overall ice growth rate or freezing capacity (defined as kilogram of ice per unit of time and employed cooling surface area) are in the range of 10^{-4}–10^{-3} kg/m^{2}s. Furthermore, layer FC progressed with somewhat faster freezing kinetics, as the values of overall growth rates of layer FC are noticeably higher when compared to those obtained by suspension FC. Nevertheless, for both FC methods, the overall growth rate decreased for the more concentrated solutions of 6 wt.% [DBNH][OAc], which means that [DBNH][OAc] decreased the ice-growth kinetics.

5. Conclusions

In this study, the solid–liquid equilibria of a binary 1,5-diazabicyclo[4.3.0]non-5-enium acetate ([DBNH][OAc])–water compound system as well as layer and suspension freeze crystallization as a concentration method were investigated. The main conclusions that can be drawn from this study are as follows:

- Based on obtained solid–liquid equilibria, it was concluded that melt crystallization can be employed as a concentration method for aqueous [DBNH][OAc] solutions with water content higher than 54.3 wt.%.
- Water activity coefficient results calculated in mole fraction range between 0 and 0.08 showed that [DBNH][OAc] solutions are non-ideal solutions with an attractive interaction between [DBNH][OAc] and water molecules.
- Single-step suspension freeze crystallization is a more suitable concentration method for aqueous [DNBH][OAc] solutions than single-step layer freeze crystallization, based on the comparison between ice crystal yield and ice purity of these two freeze crystallization methods.

Supplementary Materials: The following are available online at http://www.mdpi.com/2073-4352/10/3/147/s1, Figure S1: Schematic diagram of extracting thermal data from a DSC curve, Table S1: Liquidus and glass transition (or eutectic temperature) temperatures extracted from DSC curves, Table S2: Layer freeze crystallization data of aqueous 3 wt.% [DBNH][OAc] solutions with a freezing time of 40 min, Table S3: Layer freeze crystallization data of aqueous 3 wt.% [DBNH][OAc] solutions with a freezing time of 60 min, Table S4: Layer freeze crystallization data of aqueous 6 wt.% [DBNH][OAc] solutions with a freezing time of 40 min, Table S5: Layer freeze crystallization data of aqueous 6 wt.% [DBNH][OAc] solutions with a freezing time of 60 min, Table S6: Suspension freeze crystallization data of aqueous 3 wt.% [DBNH][OAc] solutions with a freezing time of 40 min, Table S7: Suspension freeze crystallization data of aqueous 3 wt.% [DBNH][OAc] solutions with a freezing time of 60 min, Table S8: Suspension freeze crystallization data of aqueous 6 wt.% [DBNH][OAc] solutions with a freezing time of 40 min, Table S9: Suspension freeze crystallization data of aqueous 6 wt.% [DBNH][OAc] solutions with a freezing time of 60 min.

Author Contributions: Conceptualization of freeze crystallization experiments, N.O. and M.L.-K.; conceptualization of DSC analysis, L.Y. and H.L.; methodology, validation, and investigation of freeze crystallization, N.O.; methodology, validation, and investigation of DSC analysis L.Y.; writing—original draft,

N.O.; writing—review and editing, N.O., L.Y., and M.L.-K.; visualization, N.O. and L.Y.; supervision, M.L.-K. and H.L. All authors have read and agreed to the published version of the manuscript.

Funding: The first author is thankful to Aalto CHEM's funding for financial support.

Acknowledgments: The authors are grateful to Han Bing and Peter Schulze for their valuable assistance during experimental work as well as to Sanna Hellsten for providing synthetized ionic liquid.

Conflicts of Interest: The authors declare no conflict of interest. The funders had no role in the design of the study; in the collection, analyses, or interpretation of data; in the writing of the manuscript; or in the decision to publish the results.

References

1. Ulrich, J. Introduction. In *Melt Crystallization: Fundamentals, Equipment and Applications*; Ulrich, J., Glade, H., Eds.; Berichte aus der Verfahrenstechnik; Shaker: Aachen, Germany, 2003; pp. 1–6.
2. Parviainen, A.; Wahlström, R.; Liimatainen, U.; Liitiä, T.; Rovio, S.; Helminen, J.K.J.; Hyväkkö, U.; King, A.W.T.; Suurnäkki, A.; Kilpeläinen, I. Sustainability of Cellulose Dissolution and Regeneration in 1,5-Diazabicyclo[4.3.0]Non−5-Enium Acetate: A Batch Simulation of the IONCELL-F Process. *RSC Adv.* **2015**, *5*, 69728–69737. [CrossRef]
3. Wendler, F.; Todi, L.-N.; Meister, F. Thermostability of Imidazolium Ionic Liquids as Direct Solvents for Cellulose. *Thermochim. Acta* **2012**, *528*, 76–84. [CrossRef]
4. Vigier, K.D.O.; Chatel, G.; Jérôme, F. Contribution of Deep Eutectic Solvents for Biomass Processing: Opportunities, Challenges, and Limitations. *ChemCatChem* **2015**, *7*, 1250–1260. [CrossRef]
5. Freire, M.G.; Neves, C.M.S.S.; Marrucho, I.M.; Coutinho, J.A.P.; Fernandes, A.M. Hydrolysis of Tetrafluoroborate and Hexafluorophosphate Counter Ions in Imidazolium-Based Ionic Liquids. *J. Phys. Chem. A* **2010**, *114*, 3744–3749. [CrossRef] [PubMed]
6. Meine, N.; Benedito, F.; Rinaldi, R. Thermal Stability of Ionic Liquids Assessed by Potentiometric Titration. *Green Chem.* **2010**, *12*, 1711. [CrossRef]
7. Ahmad, W.; Ostonen, A.; Jakobsson, K.; Uusi-Kyyny, P.; Alopaeus, V.; Hyväkkö, U.; King, A.W.T. Feasibility of Thermal Separation in Recycling of the Distillable Ionic Liquid [DBNH][OAc] in Cellulose Fiber Production. *Chem. Eng. Res. Des.* **2016**, *114*, 287–298. [CrossRef]
8. Kakko, T.; King, A.W.T.; Kilpeläinen, I. Homogenous Esterification of Cellulose Pulp in [DBNH][OAc]. *Cellulose* **2017**, *24*, 5341–5354. [CrossRef]
9. Hanabusa, H.; Izgorodina, E.I.; Suzuki, S.; Takeoka, Y.; Rikukawa, M.; Yoshizawa-Fujita, M. Cellulose-Dissolving Protic Ionic Liquids as Low Cost Catalysts for Direct Transesterification Reactions of Cellulose. *Green Chem.* **2018**, *20*, 1412–1422. [CrossRef]
10. Ulrich, J.; Bülau, H.C. Melt Crystallization. In *Handbook of Industrial Crystallization*, 2nd ed.; Myerson, A.S., Ed.; Elsevier Science and Technology: Oxford, UK, 2001; pp. 161–180, ProQuest Ebook Central; Available online: https://ebookcentral.proquest.com/lib/aalto-ebooks/detail.action?docID=317222 (accessed on 31 January 2020).
11. Lewis, A.; Seckler, M.; Kramer, H.J.M.; van Rosmalen, G. *Industrial Crystallization: Fundamentals and Applications*; Cambridge University Press: Cambridge, UK, 2015; pp. 261–281.
12. Liu, Y.; Meyer, A.S.; Nie, Y.; Zhang, S.; Thomsen, K. Low Energy Recycling of Ionic Liquids via Freeze Crystallization during Cellulose Spinning. *Green Chem.* **2018**, *20*, 493–501. [CrossRef]
13. Parviainen, A.; King, A.W.T.; Mutikainen, I.; Hummel, M.; Selg, C.; Hauru, L.K.J.; Sixta, H.; Kilpeläinen, I. Predicting Cellulose Solvating Capabilities of Acid-Base Conjugate Ionic Liquids. *ChemSusChem* **2013**, *6*, 2161–2169. [CrossRef] [PubMed]
14. Stepan, A.M.; Michud, A.; Hellstén, S.; Hummel, M.; Sixta, H. IONCELL-P&F: Pulp Fractionation and Fiber Spinning with Ionic Liquids. *Ind. Eng. Chem. Res.* **2016**, *55*, 8225–8233. [CrossRef]
15. Hauru, L.K.J.; Hummel, M.; Michud, A.; Sixta, H. Dry Jet-Wet Spinning of Strong Cellulose Filaments from Ionic Liquid Solution. *Cellulose* **2014**, *21*, 4471–4481. [CrossRef]
16. Ma, Y.; Stubb, J.; Kontro, I.; Nieminen, K.; Hummel, M.; Sixta, H. Filament Spinning of Unbleached Birch Kraft Pulps: Effect of Pulping Intensity on the Processability and the Fiber Properties. *Carbohydr. Polym.* **2018**, *179*, 145–151. [CrossRef] [PubMed]

17. Asaadi, S.; Hummel, M.; Ahvenainen, P.; Gubitosi, M.; Olsson, U.; Sixta, H. Structural Analysis of Ioncell-F Fibres from Birch Wood. *Carbohydr. Polym.* **2018**, *181*, 893–901. [CrossRef] [PubMed]
18. Hauru, L.K.J.; Hummel, M.; Nieminen, K.; Michud, A.; Sixta, H. Cellulose Regeneration and Spinnability from Ionic Liquids. *Soft Matter* **2016**, *12*, 1487–1495. [CrossRef] [PubMed]
19. Lê, H.Q.; Sixta, H.; Hummel, M. Ionic Liquids and Gamma-Valerolactone as Case Studies for Green Solvents in the Deconstruction and Refining of Biomass. *Curr. Opin. Green Sustain. Chem.* **2019**, *18*, 20–24. [CrossRef]
20. Gabbott, P. A Practical Introduction to Differential Scanning Calorimetry. In *Principles and Applications of Thermal Analysis*; Gabbott, P., Ed.; Blackwell Publishing Ltd.: Oxford, UK, 2008; pp. 1–50. [CrossRef]
21. Gmehling, J.; Kolbe, B.; Kleiber, M.; Rarey, J. *Chemical Thermodynamics for Process. Simulation*; Wiley-VCH Verlag & Co. KGaA: Weinheim, Germany, 2012; pp. 405–437.
22. Sippola, H.; Taskinen, P. Activity of Supercooled Water on the Ice Curve and Other Thermodynamic Properties of Liquid Water up to the Boiling Point at Standard Pressure. *J. Chem. Eng. Data* **2018**, *63*, 2986–2998. [CrossRef] [PubMed]

© 2020 by the authors. Licensee MDPI, Basel, Switzerland. This article is an open access article distributed under the terms and conditions of the Creative Commons Attribution (CC BY) license (http://creativecommons.org/licenses/by/4.0/).

Article

Effects of Synthesis Parameters on Crystallization Behavior of K-MER Zeolite and Its Morphological Properties on Catalytic Cyanoethylation Reaction

Ying-Wai Cheong [1], Ka-Lun Wong [2], Boon Seng Ooi [3], Tau Chuan Ling [4], Fitri Khoerunnisa [5] and Eng-Poh Ng [1,*]

1. School of Chemical Sciences, Universiti Sains Malaysia, Penang 11800 USM, Malaysia; thia911@hotmail.com
2. School of Energy and Chemical Engineering, Xiamen University Malaysia, Jalan Sunsuria, Bandar Sunsuria, Sepang 43900, Selangor, Malaysia; kalun.wong@xmu.edu.my
3. School of Chemical Engineering, Engineering Campus, Universiti Sains Malaysia, Seri Ampangan, Nibong Tebal 14300, Penang, Malaysia; chobs@usm.my
4. Institute of Biological Sciences, Faculty of Science, University of Malaya, Kuala Lumpur 50603, Malaysia; tcling@um.edu.my
5. Chemistry Education Department, Universitas Pendidikan Indonesia, Jl. Setiabudhi 258, Bandung 40514, Indonesia; fitri.khoerunnisa@gmail.com
* Correspondence: epng@usm.my

Received: 24 December 2019; Accepted: 19 January 2020; Published: 23 January 2020

Abstract: MER-type zeolite is an interesting microporous material that has been widely used in catalysis and separation. By carefully controlling the synthesis parameters, a procedure to synthesize K-MER zeolite crystals with various morphologies has been developed. The silica, water and mineralizer content in the synthesis gel, as well as crystallization time and temperature, have a profound impact on the crystallization kinetics, resulting in zeolite solids with various degrees of crystallinity, crystal sizes and shapes. K-MER zeolite crystals with nanorod, bullet-like, prismatic and wheatsheaf-like morphologies have been successfully obtained. The catalytic performances of the K-MER zeolites in cyanoethylation of methanol, under novel non-microwave instant heating, have been investigated. The zeolite in nanosize form shows the best catalytic performance (94.1% conversion, 100% selectivity) while the bullet-like zeolite gives poorest catalytic performance (44.2% conversion, 100% selectivity).

Keywords: K-MER zeolite; synthesis parameter; morphology; cyanoethylation of methanol; catalyst

---

1. Introduction

Zeolites are crystalline microporous aluminosilicates formed by a network of $[SiO_4]^{4-}$ and $[AlO_4]^{5-}$ tetrahedrals [1]. These materials have unique properties, such as their uniform microporous structure, hydrophilic surfaces, adjustable framework composition and strong chemical interactions with guest molecules. These properties have made them suitable candidates for widespread applications, including gas separation, adsorption, ion-exchange and catalysis [2–8]. The search for new materials is important especially for the petrochemical and pharmaceutical industries. In particular, potassium containing MER zeolite (K-MER), a zeolite which has framework topology similar to the mineral merlinoite; has drawn researchers' attention due to its three-dimensional pore channel system with medium pore sizes and high hydrophilicity. Thus, it is a promising material for adsorption [9], separation [10], and catalytic applications [11,12].

The physicochemical properties of zeolites are greatly dependent on their framework structures. Furthermore, the morphological properties such as crystal shape and size of a zeolite also have great impact on their physicochemical properties and applications [13,14]. For example, the shape of zeolite crystals has been shown to exert a significant effect in adsorption and separation applications since it modulates molecular diffusion, accessibility, interfacial energy, molecular separation and inclusion properties [15,16]. Recently, several techniques to control the morphological properties of zeolites were reported [17]. For example, the use of different organic templates has shown to produce AlPO-5 zeolite-like materials with aggregated sphere, plate, rod, prism and barrel shapes [18–20]. The employment of different heating methods such as microwave and ultrasonic radiations exhibit significant effects on the overall crystal size and shape of the zeolite products [21,22]. In addition, the influence of the hydrothermal synthesis parameters, such as silica and alumina contents, alkalinity, amount of water, type of organic template, type of mineralizer, crystallization time and temperature, on the morphological and other properties of zeolites are also reported [23–25]. Nevertheless, knowledge about the impact of these synthesis parameters on the morphological behavior, crystallization kinetics and formation process of K-MER zeolite is still not well understood.

Therefore, we have studied the influence of synthesis parameters on the formation of K-MER zeolite (e.g. crystallization kinetics, structure and purity, morphology and crystal size). In addition, the influence of crystal morphologies on the catalytic behavior of K-MER zeolite in the cyanoethylation of methanol, under novel non-microwave instant heating, is also presented in this paper.

2. Experimental

2.1. Synthesis of MER Zeolite

The synthesis of MER zeolite (W-3) was carried out as follows. Typically, Al(OH)$_3$ (1.448 g, extra pure, Acros Organics) and KOH (4.290 g, 85%, QRëC) were mixed in distilled water (11.876 g). The mixture was magnetically stirred at 100 °C for 16 h. The clear aluminate solution was then slowly introduced into the silicate solution comprising HS-40 (9.763 g, 40% SiO$_2$, Sigma–Aldrich) and distilled water (15.022 g). The resulting hydrogel with a molar composition of 1Al$_2$O$_3$:7SiO$_2$:3.5K$_2$O:196H$_2$O was stirred for another 10 min before crystallization at 180 °C for 14 h. The solid product obtained was purified with distilled water using high speed centrifugation (10,000 rpm, 10 min) until pH 7 and the sample was freeze-dried. Other samples were also prepared by varying the synthesis conditions as summarized in Table 1 using the same procedure. The samples were labelled as W-n where n was the number of the sample.

2.2. Characterization

The XRD patterns of the samples were recorded using a Bruker AXS D8 (MA, USA) diffractometer with Cu-Kα radiation (λ = 1.5418 Å) at 2θ = 3–50°. The morphology of solids was studied with a FESEM microscope (Leo Supra 50VP, Oberkochen, Germany) operating at 20 kV. The average crystal size of solids was determined using ImageJ software by counting 50 crystals randomly throughout the FESEM images. The Si/Al and K/Al ratios of the solids were determined by using a Perkin Elmer's atomic absorption spectrometer (AAS, AAnalyst 400). Prior to analysis, the sample was dissolved in hydrofluoric acid solution (0.5 M) where boric acid was also added to minimize the fluoride interference. Meanwhile, five standard solutions of each Al, Si and K elements were also prepared for calibration study. The surface area of the samples was calculated by using the BET equation where the monolayer volume (V_{mono}) was first obtained from the nitrogen adsorption isotherm; the samples were first degassed (6 h, 300 °C) before the adsorption isotherms were recorded from a Micrometrics ASAP 2010 analyzer (Norcross, USA) at −196 °C. The surface basicity of MER zeolite samples was analyzed using a BELCAT-B temperature programmed desorption (TPD) instrument (Osaka, Japan). Initially, the solid sample (ca. 100 mg) was outgassed at 450 °C overnight before CO$_2$ gas was introduced for adsorption. The excess CO$_2$ was then evacuated at room temperature and CO$_2$ desorption was performed from 40 to 500 °C at a heating rate of 10 °C/min. The TPD profile was plotted as TCD signal versus desorption temperature.

2.3. Catalytic Study

Catalytic cyanoethylation of methanol was performed using the following procedure. First, K-MER zeolite (0.500 g), methanol (28 mmol, Merck) and acrylonitrile (7 mmol, Merck) were loaded into a 10-mL quartz vial. The vial was sealed with a silicon cap, heated (140 °C) and magnetically stirred (800 rpm) in a non-microwave instant heating reactor (Anton Paar's Monowave 50, Graz, Austria) for several specific heating times (0-90 min). After cooling down, the reaction solution was isolated from the solid catalyst and injected into a GC–MS (Perkin-Elmer Clarus 500 system, Massachusetts, USA) and a GC–FID (Agilent/HP 6890 GC, Califonia, USA) for identification and quantitative analysis, respectively.

3. Results and Discussion

3.1. Effect of Heating Time

The crystallization of K-MER zeolite was first studied by varying the hydrothermal heating times at 180 °C. An amorphous solid product was formed at 0 h according to the XRD analysis where a strong broad XRD hump at 2θ = 22.3° was detected (Figure 1a: W-1). The XRD data was supported by FESEM observation of nanoparticles (ca. 58 nm) with coral-like structure (Figure 2a). With a heating time of 10 h, a significant drop in the Si/Al ratio from 7.81 to 3.73 was observed in the solids (Table 1). Yet, no crystalline phase was revealed by XRD technique at this time. Nevertheless, the amorphous hump became weaker and shifted to 2θ = 27.8° (Figure 1b: W-2). Both XRD and AAS elemental analyses thus revealed the occurrence of amorphous phase reorganization into secondary more reactive amorphous solid at 10 h [26]. This amorphous-amorphous phase transformation was also confirmed by FESEM study where bulkier amorphous entities (ca. 180 nm) were formed (Figure 2b). Interestingly, the appearance of particles with more well-defined nanorod morphology (ca. 33 × 4 nm^2) was detected randomly in W-2 sample via microscopic investigation. This indicated that the nucleation process of K-MER zeolite had occurred. These nanorods were agglomerated into bundle-like secondary particles (ca. 650 nm) and were grown on the surface of an amorphous particle.

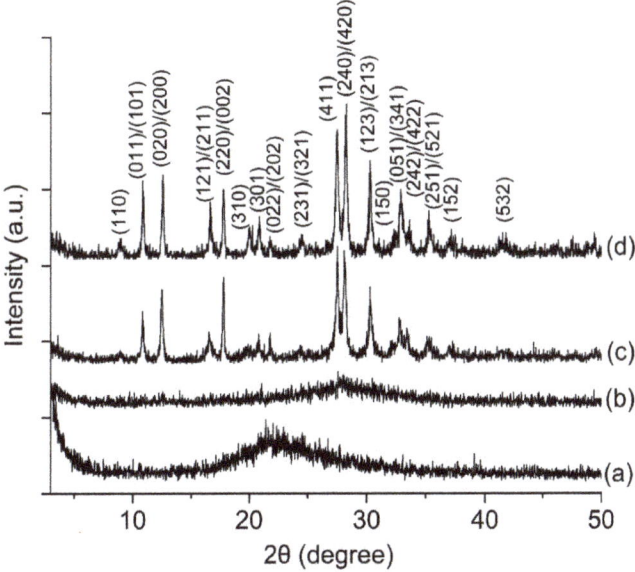

Figure 1. XRD patterns of (**a**) W-1, (**b**) W-2, (**c**) W-3 and (**d**) W-4 samples heated for 0, 10, 14 and 20 h, respectively.

Figure 2. FESEM images of (**a**) W-1, (**b**) W-2, (**c**) W-3 and (**d**) W-4 samples heated for 0, 10, 14 and 20 h, respectively.

Table 1. The chemical compositions of precursor hydrogels and their respective synthesis conditions.

Parameters	Samples	Gel Molar Composition				T (°C)	t (h)	Si/Al Ratio	Phase(s) [a]
		SiO_2	Al_2O_3	K_2O	H_2O				
Time	W-1	7	1	3.5	196	180	0	7.81	Am.
	W-2						10	3.73	Am.>>MER
	W-3						14	2.29	MER
	W-4						20	2.28	MER
Temperature	W-5	7	1	3.5	196	120	14	3.29	Am.
	W-6					140		2.63	MER
	W-7					160		2.61	MER
	W-3					180		2.29	MER
K_2O	W-3	7	1	3.5	196	180	14	2.29	MER
	W-8			5.0				2.29	MER
	W-9			7.0				2.28	MER
SiO_2	W-10	1.5	1	3.5	130	180	14	1.22	EDI
	W-11	5						2.53	MER
	W-12	7						2.73	LTL<MER
	W-13	10						3.05	LTL
H_2O	W-14	7	1	3.5	100	180	14	2.82	LTL<MER
	W-12				130			2.73	LTL<MER
	W-3				196			2.29	MER
	W-16				280			2.31	MER

[a] Am. = Amorphous.

When the heating time was prolonged to 14 h, the amorphous solids were completely consumed as nutrient and subsequently crystalline K-MER zeolite nanocrystals were produced (Figure 1c: W-3). At this time, the Si/Al ratio became nearly constant, with a value of 2.29. As shown in Figure 2c, the crystalline primary (ca. 40 nm) and secondary (ca. 1.3 μm) particles had grown to a larger size due to simultaneous occurrence of nucleation and crystal growth processes. The XRD analysis also supported this conclusion, as no amorphous hump was seen in the XRD pattern (Figure 1c). Indeed, the pattern showed major peaks at 2θ = 8.96°, 10.84°, 12.46°, 16.58°, 17.80°, 27.50°, 28.10°, 30.28° and 32.88° which were characteristics of the MER framework topology [27]. Further increasing the heating time to 20 h showed no change in the framework composition and framework type but the XRD peaks with higher intensity and narrower peaks were recorded (Figure 1d: W-4), indicating Ostwald ripening and crystal growth were dominating the crystallization process [28]. This was supported by the FESEM microscopy showing that the crystalline K-MER zeolite nanorods further agglomerated and transformed into larger MER crystals (>1 µm) with bullet-shape morphology.

3.2. Effect of Heating Temperature

Heating temperature plays a very crucial role in zeolite crystallization process as it provides energy to overcome the activation energy of the reactions (polycondensation, induction, nucleation, crystal growth, etc.) [29,30]. Hence, the hydrogel with the same molar composition ($1Al_2O_3:7SiO_2:3.5K_2O:196H_2O$) was subjected to hydrothermal treatment at 120, 140, 160 and 180 °C for 14 h. The W-5 solid product appeared to be amorphous at 120 °C and the Si/Al ratio of the solid was 3.29 (Figures 3a and 4a, Table 1). The chemical composition reached nearly 2.60 when the synthesis temperature was raised to 140 °C. At this temperature, K-MER zeolite nanorods (ca. 28 nm) were obtained which tend to form secondary agglomerated particles of ca. 240 nm (Figures 3b and 4b: W-6). Upon increasing the temperature to 160 °C (W-7) and 180 °C (W-3), no change in the framework chemical composition (Si/Al ratio) and crystalline phase were detected but the size of primary and secondary particles became larger due to further crystals growth at higher temperature (Figure 3c,d and Figure 4c,d). The results indicate that the crystallization of MER zeolite is a thermally activated reaction.

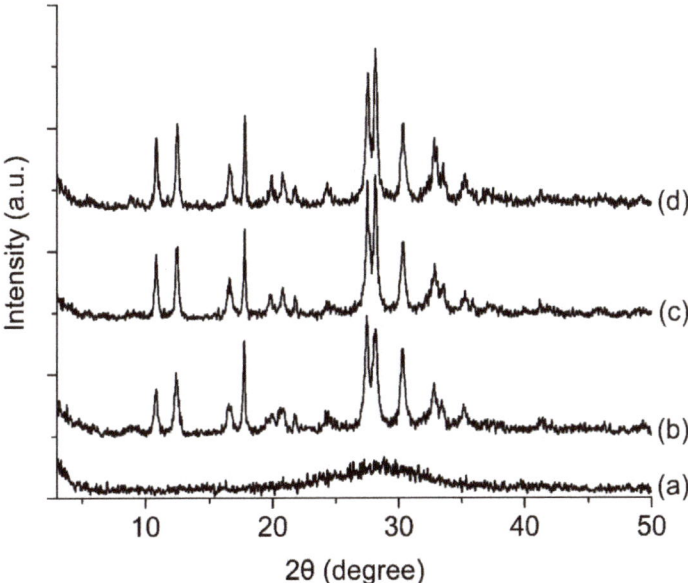

Figure 3. XRD patterns of (**a**) W-5, (**b**) W-6, (**c**) W-7 and (**d**) W-3 samples upon heating at 120, 140, 160 and 180 °C for 14 h, respectively.

Figure 4. FESEM images of (**a**) W-5, (**b**) W-6, (**c**) W-7 and (**d**) W-3 samples upon heating at 120, 140, 160 and 180 °C for 14 h, respectively.

3.3. Effect of K_2O Content

Mineralizer is an essential chemical component in the synthesis of zeolites as it increases the overall solubility of inorganic species in the precursor hydrogels by providing accessibility to useful species at a level needed for the nucleation and crystal growth of zeolites. Hence, a precursor hydrogel with a composition of $7SiO_2: 1Al_2O_3: xK_2O: 196H_2O$ (x = 3.5, 5.0 and 7.0) was heated at 180 °C for 14 h to study the effect of K_2O mineralizer. From the XRD data, the framework chemical composition and the purity of crystallized products remained intact with no other competing phases detected in the K_2O range studied (Table 1, Figure 5). However, a change in the crystal morphology was observed upon varying the K_2O content. At x = 3.5 (W-3), the obtained crystals were in nanorod morphology, while a mixing of nanorods and prismatic crystals were observed at x = 5.0 (W-8), demonstrating the significant effect of mineralizer content on the morphological properties of K-MER zeolite (Figure 6a). A similar observation was also reported by Zhang et al. where the addition of a higher amount of NaOH as mineralizer promotes the crystal growth at a specific axis, resulting in the crystallization of silicalite-1 zeolite with spherical shape instead of common coffin shape [31]. Further increasing the K_2O content (x = 7.0: W-9) led to the formation of K-MER zeolite with only pure prismatic-shape crystals (Figure 6c). As seen, the average size of prismatic shape crystals also reduced from 125 × 167 nm^2 to 114 × 160 nm^2 when the K_2O content increased due to the enhancement of the solubility of aluminosilicate species as a result of higher alkalinity in the hydrogel solution [32].

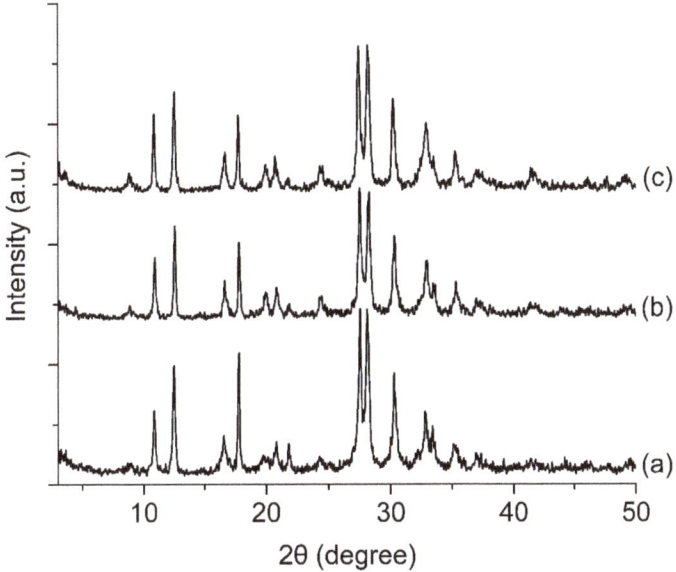

Figure 5. XRD patterns of samples prepared from an aluminosilicate gel precursor with a composition of 7SiO$_2$: 1Al$_2$O$_3$: xK$_2$O: 196H$_2$O with x = (**a**) 3.5 (W-3), (**b**) 5 (W-8) and (**c**) 7 (W-9). All samples were heated at 180 °C for 14 h.

Figure 6. FESEM images of (**a**) W-3, (**b**) W-8 and (**c**) W-9 samples prepared from an aluminosilicate gel precursor with the composition of 7SiO$_2$: 1Al$_2$O$_3$: xK$_2$O: 196H$_2$O with x = 3.5, 5 and 7, respectively. The arrows shown in (**b**) indicate the existence of K-W nanorod crystals in midst of K-W prismatic crystals.

3.4. Effect of SiO$_2$ Content

The effect of SiO$_2$ content was also studied ranging from SiO$_2$/Al$_2$O$_3$ = 1.5 to 10. The results showed that the phase purity was found to be more sensitive by varying the SiO$_2$ content as compared to the K$_2$O content and heating temperature (Table 1). Crystalline W-10 with a high silica content (Si/Al ratio = 1.22) and cubic morphology was obtained when the hydrogel with low silica content (SiO$_2$/Al$_2$O$_3$ ratio = 1.5) was used. The solid was proven to be an EDI-type zeolite according to XRD and SEM analyses (Figures 7a and 8a). At a SiO$_2$/Al$_2$O$_3$ molar ratio of 5.0, W-11 (Si/Al ratio = 2.53) with a pure MER crystalline phase was produced (Figures 7b and 8b). Further increasing the silica content led to the co-crystallization of MER- and LTL-type zeolites before single phase of LTL-type zeolite product was crystallized at a SiO$_2$/Al$_2$O$_3$ molar ratio of 10 (Figure 7c,d and Figure 8c,d). As shown, the LTL-type zeolite (W-13), having a one-dimensional pore structure, possessed a higher framework silica content (Si/Al ratio = 3.05) and exhibited a novel spinning top-like shape instead of conventional cylindrical structure [33] owing to the silica source and the precursor molar composition used.

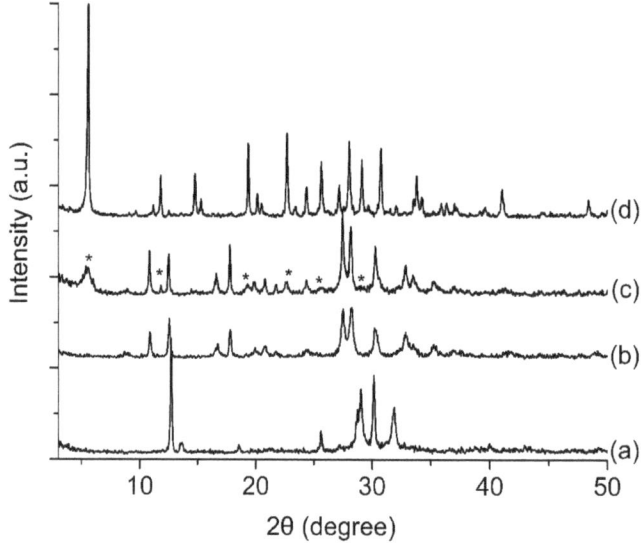

Figure 7. XRD patterns of (**a**) W-10, (**b**) W-11, (**c**) W-12 and (**d**) W-13 samples crystallized at 180 °C for 14 h using an aluminosilicate gel precursor of $y\text{SiO}_2:1\text{Al}_2\text{O}_3:3.5\text{K}_2\text{O}:130\text{H}_2\text{O}$ where y = 1.5, 5, 7 and 10, respectively. The * marks in (**c**) indicate the presence of the LTL crystalline phase.

Figure 8. FESEM images of solids prepared using a precursor hydrogel of $y\text{SiO}_2:1\text{Al}_2\text{O}_3:3.5\text{K}_2\text{O}:130\text{H}_2\text{O}$ at 180 °C for 14 h where y = (**a**) 1.5 (W-10), (**b**) 5 (W-11), (**c**) 7 (W-12) and (**d**) 10 (W-13).

3.5. Effect of Water Content

Water content plays an important role in the hydrothermal synthesis of zeolites. It not only serves as a solvent during the crystallization process, but also regulates the formation of possible precursors of zeolite frameworks [34]. Hence, the crystallization of K-MER zeolite was studied at 180 °C for 14 h by varying the molar ratio of water from 100 to 280 (Table 1). At low water content (H_2O/Al_2O_3 = 100: W-14), LTL crystalline phase was co-crystallized with the MER crystalline phase (Figure 9a). This observation could also be confirmed by the AAS spectroscopy results where the Si/Al ratio of solid was 2.82 indicating the mixing of both high silica (LTL) and low silica (MER) zeolites. The results were further supported by FESEM analysis where cylindrical nanorods of ca. 350 × 970 nm^2, which were characteristic of LTL-type zeolite, were observed and grown together with the K-MER nanocrystals (Figure 10a). The LTL crystalline phase, however, slowly disappeared with increasing water content, indicating the direct influence of water on the crystallization of LTL-type zeolite. At higher H_2O/Al_2O_3 ratios (196–280), zeolites of pure MER crystalline phase with different crystallite sizes and morphologies were obtained. As shown, the bundle-like MER-type zeolite particles made up of nanorod primary crystals (205 × 40 nm^2) were captured at H_2O/Al_2O_3 = 196 (Figures 9c and 10c: W-3). The nanocrystals grew further, and larger secondary crystals with wheatsheaf morphology were formed when the water content was further increased to H_2O/Al_2O_3 = 280 (Figures 9d and 10d: W-16). The change of morphology and increment in particle size could be explained by the dilution of nutrient when the water content increased. As a result, the low concentration of nutrients in the synthesis medium favors the crystal growth more than nucleation process. Hence, particles with larger size and different morphology are formed [35].

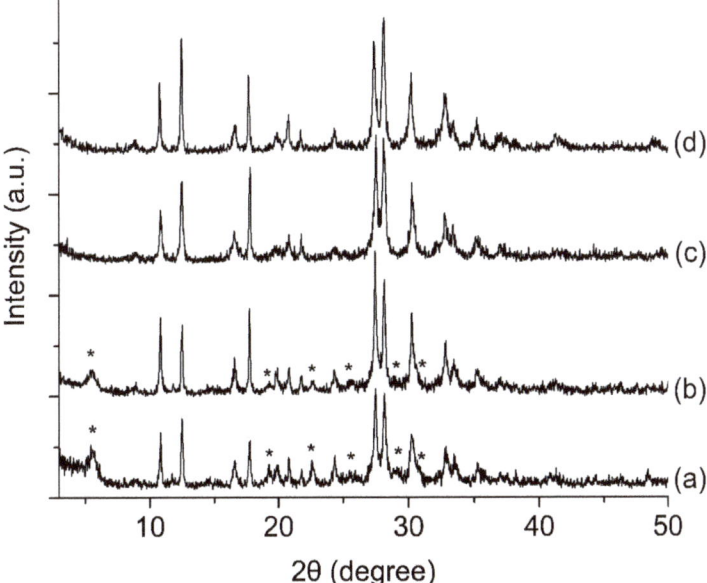

Figure 9. XRD patterns of (**a**) W-14, (**b**) W-12, (**c**) W-3 and (**d**) W-16 samples prepared from an aluminosilicate precursor hydrogel with a composition of $7SiO_2$: $1Al_2O_3$: $3.5K_2O$: zH_2O with z = 100, 130, 196 and 280, respectively. All samples were heated at 180 °C for 14 h. The * marks in (**a**,**b**) indicate the presence of the LTL crystalline phase.

Figure 10. FESEM images of (**a**) W-14, (**b**) W-12, (**c**) W-3 and (**d**) W-16 samples solids prepared using a gel precursor of 7SiO$_2$:1Al$_2$O$_3$:3.5K$_2$O: zH$_2$O at 180 °C for 14 h where z = 100, 130, 196 and 280, respectively.

3.6. Morphological Effects on the Catalytic Behavior of K-MER Zeolite

By changing the synthesis parameters, K-MER zeolites with four distinct morphologies, namely nanorod (W-3), bullet-like (W-4), prismatic (W-9) and wheatsheaf-like (W-16) shapes, were obtained. Nevertheless, from the elemental analysis, all the samples exhibited nearly similar Si/Al ratio (ca. 2.29), which was close to the theoretical one (2.12) (Table 2) [36]. Furthermore, the K/Al ratio of all the samples was found to be near unity due to the fact that the positive charge of each K$^+$ non-framework cation has to be counter-balanced by a negative charge contributed by an Al atom in the [Si-O-Al]$^-$ form [25]. On the other hand, the total surface area of the samples was measured with the N$_2$ adsorption isotherm analysis. Note that the total surface area determined was actually contributed only by the external surfaces because the size of N$_2$ molecules is too large to probe the micropores of the K-MER zeolite [27,37]. The results indicated that the external surface area had positive correlation with the crystallite size of K-MER zeolite. For instance, K-MER zeolite with nanorod shape had the highest external surface area (39.57 m^2/g) and the external surface area generally reduced as the crystallite size increased.

The surface basicity of these four zeolite samples was also characterized by using CO$_2$-TPD. Upon CO$_2$ adsorption and desorption from all samples, four deconvoluted signals with different intensities were observed indicating that the morphology had considerable effects on the basic strengths (weak basic sites: ca. 105 and 175 °C; medium basic sites: ca. 260 °C, medium-strong: 330 °C) (Figure 11). The number of active sites with different basic strengths was also quantified based on the amount of CO$_2$ sorbed per gram of zeolite (Table 2). It was found that the number of total active sites (weak, medium and medium-strong basic strengths) was linearly proportional to the surface area of K-MER zeolite (R^2 = 0.954). In fact, this is not surprising because most of the accessible basic sites, namely [Si-O-Al]$^-$K$^+$, are located at the external surfaces of the zeolite particles. As a result, K-MER zeolite with smaller crystallite size exhibited a larger number of basic sites [38]. In addition, the low Si/Al ratio of the zeolite framework also contributed to the basicity of K-MER zeolite because when the Si/Al ratio is low, more K$^+$ cations are needed by the zeolite for surface charge counter-balance, which leads to the enhancement of zeolite basicity. Nanorod-shaped K-MER zeolite appeared to have the largest number of medium-strong basic sites (2.03 mmol/g) followed by prismatic (0.94 mmol/g), wheatsheaf-like (0.65 mmol/g) and bullet-like (0.27 mmol/g) K-MER zeolite.

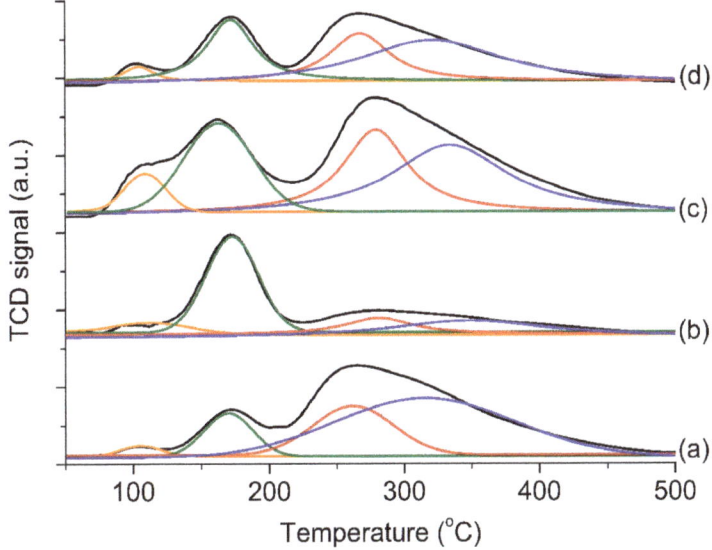

Figure 11. TPD-CO$_2$ profiles of (**a**) nanorod (W-3), (**b**) bullet-like (W-4), (**c**) prismatic (W-9), and (**d**) wheatsheaf-like K-MER zeolites.

Table 2. TPD-CO$_2$ basicity of K-MER zeolites with various morphologies.

Sample	Morphology	Si/Al Ratio	K/Al Ratio	Surface Area (m^2/g) [a]	TPD-CO$_2$ Basicity (mmol/g)			
					Weak	Medium	Medium-Strong	Total
W-3	Nanorod	2.29	1.04	39.57	0.08	1.99	2.03	3.10
W-4	Bullet	2.28	1.07	12.24	0.18	0.95	0.27	1.39
W-9	Prismatic	2.28	0.97	26.63	0.24	1.53	0.94	2.71
W-16	Wheatsheaf	2.31	1.06	9.01	0.04	0.56	0.65	1.25

[a] Equivalent to external surface area because micropore surface area was not measureable due to small micropore size of K-MER zeolite.

To study the morphological influences on the catalytic properties, the K-MER zeolites were tested in a model base-catalyzed reaction, i.e., cyanoethylation of methanol. The cyanoethylation of methanol with acrylonitrile was carried out under non-microwave instant heating where K-MER zeolites with different morphologies (W-3, W-4, W-9 and W-16) were used as the base catalysts (Figure 12). In general, the catalytic reactivity had a strong correlation with the morphology of zeolite catalysts. Remarkably, K-MER zeolite nanorods (W-3) exhibited superior catalytic activity with 94.1% of methanol conversion (100% selectivity) within 45 min of reaction at 140 °C, which could be explained by the largest number of accessible basic sites (particularly medium-strong basic sites) at its external surface. In contrast, bullet-like K-MER zeolite catalyst (W-4), which had the largest crystallite size and the lowest number of basic sites, showed the lowest catalytic conversion (44.2%) among the four K-MER zeolites studied. Hence, the results showed that the morphological properties had a direct influence on the catalytic activity of a zeolite whereby the morphology is directly associated with the number of accessible catalytic active sites [39]. Comprehensive work on the aspects of molecular diffusion on K-MER zeolites with different morphologies is in progress.

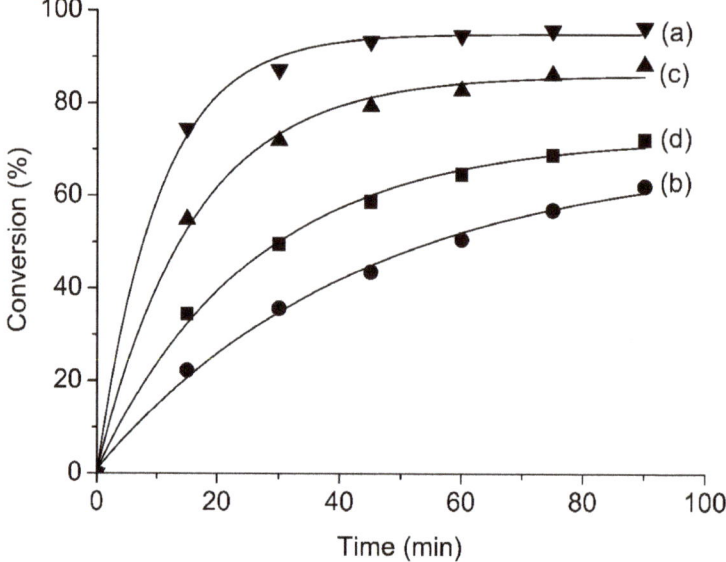

Figure 12. Catalytic performance of (**a**) nanorod (W-3), (**b**) bullet-like (W-4), (**c**) prismatic (W-9), and (**d**) wheatsheaf-like K-MER zeolites on cyanoethylation of methanol at 140 °C.

4. Conclusions

In conclusion, the effects of synthesis parameters on the crystallization profiles of K-MER zeolite have been investigated. The results reveal that the chemical composition of hydrogel (silica content, water concentration, and mineralizer loading), hydrothermal synthesis time and temperature are found to have many profound effects on the physico-chemical properties of the solid samples such as nucleation and crystallization rates, crystallite size, crystalline phase, purity and morphological properties. By carefully tuning the synthesis conditions, K-MER zeolites with four distinct morphologies, namely nanorod, bullet-like, prismatic and wheatsheaf-like shapes, have been successfully prepared. Furthermore, the influence of morphological properties of K-MER zeolites on their catalytic behavior has also been investigated. The results reveal that K-MER zeolite with nanorod shape gives the best catalytic performance in the cyanoethylation of methanol (94.1%) where its catalytic activity is associated with its stronger and higher number of accessible basic sites located at the external surface. Hence, from a material engineering point of view, this work not only provides an insight on the crystallization process of zeolite but also provides insights on designing zeolites with specific morphologies for advanced applications such as high throughput membranes for catalysis and selective gas separation.

Author Contributions: Y.-W.C.: Performed most of the experiments, wrote the manuscript; F.K., T.C.L. and E.-P.N.: wrote and proof-read the manuscript supervised; Y.-W.C.; conceived the entire project, applied for research funding; K.-L.W. and B.S.O.: ran XRD, SEM investigation, proofread the manuscript. All authors have read and agreed to the published version of the manuscript.

Funding: The financial support from a FRGS (203/PKIMIA/6711642) grant is gratefully acknowledged. Y.-W. Cheong would also like to thank the USM Fellowship for the scholarship provided.

Conflicts of Interest: The authors declare no conflict of interest.

References

1. Davis, M.E. Ordered porous materials for emerging applications. *Nature* **2002**, *417*, 813–821. [CrossRef] [PubMed]
2. Saqib, N.U.; Adnan, R.; Shah, I. Zeolite supported TiO_2 with enhanced degradation efficiency for organic dye under household compact fluorescent light. *Mater. Res. Exp.* **2019**, *6*, 095506. [CrossRef]
3. Derakhshankhah, H.; Hajipour, M.J.; Barzegari, E.; Lotfabadi, A.; Ferdousi, M.; Saboury, A.A.; Ng, E.-P.; Raoufi, M.; Awala, H.; Mintova, S.; et al. Zeolite nanoparticles inhibit Aβ–fibrinogen interaction and formation of a consequent abnormal structural clot. *ACS Appl. Mater. Interfaces* **2016**, *8*, 30768–30779. [CrossRef] [PubMed]
4. Majano, G.; Ng, E.-P.; Lakiss, L.; Mintova, S. Nanosized molecular sieves utilized as an environmentally friendly alternative to antioxidants for lubricant oils. *Green Chem.* **2011**, *13*, 2435–2440. [CrossRef]
5. Balkus, K.J., Jr.; Shi, J. A study of suspending agents for gadolinium (III)-exchanged hectorite. An oral magnetic resonance imaging contrast agent. *Langmuir* **1996**, *12*, 6277–6281. [CrossRef]
6. Adam, F.; Appaturi, J.N.; Ng, E.-P. Halide aided synergistic ring opening mechanism of epoxides and their cycloaddition to CO_2 using MCM-41-imidazolium bromide catalyst. *J. Mol. Catal. A Chem.* **2014**, *386*, 42–48. [CrossRef]
7. Ng, E.-P.; Lim, G.K.; Khoo, G.-L.; Tan, K.-H.; Ooi, B.S.; Adam, F.; Ling, T.C.; Wong, K.-L. Synthesis of colloidal stable Linde Type J (LTJ) zeolite nanocrystals from rice husk silica and their catalytic performance in Knoevenagel reaction. *Mater. Chem. Phys.* **2015**, *155*, 30–35. [CrossRef]
8. Ng, E.-P.; Nur, H.; Wong, K.-L.; Muhid, M.N.M.; Hamdan, H. Generation of Brönsted acidity in AlMCM-41 by sulphation for enhanced liquid phase tert-butylation of phenol. *Appl. Catal. A Gen.* **2007**, *323*, 58–65. [CrossRef]
9. Kakutani, Y.; Weerachawanasak, P.; Hirata, Y.; Sano, M.; Suzuki, T.; Miyake, T. Highly effective K-Merlinoite adsorbent for removal of Cs^+ and Sr^{2+} in aqueous solution. *RSC Adv.* **2017**, *7*, 30919–30928. [CrossRef]
10. Mirfendereski, S.M. Synthesis and application of high-permeable MER zeolite membrane for separation of carbon dioxide from methane. *J. Aust. Ceram. Soc.* **2019**, *55*, 103–114. [CrossRef]
11. Seo, Y.H.; Prasetyanto, E.A.; Jiang, N.; Oh, S.M.; Park, S.E. Catalytic dehydration of methanol over synthetic K-MER zeolite. *Microporous Mesoporous Mater.* **2010**, *128*, 108–114. [CrossRef]
12. Cheong, Y.-W.; Wong, K.-L.; Ling, T.C.; Ng, E.-P. Rapid synthesis of nanocrystalline zeolite W with hierarchical mesoporosity as an efficient solid basic catalyst for nitroaldol Henry reaction of vanillin with nitroethane. *Mater. Express* **2018**, *8*, 463–468. [CrossRef]
13. Liu, X.-D.; Wang, Y.-P.; Cui, X.-M.; He, Y.; Mao, J. Influence of synthesis parameters on NaA zeolite crystals. *Powder Technol.* **2013**, *243*, 184–193. [CrossRef]
14. Sivalingam, S.; Sen, S. Optimization of synthesis parameters and characterization of coal fly ash derived microporous zeolite X. *Appl. Surf. Sci.* **2018**, *455*, 903–910. [CrossRef]
15. Quan, Y.; Li, S.; Wang, S.; Li, Z.; Dong, M.; Qin, Z.; Chen, G.; Wei, Z.; Fan, W.; Wang, J. Synthesis of chainlike ZSM-5 zeolites: Determination of synthesis parameters, mechanism of chainlike morphology formation, and their performance in selective adsorption of xylene isomers. *ACS Appl. Mater. Interfaces* **2017**, *9*, 14899–14910. [CrossRef]
16. Au, L.T.Y.; Yeung, K.L. An investigation of the relationship between microstructure and permeation properties of ZSM-5 membranes. *J. Membr. Sci.* **2001**, *194*, 33–55. [CrossRef]
17. Li, S.; Li, J.; Dong, M.; Fan, S.; Zhao, T.; Wang, J.; Fan, W. Strategies to control zeolite particle morphology. *Chem. Soc. Rev.* **2019**, *48*, 885–907. [CrossRef]

18. Jhung, S.H.; Chang, J.-S.; Hwang, Y.K.; Park, S.-E. Crystal morphology control of AFI type molecular sieves with microwave irradiation. *J. Mater. Chem.* **2004**, *14*, 280–285. [CrossRef]
19. Khoo, D.Y.; Kok, W.-M.; Mukti, R.R.; Mintova, S.; Ng, E.-P. Ionothermal approach for synthesizing AlPO-5 with hexagonal thin-plate morphology influenced by various parameters at ambient pressure. *Solid State Sci.* **2013**, *25*, 63–69. [CrossRef]
20. Ng, E.-P.; Ng, D.T.-L.; Awala, H.; Wong, K.-L.; Mintova, S. Microwave synthesis of colloidal stable AlPO-5 nanocrystals with high water adsorption capacity and unique morphology. *Mater. Lett.* **2014**, *132*, 126–129. [CrossRef]
21. Ou, X.; Xu, S.; Warnett, J.M.; Holmes, S.M.; Zaheer, A.; Garforth, A.A.; Williams, M.A.; Jiao, Y.; Fan, X. Creating hierarchies promptly: Microwave-accelerated synthesis of ZSM-5 zeolites on macrocellular silicon carbide (SiC) foams. *Chem. Eng. J.* **2017**, *312*, 1–9. [CrossRef]
22. Askari, S.; Halladj, R. Ultrasonic pretreatment for hydrothermal synthesis of SAPO-34 nanocrystals. *Ultrason. Sonochem.* **2012**, *19*, 554–559. [CrossRef] [PubMed]
23. Ng, E.-P.; Chow, J.-H.; Mukti, R.R.; Muraza, O.; Ling, T.C.; Wong, K.-L. Hydrothermal synthesis of zeolite a from bamboo leaf biomass and its catalytic activity in cyanoethylation of methanol under autogenic pressure and air conditions. *Mater. Chem. Phys.* **2017**, *201*, 78–85. [CrossRef]
24. Ginés-Molina, M.J.; Ahmad, N.H.; Mérida-Morales, S.; García-Sancho, C.; Mintova, S.; Ng, E.-P.; Maireles-Torres, P. Selective Conversion of Glucose to 5-Hydroxymethylfurfural by Using L-Type Zeolites with Different Morphologies. *Catalysts* **2019**, *9*, 1073. [CrossRef]
25. Ghrear, T.M.A.; Rigolet, S.; Daou, T.J.; Mintova, S.; Ling, T.C.; Tan, S.H.; Ng, E.-P. Synthesis of Cs-ABW nanozeolite in organotemplate-free system. *Microporous Mesoporous Mater.* **2019**, *277*, 78–83. [CrossRef]
26. Wong, S.-F.; Awala, H.; Vincente, A.; Retoux, R.; Ling, T.C.; Mintova, S.; Mukti, R.R.; Ng, E.-P. KF zeolite nanocrystals synthesized from organic-template-free precursor mixture. *Microporous Mesoporous Mater.* **2017**, *249*, 105–110. [CrossRef]
27. IZA-SC Database of Zeolite Structures. Available online: http://www.iza-structure.org/databases/ (accessed on 1 December 2019).
28. Wong, S.-F.; Deekomwong, K.; Wittayakun, J.; Ling, T.C.; Muraza, O.; Adam, F.; Ng, E.-P. Crystal growth study of KF nanozeolite and its catalytic behavior in Aldol condensation of benzaldehyde and heptanal enhanced by microwave heating. *Mater. Chem. Phys.* **2017**, *196*, 295–301. [CrossRef]
29. Ng, E.-P.; Sekhon, S.S.; Mintova, S. Discrete MnAlPO-5 nanocrystals synthesized by an ionothermal approach. *Chem. Commun.* **2009**, 1661–1663. [CrossRef]
30. Ng, E.-P.; Awala, H.; Ghoy, J.P.; Vicente, A.; Ling, T.C.; Ng, Y.H.; Mintova, S.; Adam, F. Effects of ultrasonic irradiation on crystallization and structural properties of EMT-type zeolite nanocrystals. *Mater. Chem. Phys.* **2015**, *159*, 38–45. [CrossRef]
31. Zhang, J.; Lu, X.; Wang, Z. Control of crystallization rate and morphology of zeolite silicalite-1 in solvent-free synthesis. *Microporous Mesoporous Mater.* **2019**, *283*, 14–24. [CrossRef]
32. Ng, E.-P.; Rigolet, S.; Daou, T.J.; Mintova, S.; Ling, T.C. Micro-and macroscopic observations of the nucleation process and crystal growth of nanosized Cs-pollucite in an organotemplate-free hydrosol. *New J. Chem.* **2019**, *43*, 17433–17440. [CrossRef]
33. Gomez, A.G.; de Silveira, G.; Doana, H.; Cheng, C.-H. A facile method to tune zeolite L crystals with low aspect ratio. *Chem. Commun.* **2011**, *47*, 5876–5878. [CrossRef] [PubMed]
34. Ng, E.-P.; Goh, J.-Y.; Ling, T.C.; Mukti, R.R. Eco-friendly synthesis for MCM-41 nanoporous materials using the non-reacted reagents in mother liquor. *Nanoscale Res. Lett.* **2013**, *8*, 120. [CrossRef] [PubMed]
35. Ghrear, T.M.A.; Cheong, Y.-W.; Lim, G.K.; Chateigner, D.; Ling, T.C.; Tan, S.H.; Ng, E.-P. Fast, low-pressure, low-temperature microwave synthesis of ABW cesium aluminosilicate zeolite nanocatalyst in organotemplate-free hydrogel system. *Mater. Res. Bull.* **2020**, *122*, 110691. [CrossRef]
36. Barrett, P.A.; Valencia, S.; Camblor, M.A. Synthesis of a merlinoite-type zeolite with an enhanced Si/Al ratioviapore filling with tetraethylammonium cations. *J. Mater. Chem.* **1998**, *8*, 2263–2268. [CrossRef]
37. Mohammad, S.A.G.; Khoerunnisa, F.; Rigolet, S.; Daou, T.J.; Ling, T.C.; Ng, E.-P. Hierarchical Cs–Pollucite Nanozeolite Modified with Novel Organosilane as an Excellent Solid Base Catalyst for Claisen–Schmidt Condensation of Benzaldehyde and Acetophenone. *Processes* **2020**, *8*, 96. [CrossRef]

38. Choo, M.-Y.; Juan, J.C.; Oi, L.E.; Ling, T.C.; Ng, E.-P.; Noorsaadah, A.R.; Centi, G.; Lee, K.T. The role of nanosized zeolite Y in the H 2-free catalytic deoxygenation of triolein. *Catal. Sci. Technol.* **2019**, *9*, 772–782. [CrossRef]
39. Hargreaves, J.S.J.; Hutchings, G.J.; Joyner, R.W.; Kiely, C.J. The relationship between catalyst morphology and performance in the oxidative coupling of methane. *J. Catal.* **1992**, *135*, 576–595. [CrossRef]

 © 2020 by the authors. Licensee MDPI, Basel, Switzerland. This article is an open access article distributed under the terms and conditions of the Creative Commons Attribution (CC BY) license (http://creativecommons.org/licenses/by/4.0/).

Article

Solvent Effects on Catechol Crystal Habits and Aspect Ratios: A Combination of Experiments and Molecular Dynamics Simulation Study

Dan Zhu [1], Shihao Zhang [1], Pingping Cui [1], Chang Wang [1], Jiayu Dai [1], Ling Zhou [1,*], Yaohui Huang [1], Baohong Hou [1,2], Hongxun Hao [1,2], Lina Zhou [1,2] and Qiuxiang Yin [1,2,3]

- [1] School of Chemical Engineering and Technology, Tianjin University, Tianjin 300072, China; sallyzhudan@tju.edu.cn (D.Z.); zshihao@tju.edu.cn (S.Z.); pp_cui@tju.edu.cn (P.C.); changwang@tju.edu.cn (C.W.); daijiayu@tju.edu.cn (J.D.); huangyaohui@tju.edu.cn (Y.H.); houbaohong@tju.edu.cn (B.H.); hongxunhao@tju.edu.cn (H.H.); linazhou@tju.edu.cn (L.Z.)
- [2] Collaborative Innovation Center of Chemical Science and Engineering (Tianjin), Tianjin 300072, China
- [3] State Key Laboratory of Chemical Engineering, Tianjin University, Tianjin 300072, China; qxyin@tju.edu.cn
- * Correspondence: zhouling@tju.edu.cn

Received: 29 March 2020; Accepted: 16 April 2020; Published: 18 April 2020

Abstract: This work could help to better understand the solvent effects on crystal habits and aspect ratio changes at the molecular level, which provide some guidance for solvent selection in industrial crystallization processes. With the catechol crystal habits acquired using both experimental and simulation methods in isopropanol, methyl acetate and ethyl acetate, solvent effects on crystal morphology were explored based on the modified attachment energy model. Firstly, morphologically dominant crystal faces were obtained with the predicted crystal habit in vacuum. Then, modified attachment energies were calculated by the molecular dynamics simulation to modify the crystal shapes in a real solvent environment, and the simulation results were in agreement with the experimental ones. Meanwhile, the surface properties such as roughness and the diffusion coefficient were introduced to analyze the solvent adsorption behaviors and the radial distribution function curves were generated to distinguish diverse types of interactions like hydrogen bonds and van der Waals forces. Results show that the catechol crystal habits were affected by the combination of the attachment energy, surface structures and molecular interaction types. Moreover, the changing aspect ratios of catechol crystals are closely related to the existence of hydrogen bonds which contribute to growth inhibition on specific faces.

Keywords: solvent effect; crystal habit; aspect ratio; molecular dynamics (MD); surface structure

1. Introduction

Crystal size and shape have essential effects on downstream processing such as filtration, washing and drying for solution crystallization [1–3]. Meanwhile, crystal properties such as flowability and bulk density [4,5] are closely associated with crystal morphology. Therefore, it has been very crucial to explore the possible crystal habits under different conditions to select the suitable crystals for industrial operations. Many factors contribute to affect crystal shapes, such as solvents [6], additives [7], temperature [8], supersaturation [9] and even stirring rate [10], which can be modified and controlled to obtain the morphology optimal for industrial applications. For solution crystallization, the effect of solvents is one of the most primary factors that affects final crystal habits. As indicated by many researches [11–13], the interactions between solvent molecules and crystal faces are quite essential to crystal growth. Hence, in order to obtain desired crystal shapes, it is necessary to investigate solvent effects on crystal habits and aspect ratio changes.

Although real and effective crystal morphology could be acquired by experimental methods, it is labor-intensive and time-consuming to conduct a large number of experiments just to find the optimal one that benefits production. In recent years, computer-aided methods such as molecular simulations and first principles (FP) simulations have been explored to explain and predict crystal habits [14–17], and the gradually developing simulation methods have provided a broader perspective for crystal morphology research at a molecular level.

Molecular dynamics (MD) is one of the molecular simulation methods that helps to investigate the directions of crystal growth with molecular information, reveal the interactions between solvent molecules and crystal surfaces and provide more microscopic details for experiments. The MD method has been successfully applied to simulate crystal morphology in many cases [18–20]. Wang et al. [21] utilized a surface docking model for a crystal growth simulation of cefaclor dihydrate and proposed a competitive relationship of surface adsorption by the solute and solvent molecules based on pure solvent and solution models. Zhang et al. [22] simulated the crystal morphology of ibuprofen obtained from four different solvents using the modified attachment energy (MAE) model and found that the crystal aspect ratios were sensitive to the relative polarity of the solvents. They regarded the method as a promising way for the computer-aided design of desirable pharmaceutical crystal habits with rapid solvent screening. Poornachary et al. [23] investigated the mechanism of additive inhibition on naproxen crystal growth by modeling the intermolecular interactions between the polymeric additive and the crystal surface and ascribed the phenomenon to solute diffusion control.

However, the exploration of solvent–crystal interactions from the thermodynamic perspective is just a partial picture, and discussions are not sufficient on the differences in crystal aspect ratios because of solvent system distinctions, which makes the crystal morphology selection in industrial application still based on experiences.

Catechol is a common fine chemical raw material, which is mainly used as an intermediate for polymerization inhibitor synthesis and pesticide production. In addition, it is also consumed as a precursor to fragrances, pharmaceuticals and dyes [24]. This organic chemical commodity has been in industrial production for over forty years, and the research emphasis is usually focused on its synthesis techniques [25,26]. Further experimental and simulation investigations on catechol crystal morphology are needed to get a deeper insight into the crystal growth and habits in solvents, which helps to select optimal crystal shapes and solvents to avoid agglomeration and improve product quality for industrial production.

In this paper, the catechol crystal morphology in isopropanol, methyl acetate and ethyl acetate were simulated using the MAE model, which helped to quantify the interactions between the crystal surfaces and solvent molecules. Experiments of cooling crystallization were conducted at the same time for model verification. This study aimed to describe the solvent effects on crystal habits from the perspective of crystal–solvent interactions and surface properties, which may favor a better understanding of crystal habit distinctions, especially the aspect ratio changes of crystal shapes in solvents.

2. Calculation Methodology

2.1. Theory

The Bravais–Friedel–Donnay–Harker (BFDH) model [27] is one of the models which are initially applied to predict crystal habits, but the model lacks precision because it simulates possible crystal facets merely according to geometric factors without considering the actual chemical environment, so it was soon developed into the attachment energy (AE) model by Hartman and Bennema [28–30], taking into account the energies of the system on the basis of the period bond chain (PBC) theory [31]. This kind of model is based on the attachment energy (E_{att}), which is defined as the released energy when a growth slice attaches on to a growing crystal surface. The relative growth rate (R_{hkl}) of each crystal face (h k l) is assumed to be proportional to the absolute value of the attachment energy [28]:

$$R_{hkl} \propto |E_{att}|. \tag{1}$$

The attachment energy (E_{att}) is calculated as

$$E_{att} = E_{latt} - E_{slice}, \tag{2}$$

where E_{latt} is the lattice energy of the crystal and E_{slice} is the energy of a growth slice with a thickness of d_{hkl}.

However, the AE model was put forward to simulate ideal crystal morphology in the premise of a vacuum environment, leading to a precision loss when the model was applied to solution crystallization. As a matter of fact, solvent molecules adsorb on the crystal surfaces as solute molecules do [32], which means that the growth of the crystal faces may be impeded by the solvent molecule adsorption. In order to cover the effects of the solvent on crystal growth, a modified attachment energy (MAE) model was developed. The modified model introduced an energy correction term E_s for the initial E_{att} to take into consideration the effects of the solvent molecules on crystal faces, and the modified attachment energy E'_{att} is calculated in the following formula:

$$E'_{att} = E_{att} - E_s, \tag{3}$$

where E_s represents the binding energy between the solvent molecules and the crystal face (h k l), which it can be obtained using the follow equation:

$$E_s = E_{int} \times \frac{A_{acc}}{A_{box}}, \tag{4}$$

where A_{box} is the total crystal surface area of the simulated supercell along the (h k l) direction, and A_{acc} represents the solvent-accessible area of the crystal face in the unit cell, which can be approximated by the Connolly surface algorithm [33]. E_{int} was defined as the interaction energy between a specific crystal face and the corresponding solvent layer, which can be calculated as follows:

$$E_{int} = E_{tot} - (E_{sur} + E_{sol}). \tag{5}$$

In Equation (5), E_{tot} means the total energy of the entire simulation box including all crystal and solvent molecules, and E_{sur} and E_{sol} represent the energy of the isolated crystal surface and solvent layer, respectively.

After correction, the relative growth rate R'_{hkl} in the MAE model was still in proportion to the absolute value of the modified attachment energy $|E'_{att}|$:

$$R'_{hkl} \propto |E'_{att}|. \tag{6}$$

2.2. Simulation Details

The initial crystal structure of catechol (identifier CATCOL17) was obtained from CSD (Cambridge Structural Database), derived from experiments by Clegg and Scott, with the space group of P2$_1$/n (a = 9.8326 Å, b = 5.5910 Å, c = 10.467 Å, $\alpha = \gamma = 90°$, $\beta = 114.988°$) [34]. As shown in Figure 1, there are four catechol molecules in the unit cell.

The entire geometry optimization and MD simulation process were implemented using the program Materials Studio 6.0 (Accelrys, San Diego, CA, USA) [35] with the COMPASS force field, which is a powerful ab initio force field with many accurate predictions of structures [36]. After the initial downloaded crystal was optimized using the COMPASS force field, the similar lattice parameters of the two crystals listed in Table 1, proved that it was a suitable choice to use the COMPASS field since the relative error of each parameter was less than 5%.

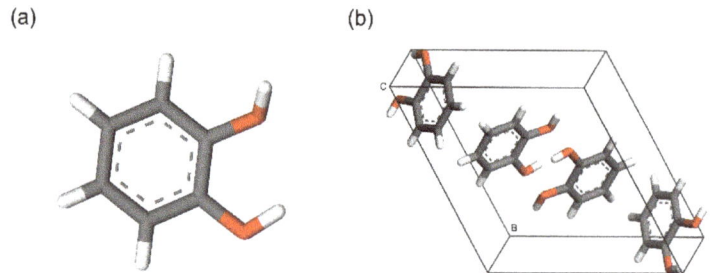

Figure 1. The molecular structure (**a**) and the unit cell (**b**) of catechol.

Table 1. The comparison of the initial and the optimized lattice parameters of catechol.

Lattice Parameter	a (Å)	b (Å)	c (Å)	α (°)	β (°)	γ (°)
Initial	10.941	5.509	10.069	90.000	119.000	90.000
COMPASS optimized	10.969	5.555	9.588	90.000	121.403	90.000
Relative error/%	0.26	0.83	4.78	0	2.02	0

After geometry optimization, the crystal habit of catechol in vacuum was predicted with the growth morphology method in the morphology module, providing several potential growing crystal faces that possibly existed in the solution environment. Based on the acquired morphologically important faces (h k l), the molecular models of catechol crystal surfaces were built by cleaving the corresponding (h k l) faces from the crystal cell with a depth of 2–4 d_{hkl} and then extending the surface to supercells containing m × n unit cells (specific data for each crystal face listed in Supplementary Materials Table S1). The solvent effects were modeled using the Amorphous Cell module by setting up a three-dimensional solvent box containing 200 randomly distributed solvent molecules and the size of the solvent box was consistent with the crystal supercell and the thickness determined by the solvent density.

Subsequently, a double-layer crystal–solvent interface model was constructed, and then a vacuum layer with a thickness of 50 Å was added onto the model box to avoid any additional effects of periodic boundaries. After the interface model was geometrically optimized with catechol molecules fixed in the crystal, the MD simulation was operated with NVT ensemble at 298 K, controlled by the Andersen thermostat. The total duration of the simulation was 200 ps with a time step of 1 fs, and the data were collected every 100 steps. In the non-bonding interaction calculation procedure, the standard Ewald method was used to calculate the electrostatic interactions with an accuracy of 0.0001 kcal mol^{-1} while the atom-based summation method was applied to calculate van der Waals interactions with a cut-off distance (d_c) of 15.5 Å. The cleaving and extending operations were implemented to balance the accuracy and simulation time [37], assuring that the length and width of the crystal supercell were both no less than $2d_c$, while the thickness of the solvent layer and crystal layer were greater than d_c [38].

3. Experiment and Characterization

3.1. Materials

Commercial catechol (99% purity) was supplied by Tianjin Damao Chemical Reagent Factory (Tianjin, China). Isopropanol, ethyl acetate and methyl acetate were provided by Tianjin Jiangtian Chemical Technology Co., Ltd. (Tianjin, China). All reagents were of analytical grade purity and used without further purification.

3.2. Cooling Crystallization Experiments

Firstly, saturated solutions in isopropanol, methyl acetate and ethyl acetate at 298 K were prepared based on the solubility data of catechol [39]. Then, the saturated solution was filtered into a double jacketed crystallizer preheated to 298 K. Afterwards, the solution was slowly cooled to 288 K at a cooling rate of 5 K per hour, using a thermostat (CF41, Julabo, Seelbach, Germany) with a ±0.02 K temperature stability. Afterwards, the crystals were obtained and dried for further solid characterization.

3.3. Crystal Characterization

Firstly, the crystal habits of the experimental catechol samples were observed by optical microscopy (BX51, Olympus, Tokyo, Japan), and ten crystals in each solvent were randomly selected to measure the aspect ratios. Then, the crystal forms of the dried products were detected by a powder X-ray diffractometer (D/MAX-2500, Rigaku, Tokyo, Japan) at ambient temperature with a scanning rate of 8 degrees per minute. The data were collected using Cu Kα radiation ($\lambda K\alpha$ = 0.15406 nm) and the 2θ scan range was 2°–40°.

4. Results and Discussion

4.1. Polymorph Identification

In general, changes in crystal morphology are probable to occur when polymorphism exists. Numerous cases [40–42] indicated various crystal forms with diverse habits in different solvent environments. Therefore, it is necessary to examine the crystal polymorphism of the products obtained from different solutions. As illustrated in Figure 2, the catechol products crystallized from isopropanol, methyl acetate and ethyl acetate have similar characteristic peaks with the raw material, proving that all the samples were in the same crystal form as reported [43]. In addition, the simulated powder X-ray diffraction (PXRD) pattern of the catechol obtained from CSD was consistent with the patterns of the experimental crystals (Supplementary Materials Figure S1). The results suggest that no polymorphic phenomenon occurred in our study, and it is reasonable to simulate crystal habits using the structure obtained from CSD.

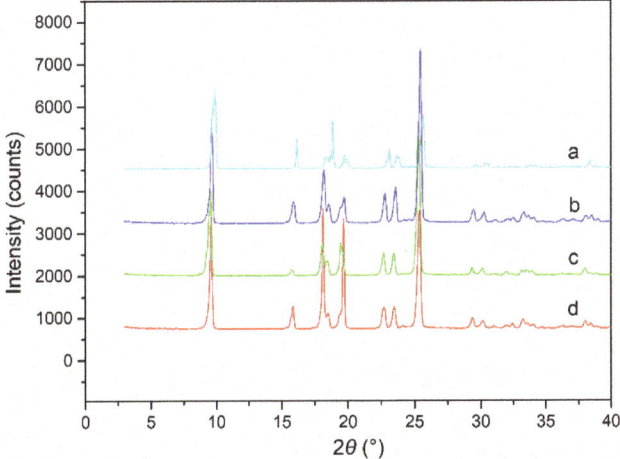

Figure 2. Powder X-ray diffraction (PXRD) patterns of the catechol crystals: (**a**) the raw material and the products crystallized from (**b**) ethyl acetate, (**c**) methyl acetate, and (**d**) isopropanol.

4.2. Prediction of Catechol Crystal Morphology in Vacuum

Figure 3a depicts the crystal morphology of catechol in vacuum predicted by the AE model. The prismatic crystal had six dominant faces, owing to not only crystallography geometry but also intermolecular interactions. Figure 3b visualizes the interactions between the catechol molecules calculated and generated by the Crystal Graph in the morphology module, with the blue and red lines representing strong and weak interactions, respectively. Previous studies indicated that crystals grow faster along the direction with strong molecular interactions [44]. Therefore, catechol crystals grow faster along the blue line direction and the surfaces with fast growth rate may disappear, leaving the slowly growing surface appear in the final morphology with six important faces, (1 0 −1), (1 0 1), (0 1 1), (1 1 0), (0 0 2) and (1 1 −1). In the AE model, the surface with higher absolute value of E_{att} has stronger ability to adsorb catechol molecules, which means a relatively faster growing rate of the crystal face, and vice versa. As listed in Table 2, the most important face is (1 0 −1) occupying more than 48% of the total habit facet area, followed by (1 1 −1) and (1 0 1) with 20% and 19% in total areas, respectively. Therefore, (1 1 0), (0 0 2) and (0 1 1) faces grow faster with relatively larger $|E_{att}|$ among all the crystal faces, which are of lower morphological importance and are more likely to disappear.

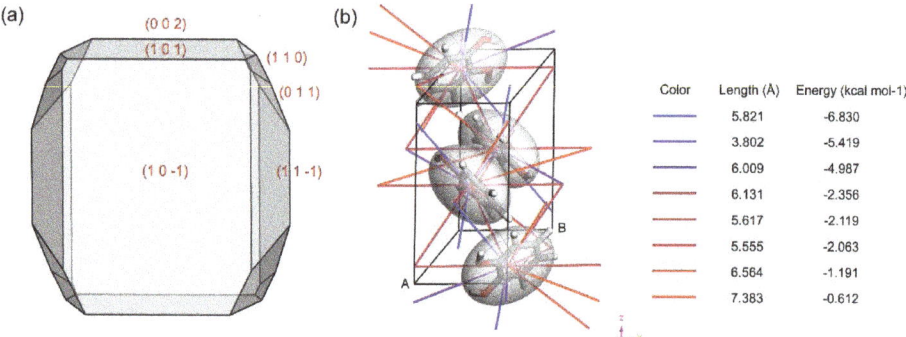

Figure 3. The predicted catechol crystal morphology in the vacuum (**a**) and the intermolecular interactions in the catechol unit cell calculated by the Crystal Graph (**b**).

Table 2. The parameters of the catechol crystal faces in the vacuum predicted via the attachment energy (AE) model.

Faces (h k l)	Multiplicity	d_{hkl} (Å)	E_{att} (kcal mol^{-1})	Area (%)
(1 0 −1)	2	8.19	−18.09	48.55
(1 0 1)	2	5.47	−43.45	19.10
(0 1 1)	4	4.78	−46.15	10.97
(1 1 0)	4	4.71	−50.50	1.25
(0 0 2)	2	4.68	−47.41	0.10
(1 1 −1)	4	4.60	−37.64	20.03

4.3. Modified Results of Catechol Crystal Morphology in Solvent Systems

Consistent with the experiments, the MD simulation based on the MAE model was conducted in three kinds of solvent systems: isopropanol, methyl acetate and ethyl acetate. Table 3 lists the simulated results of the six significant crystal faces in the different solvents. The negative values of E_{int}, which indicate the interaction energies between the solvents and the catechol crystal faces, revealed that the solvent molecule adsorption was spontaneous because the process was exothermic. The diverse absolute values of E_s for different crystal faces in the solvents implied that the solvents had different effects on each crystal face due to their distinct characteristics. However, ordered from largest to smallest, the $|E_s|$ values of the (1 0 1), (1 1 0), and (0 1 1) face were the top three in all three solvents, which means that the interactions between the solvents and crystal faces were relatively strong on the (1 1 0), (0 1 1) and (1 0 1) faces. After correction, the absolute values of the modified attachment energy $|E'_{att}|$ were sorted as follows: (0 0 2) > (1 1 0) > (0 1 1) > (1 1 −1) > (1 0 1) > (1 0 −1) in isopropanol, (0 0 2) > (0 1 1) > (1 0 1) > (1 1 0) > (1 1 −1) > (1 0 −1) in methyl acetate, and (0 0 2) > (1 0 1) > (0 1 1) > (1 1 0) > (1 1 −1) > (1 0 −1) in ethyl acetate. Although the orders were not identical, the relatively most fast-growing (0 0 2) face disappeared in all three solvent systems with the largest $|E'_{att}|$. Meanwhile, the (1 0 −1) face remained to take up the largest percentage of the crystal facet areas compared to the crystal morphology in vacuum: 43.87% in isopropanol and more than 70% in the other two solvents. For the crystal in methyl acetate, the (1 1 0) face had more area proportion than the (1 0 1) face due to its slower growth rate with a smaller $|E'_{att}|$ value.

Table 3. Simulated results of the dominant crystal faces in isopropanol, methyl acetate and ethyl acetate [1].

Solvent	Faces	E_{tot}	E_{sur}	E_{sol}	E_{int}	E_s	E'_{att}	R'_{hkl}	Area/%
Isopropanol	(1 0 −1)	−11,206.93	−3093.73	−7916.14	−197.05	−13.48	−4.61	1.00	43.87
	(1 0 1)	−10,774.85	−2646.98	−7913.94	−213.93	−37.01	−6.44	1.40	35.57
	(0 1 1)	−12,056.38	−3950.79	−7947.46	−158.13	−26.70	−19.45	4.21	-
	(1 1 0)	−12,186.14	−4017.98	−7954.85	−213.31	−30.16	−20.35	4.41	-
	(0 0 2)	−10,824.04	−2703.56	−7877.07	−243.41	−17.44	−29.97	6.50	-
	(1 1 −1)	−11,326.17	−3180.69	−7990.31	−155.17	−25.20	−12.44	2.70	20.56
Methyl acetate	(1 0 −1)	−6423.31	−3093.60	−3046.64	−283.06	−19.36	1.27	1.00	70.81
	(1 0 1)	−6027.75	−2646.91	−3052.75	−328.10	−56.76	13.31	10.49	2.50
	(0 1 1)	−7330.46	−3950.59	−2990.34	−389.53	−65.77	19.63	15.47	-
	(1 1 0)	−7395.82	−4017.81	−2971.41	−406.60	−57.48	6.98	5.50	14.97
	(0 0 2)	−6059.46	−2703.51	−3024.78	−331.17	−23.72	−23.68	18.67	-
	(1 1 −1)	−6504.15	−3180.63	−3062.49	−261.03	−42.39	4.75	3.75	11.72
Ethyl acetate	(1 0 −1)	−8963.16	−3093.20	−5609.78	−260.18	−17.80	−0.30	1.00	71.14
	(1 0 1)	−8479.98	−2646.54	−5548.65	−284.79	−49.27	5.82	19.65	3.98
	(0 1 1)	−9801.90	−3950.16	−5555.32	−296.41	−50.05	3.90	13.17	-
	(1 1 0)	−9927.83	−4017.27	−5527.68	−382.88	−54.13	3.63	12.24	-
	(0 0 2)	−8487.97	−2703.17	−5532.21	−252.59	−18.10	−29.31	98.96	-
	(1 1 −1)	−9038.79	−3180.23	−5633.03	−225.53	−36.62	−1.01	3.42	24.88

[1] All energies are in kcal mol^{-1}.

As two or three crystal faces disappeared because of their relatively fast growth rate, the crystal morphology of catechol obviously changed in the three solvent systems, which powerfully supported the non-negligible effects of the solvents on crystal habits. In the results shown in Figure 4, the simulated crystals basically conform with the experimental ones with prismatic, fusiform or hexagonal tubular shapes in isopropanol, methyl acetate and ethyl acetate, respectively. Here we introduced the aspect ratio of the crystal as a quantitive index to describe the differences between the crystals grown from distinct solvent systems. As can be seen in Figure 4, the aspect ratio of catechol crystal was mainly determined by the areas of the (1 0 −1) face and the (1 0 1) face, so here we defined the aspect ratio as the length along the (1 0 −1) face divided by the width along the (1 0 1) face in order to summarize the unified rules of morphology change. The calculated average aspect ratios of experimental crystals are shown in Figure 5, in which the catechol crystal in isopropanol and ethyl acetate has the smallest (1.29) and the largest aspect ratio (4.95), respectively. For the convenience of downstream processes, crystal products with small aspect ratios were preferred with high flowability and unbreakable shapes. Therefore, isopropanol may be the optimal solvent for catechol crystallization among the three solvents. The obvious differences in aspect ratios may be attributed to the relative growing rate of the crystal faces in specific solvent systems, which is fundamentally related to the effects of hydrogen bond or solvent diffusion velocity. This will be discussed in the following sections.

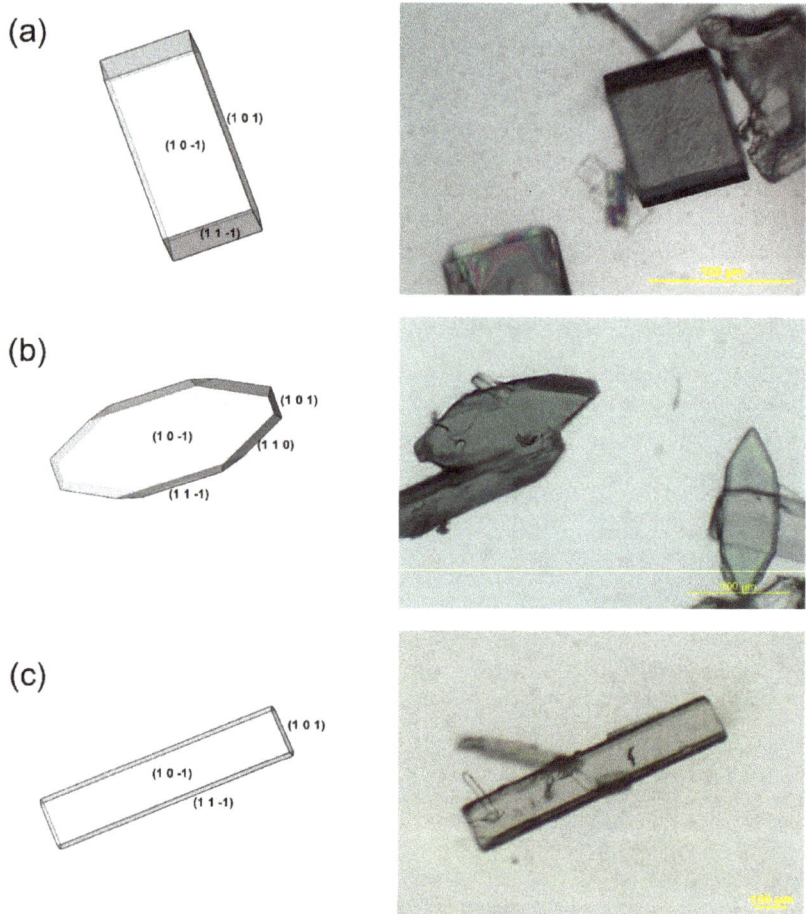

Figure 4. The modified morphology with the modified attachment energy (MAE) model (on the left) and the corresponding experimental crystal habits (on the right) of catechol in (**a**) isopropanol (IPA), (**b**) methyl acetate (MAC), and (**c**) ethyl acetate (EAC).

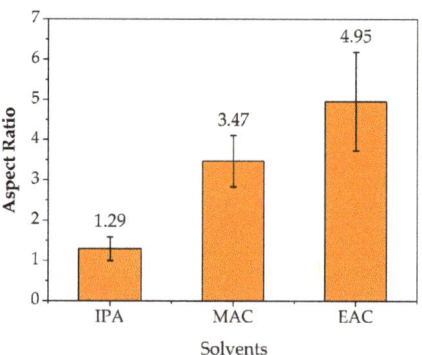

Figure 5. The aspect ratios of the catechol crystals obtained from solvents.

4.4. Solvent–Crystal Interactions on the Interface

As discussed above, the modified attachment energy indicated the interactions between the crystal layer and the solvent layer in the simulation box, which was of great significance to crystal morphology. Apart from the attachment energy from a thermodynamic point of view, further comprehension of the crystal–solvent interaction mechanism helped to optimize crystal morphology and select a suitable solvent in industrial crystallization. Therefore, considerations should be taken into account, not only for contact interface properties, but also for non-bonding interactions between crystals and solvent molecules. All these factors contributed to understanding the solvent effects entirely.

4.4.1. Molecular Alignment on Crystal Surfaces

The anisotropy of the catechol crystal structure leads to a distinct molecular alignment on each crystal face, and then gives rise to a different solvent molecular distribution at the solvent–crystal interface [45]. Such differences caused by surface structure have crucial effects on the adsorption of solvents on crystal faces, finally bringing changes to crystal growth and morphology. Here, we took the MD-simulated equilibrium configurations of the interface model of catechol and methyl acetate molecules as an example to explore how surface properties affected crystal–solvent interactions. As can be seen in Figure 6, the alignment of methyl acetate molecules differed on the six crystal faces owing to the diverse crystal surface structure. The catechol molecules on the (1 0 −1) face were arranged relatively orderly and compactly to form a relatively flat surface. However, the catechol molecules on the (0 1 1) and (0 0 2) faces were angled to the crystal plane, leading to a bumpy surface which may surround more solvent and solute molecules. In addition, the molecular alignment makes polar hydroxyl groups exposed on crystal surfaces, which was more favorable to the adsorption of solvent molecules with polar groups. Different positions and angles of the exposed groups may have formed various non-bonding interactions with distinct strength and provided different adsorption areas for solvent molecules. Numerous methyl acetate molecules adsorbed on the grooves of (1 0 1), (0 1 1), (1 1 0) and (0 0 2) faces, while for (1 0 −1) and (1 1 −1) faces, relatively flat crystal surfaces were unhelpful for molecule adsorption, which may have resulted in a smaller amount of adsorbed solvent molecules on these faces compared to the others.

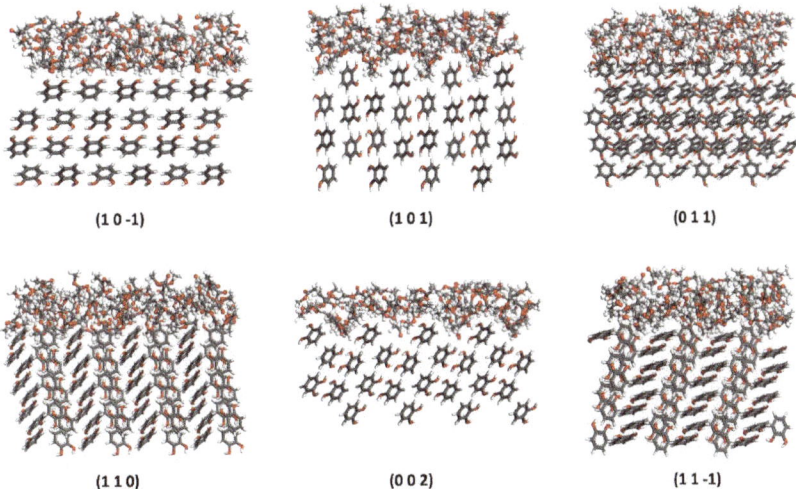

Figure 6. Methyl acetate molecular alignment on the six main possible crystal faces of catechol via a molecular dynamics (MD) simulation.

4.4.2. Roughness of Surfaces

To quantitatively describe the surface properties, a parameter S was introduced to evaluate the roughness of crystal surfaces:

$$S = \frac{A_{acc}}{A_{box}} \quad (7)$$

where A_{acc} and A_{hkl} represent the solvent-accessible area and surface area for the (h k l) crystal surface in the unit cell, respectively. A larger S means a rougher surface with larger adsorption areas for the solute and solvent molecules to interact with crystal surfaces. With the accessible areas of solvents calculated by the Connolly surface model (showed in blue in Figure 7), Table 4 lists the calculated S values of the crystal faces. Although the shapes of the solvent-accessible areas varied on each crystal face, they all had periodic fluctuations which could form grooves of more contact area with solute and solvent molecules. As shown in Table 4, the order of the roughness values for the six crystal faces was as follows: (1 0 1) > (0 1 1) > (0 0 2) > (1 1 0) > (1 1 −1) > (1 0 −1). The (1 0 −1) face with the smallest S value (1.23) provided the minimum areas for molecule incorporation, which meant it was probably difficult for solute and solvent molecules to adsorb on this face compared with the other faces. This was consistent with the relatively slow growth rate of the (1 0 −1) face, which led to a larger face area. Similarly, the (1 1 0) face grew more slowly than the (1 0 1) face and had a larger area proportion in methyl acetate. It is worth noting that the (0 0 2) faces with a moderate S value disappeared in all kinds of crystals grown from solvent systems, while the (1 0 1) faces with the largest S value remain at last. As can be seen from Table 3, the $|E_s|$ values of the (1 0 1) faces were larger than those of the (0 0 2) faces in all three solvents, which meant that the solvent molecules were more likely to adsorb on the (1 0 1) face compared to the (0 0 2) face, leading to a stronger solvent inhibition on the growth of the (1 0 1) face. As for the (0 0 2) face, less adsorbed solvent molecules provided possibilities for the continuous adsorption of the solute molecules, which indicated a fast growth rate.

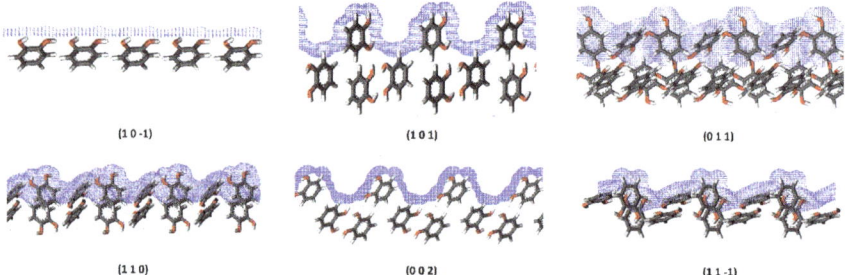

Figure 7. Solvent-accessible areas (the blue grid) of the six important faces of the catechol crystal represented by the Connolly surface.

Table 4. The roughness values (S) of the dominant crystal faces of the catechol crystal.

Faces	A_{hkl} (Å^2)	A_{acc} (Å^2)	S
(1 0 −1)	60.93	75.02	1.23
(1 0 1)	91.11	189.13	2.08
(0 1 1)	104.40	211.54	2.03
(1 1 0)	105.99	179.81	1.70
(0 0 2)	53.25	91.56	1.72
(1 1 −1)	108.51	158.59	1.46

4.4.3. Diffusion Coefficient

As mentioned above, the independent S values analysis was not entirely consistent with the crystal morphology results. Therefore, another factor, the diffusion capacity of solvent molecules, was introduced to look into the solvent diffusion effects on crystal growth. Based on the well known Einstein relationship [46], the diffusion coefficient (D) of the solvent molecules was defined as the derivative of the mean square displacement (MSD) with respect to time [47]:

$$D = \frac{1}{6} \lim_{t \to \infty} \frac{d}{dt} \sum_{i=1}^{N} \langle |r_i(t) - r_i(0)|^2 \rangle, \tag{8}$$

where N stands for the number of solvent molecules and $r_i(t)$ represents the position of the molecule i at time t.

Stronger interactions existed between the solvents and crystal faces with larger D values due to an increasing number of solvent molecules diffusing to the interface [21]. As listed in Table 5, the D values were listed in the following sequence: (1 0 −1) > (0 0 2) > (1 0 1) > (1 1 −1) > (0 1 1) > (1 1 0) in isopropanol, (1 0 −1) > (0 0 2) > (1 1 −1) > (0 1 1) > (1 1 0) > (1 0 1) in methyl acetate, and (1 0 −1) > (1 0 1) > (0 0 2) > (1 1 −1) > (1 1 0) > (0 1 1) in ethyl acetate. Compared with those of isopropanol, the D values of methyl acetate and ethyl acetate were relatively large, indicating stronger interactions with the crystal surfaces which were consistent with the values of E_{int} in the corresponding solvent systems. Obviously, the diffusion coefficient on the (1 0 −1) face was the largest one among the six crystal faces for all the three kinds of solvent molecules. This indicated that more solvent molecules gathered on the (1 0 −1) face and strong interactions existed between the solvent molecules and the crystal face although the adsorption areas provided by this face were small with the minimum S value. Thus, for the factors affecting the growth of the (1 0 −1) face, the solvent–crystal interface interaction was more dominant than the surface roughness. As a result, the (1 0 −1) face manifested a relatively slow growth of the crystal face and the largest area in real morphology. Apart from the large D value on (0 0 2) faces, the D values on crystal faces were in reasonable agreement with the experimental crystal area results, as solute molecules were more likely to adsorb on the faces with smaller D values, on which solvent molecules take less growth active sites [48]. For example, the (1 1 0) and (0 1 1) faces disappeared in ethyl acetate with smaller D values (2.11 × 10^{-9} m^2 s^{-1} and 1.94 × 10^{-9} m^2 s^{-1}, respectively), while the (1 1 0) and (0 1 1) face with small D values (0.67 × 10^{-9} m^2 s^{-1} and 0.69 × 10^{-9} m^2 s^{-1}, respectively) did not exist in the final morphology in isopropanol. Despite the high D values on (0 0 2) faces, it was easier for solute molecules to adsorb on the crystal surface than it was for the solvent molecules to form a new layer of catechol crystal because their $|E'_{att}|$ values were larger than the corresponding $|E_s|$ values.

Table 5. The diffusion coefficient (D)[1] of the solvent molecules on the different catechol crystal faces.

Faces	Isopropanol	Methyl Acetate	Ethyl Acetate
(1 0 −1)	1.09	3.00	2.53
(1 0 1)	0.75	2.25	2.28
(0 1 1)	0.69	2.57	1.94
(1 1 0)	0.67	2.52	2.11
(0 0 2)	0.91	2.88	2.21
(1 1 −1)	0.74	2.82	2.18

[1] All D values are in 10^{-9} m^2 s^{-1}.

4.4.4. Solvent Effects on Crystal Aspect Ratios

Discussions on energies and surface structures on solvent–crystal interfaces have pointed out some reasons for changes in crystal morphology. Apart from these factors, interaction types, especially hydrogen bonds, are worth discussing in our case since catechol and the three kinds of solvent molecules possess hydrogen and oxygen atoms to form hydrogen bonds.

The existence of hydrogen bonds is a significant factor affecting the types and the strength of the interactions on solvent–crystal interfaces, which play a vital role in crystal morphology [49]. Therefore, the radial distribution function (RDF), $g(r)$, was applied to explore the interactions on the solvent–crystal interfaces, which described how atom density varied as a function of the distance from the specified hydrogen or oxygen atom [50].

In general, hydrogen bonds and van der Waals interactions belong to short-range interactions whose effective intermolecular range is under 5.0 Å. The effective range of hydrogen bonds is usually defined to be within 3.1 Å, while the range for van der Waals interaction is between 3.1 Å to 5.0 Å [51]. Interactions that are effective above 5.0 Å are called long-range interactions and usually refer to electrostatic interactions [37]. As mentioned above, the aspect ratio was defined as the length along the (1 0 −1) face divided by the width along the (1 0 1) face. From Figure 4 we can conclude that the crystal length along the (1 0 −1) face was decided by the growth of the (1 0 1) face, while the crystal width along the (1 0 1) face is mainly related to the growth of the (1 1 −1) face. So here we took the examples of the RDF on the (1 0 1) face and the (1 1 −1) face in all three solvents in order to find the solvent effects on crystal aspect ratios, with the outermost layer of each face analyzed because of its proximity to solvent molecules. As shown in Figure 8, the positions of the first sharp peaks in red were all in the range of 1.70–2.00 Å, indicating that the oxygen atoms of the solvent molecules formed a strong hydrogen bond with the hydrogen atoms of the catechol molecules on the (1 0 1) face and the (1 1 −1) face. Therefore, the crystal face growth was inhibited by the solvent molecules adsorbed around the crystal surface. With different positions and strengths of the peaks, the inhibition resulted in diverse effects on the crystal face growth, which could be analyzed by the aspect ratios and face areas in final crystal habits. In particular, as shown in Figure 8a, with larger numbers of sharper peaks (r = 1.71, 2.49 and 2.87 Å), the (1 0 1) face grew at a slower rate due to the stronger solvent inhibition in isopropanol compared to those in methyl acetate and ethyl acetate, leading to a shorter crystal length compared with those in the other two solvents. In other words, compared to that in isopropanol, the crystal morphology turns longer as a hexagonal tabular shape in ethyl acetate with the relatively weaker hydrogen bonds on the (1 0 1) face. The results were consistent with the R'_{hkl} values of (1 0 1) faces in Table 3 which indicated the relative growth rates (1.40 and 19.65 in isopropanol and ethyl acetate, respectively). In addition, the crystal in methyl acetate had a fusiform-like morphology with the existence of the (1 1 0) face (the RDF curve shown in Supplementary Materials Figure S2). It is remarkable that the RDF curves of the (1 1 0) face and the (1 0 1) face showed great similarity in methyl acetate, but the (1 1 0) face was larger than the (1 0 1) face in the final morphology. Thus, the catechol crystal habit in methyl acetate was mainly related to the surface structure and the attachment energy of the crystal faces.

The similar RDF curves of the (1 1 −1) faces in methyl acetate and ethyl acetate showed the same numbers of peaks, but the peaks (r = 1.81 Å) in ethyl acetate appeared earlier compared to those in methyl acetate, which indicates the stronger inhibition of the (1 1 −1) face in ethyl acetate. The results corresponded with the R'_{hkl} values of the (1 1 −1) faces in Table 3 (3.42 and 3.75 in ethyl acetate and methyl acetate, respectively). The relatively slow growth of the (1 1 −1) face led to a shorter crystal width in the final morphology, increasing the aspect ratio of catechol crystal in ethyl acetate indirectly. Therefore, the crystals tended to be longer in ethyl acetate than those in methyl acetate, which supports the conclusion that the crystal aspect ratio was mainly dependent on the growth of the (1 0 1) and (1 1 −1) faces. Apart from the peaks within 3.1 Å, several peaks appeared in the range of 3.1–5.0 Å and above 5.0 Å in all the RDF curves, indicating the existence of strong van der Waals and electrostatic interactions between the selected atoms.

Above all, it can be concluded that the differences in the aspect ratio of the catechol crystals were attributed to distinct growth inhibition effects of the three solvents mainly on the (1 0 1) and (1 1 −1) faces. Strong hydrogen bonds exist between the hydrogen atoms of catechol and the oxygen atoms of solvents in the three solvents, which becomes a non-negligible factor in catechol morphology, especially the crystal aspect ratio.

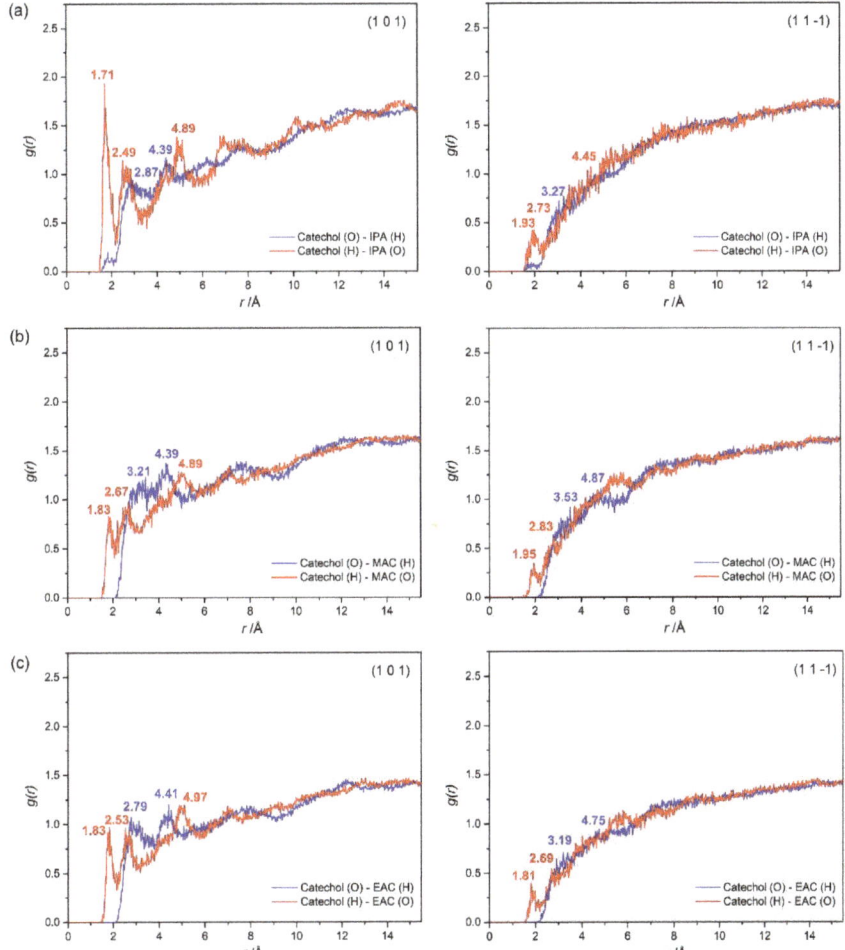

Figure 8. The radial distribution function (RDF) analysis between catechol and (**a**) isopropanol, (**b**) methyl acetate, (**c**) ethyl acetate molecules on the (1 0 1) face (left) and the (1 1 −1) face (right).

5. Conclusions

In this study, we successfully simulated the crystal habits of catechol in isopropanol, methyl acetate and ethyl acetate using the MAE model which takes solvent effects into consideration. The analysis on the calculated attachment energy indicates that the interactions on solvent−crystal interfaces had essential effects on catechol crystal morphology in all three experimental systems. Factors such as surface structure and diversity of interaction types were explored to find their synergy for crystal shapes.

The (1 0 −1) face was the most morphologically dominant crystal face because its relatively flat surface provided less adsorption sites for both solvent and solute molecules, leading to a relatively slow face growth. The molecular alignment, roughness and diffusion coefficient analysis on dominant crystal faces indicates that the (0 0 2) and (0 1 1) faces disappeared in final crystal morphology due to the competitive adsorption of molecules favorable for solutes to continuously adsorb on the surface. The RDF curves reveal that several types of interactions contributed to real crystal morphology in solvent systems, of which the hydrogen bond was a crucial factor to analyze the change in the aspect ratios of crystal. The shape distinctions of catechol crystals were mainly attributed to the attachment energy as well as the diverse strength of hydrogen bonds between solvent molecules and catechol molecules on (1 0 1) and (1 1 −1) faces. Moreover, the simulated crystal morphology of catechol was consistent with the crystallized ones obtained from all three solvents, proving the practicability of the MAE model to select optimal crystal morphology in industrial crystallization using simulation methods.

Supplementary Materials: The following are available online at http://www.mdpi.com/2073-4352/10/4/316/s1, Table S1: The cleaved depth and size of catechol crystal faces to form a simulation supercell. Figure S1: The simulated XRD pattern of catechol crystal. Figure S2: The RDF analysis between catechol and methyl acetate on the (1 1 0) face.

Author Contributions: Conceptualization, L.Z. (Ling Zhou) and D.Z.; methodology, D.Z., S.Z. and C.W.; software, S.Z., D.Z. and P.C.; validation, S.Z. and Q.Y.; formal analysis, D.Z., P.C. and C.W.; investigation, D.Z., P.C. and J.D.; data curation, D.Z., S.Z. and L.Z. (Ling Zhou) writing—original draft preparation, D.Z. and S.Z.; writing—review and editing, D.Z., S.Z., J.D., C.W., Y.H. and Q.Y.; visualization, D.Z. and P.C.; supervision, B.H. and L.Z. (Lina Zhou); project administration, Q.Y. and H.H.; funding acquisition, Q.Y. All authors have read and agreed to the published version of the manuscript.

Funding: The authors are grateful for the financial support of the National Natural Science Foundation of China (No. 21706183).

Conflicts of Interest: The authors declare no conflict of interest.

References

1. Beckmann, W. *Crystallization: Basic Concepts and Industrial Applications*; Wiley-VCH: Weinheim, Germany, 2013.
2. MacLeod, C.S.; Muller, F.L. On the Fracture of Pharmaceutical Needle-Shaped Crystals during Pressure Filtration: Case Studies and Mechanistic Understanding. *Org. Process Res. Dev.* **2012**, *16*, 425–434. [CrossRef]
3. Kim, S.; Lotz, B.; Lindrud, M.; Girard, K.; Moore, T.; Nagarajan, K.; Alvarez, M.; Lee, T.; Nikfar, F.; Davidovich, M.; et al. Control of the particle properties of a drug substance by crystallization engineering and the effect on drug product formulation. *Org. Process Res. Dev.* **2005**, *9*, 894–901. [CrossRef]
4. Rasenack, N.; Müller, B.W. Ibuprofen crystals with optimized properties. *Int. J. Pharm.* **2002**, *245*, 9–24. [CrossRef]
5. Širola, I.; Lukšić, J.; Šimunić, B.; Kujundžić, N. Effect of crystal size and shape on bulk density of pharmaceutical powders. *J. Cryst. Growth* **1997**, *181*, 403–409. [CrossRef]
6. Stoica, C.; Verwer, P.; Meekes, H.; van Hoof, P.J.C.M.; Kaspersen, F.M.; Vlieg, E. Understanding the effect of a solvent on the crystal habit. *Cryst. Growth Des.* **2004**, *4*, 765–768. [CrossRef]
7. Cashell, C.; Corcoran, D.; Hodnett, B.K. Effect of amino acid additives on the crystallization of L-glutamic acid. *Cryst. Growth Des.* **2005**, *5*, 593–597. [CrossRef]
8. Yang, D.; Hrymak, A.N. Crystal Morphology of Hydrogenated Castor Oil in the Crystallization of Oil-in-Water Emulsions: Part I. Effect of Temperature. *Ind. Eng. Chem. Res.* **2011**, *50*, 11585–11593. [CrossRef]
9. Boerrigter, S.X.M.; Cuppen, H.M.; Ristic, R.I.; Sherwood, J.N.; Bennema, P.; Meekes, H. Explanation for the supersaturation-dependent morphology of monoclinic paracetamol. *Cryst. Growth Des.* **2002**, *2*, 357–361. [CrossRef]
10. Di Renzo, F. Zeolites as tailor-made catalysts: Control of the crystal size. *Catal. Today* **1998**, *41*, 37–40. [CrossRef]
11. Blagden, N.; Davey, R.J.; Lieberman, H.F.; Williams, L.; Payne, R.; Roberts, R.; Rowe, R.; Docherty, R. Crystal chemistry and solvent effects in polymorphic systems—Sulfathiazole. *J. Chem. Soc. Faraday T* **1998**, *94*, 1035–1044. [CrossRef]
12. Urbelis, J.H.; Swift, J.A. Solvent Effects on the Growth Morphology and Phase Purity of CL-20. *Cryst. Growth Des.* **2014**, *14*, 1642–1649. [CrossRef]
13. Liu, J.; Chang, Z.; He, Y.; Yang, M.; Dang, J. Solvent effects on the crystallization of avermectin B1a. *J. Cryst. Growth* **2007**, *307*, 131–136. [CrossRef]
14. Khoshkhoo, S.; Anwar, J. Study of the effect of solvent on the morphology of crystals using molecular simulation: Application to α-resorcinol and N-n-octyl-D-gluconamide. *J. Chem. Soc. Faraday Trans.* **1996**, *92*, 1023–1025. [CrossRef]

15. de Graaf, J.; Filion, L.; Marechal, M.; van Roij, R.; Dijkstra, M. Crystal-structure prediction via the floppy-box Monte Carlo algorithm: Method and application to hard (non)convex particles. *J. Chem. Phys.* **2012**, *137*, 214101. [CrossRef]
16. Zhang, W.-B.; Chen, C.; Zhang, S.-Y. Equilibrium Crystal Shape of Ni from First Principles. *J. Phys. Chem. C* **2013**, *117*, 21274–21280. [CrossRef]
17. Dzade, N.Y.; Roldan, A.; de Leeuw, N.H. Surface and shape modification of mackinawite (FeS) nanocrystals by cysteine adsorption: A first-principles DFT-D2 study. *Phys. Chem. Chem. Phys.* **2016**, *18*, 32007–32020. [CrossRef]
18. Lynch, A.; Verma, V.; Zeglinski, J.; Bannigan, P.; Rasmuson, A. Face indexing and shape analysis of salicylamide crystals grown in different solvents. *CrystEngComm* **2019**, *21*, 2648–2659. [CrossRef]
19. Shi, W.; Chu, Y.; Xia, M.; Lei, W.; Wang, F. Crystal morphology prediction of 1,3,3-trinitroazetidine in ethanol solvent by molecular dynamics simulation. *J. Mol. Graph. Model.* **2016**, *64*, 94–100. [CrossRef]
20. Song, L.; Chen, L.; Wang, J.; Chen, F.; Lan, G. Prediction of crystal morphology of 3,4-Dinitro-1H-pyrazole (DNP) in different solvents. *J. Mol. Graph. Model.* **2017**, *75*, 62–70. [CrossRef]
21. Wang, C.; Zhang, X.; Du, W.; Huang, Y.-H.; Guo, M.-X.; Li, Y.; Zhang, Z.-X.; Hou, B.-H.; Yin, Q.-X. Effects of solvent and supersaturation on crystal morphology of cefaclor dihydrate: A combined experimental and computer simulation study. *CrystEngComm* **2016**, *18*, 9085–9094. [CrossRef]
22. Zhang, M.; Liang, Z.; Wu, F.; Chen, J.-F.; Xue, C.; Zhao, H. Crystal engineering of ibuprofen compounds: From molecule to crystal structure to morphology prediction by computational simulation and experimental study. *J. Cryst. Growth* **2017**, *467*, 47–53. [CrossRef]
23. Poornachary, S.K.; Chia, V.D.; Yani, Y.; Han, G.; Chow, P.S.; Tan, R.B.H. Anisotropic Crystal Growth Inhibition by Polymeric Additives: Impact on Modulation of Naproxen Crystal Shape and Size. *Cryst. Growth Des.* **2017**, *17*, 4844–4854. [CrossRef]
24. Fiege, H.; Voges, H.W.; Hamamoto, T.; Umemura, S.; Iwata, T.; Miki, H.; Fujita, Y.; Buysch, H.J.; Garbe, D.; Paulus, W. Phenol Derivatives. In *Ullmann's Encyclopedia of Industrial Chemistry*; Wiley-VCH: Weinheim, Germany, 2000; pp. 552–555.
25. Krab-Hüsken, L. *Production of Catechols: Microbiology and Technology*; Wageningen University: Wageningen, The Netherlands, 2002.
26. Mabrouk, A.; Erdocia, X.; Alriols, M.G.; Labidi, J. Economic analysis of a biorefinery process for catechol production from lignin. *J. Clean. Prod.* **2018**, *198*, 133–142. [CrossRef]
27. Docherty, R.; Clydesdale, G.; Roberts, K.J.; Bennema, P. Application of Bravais-Friedel-Donnay-Harker, attachment energy and Ising models to predicting and understanding the morphology of molecular crystals. *J. Phys. D: Appl. Phys.* **1991**, *24*, 89–99. [CrossRef]
28. Hartman, P.; Bennema, P. The attachment energy as a habit controlling factor: I. Theoretical considerations. *J. Cryst. Growth* **1980**, *49*, 145–156. [CrossRef]
29. Hartman, P. The attachment energy as a habit controlling factor II. Application to anthracene, tin tetraiodide and orthorhombic sulphur. *J. Cryst. Growth* **1980**, *49*, 157–165. [CrossRef]
30. Hartman, P. The attachment energy as a habit controlling factor: III. Application to corundum. *J. Cryst. Growth* **1980**, *49*, 166–170. [CrossRef]
31. Hartman, P.; Perdok, W.G. On the relations between structure and morphology of crystals. I. *Acta Crystallogr.* **1955**, *8*, 49–52. [CrossRef]
32. Weissbuch, I.; Addadi, L.; Lahav, M.; Leiserowitz, L. Molecular Recognition at Crystal Interfaces. *Science* **1991**, *253*, 637–645. [CrossRef]
33. Connolly, M. Analytical molecular surface calculation. *J. Appl. Crystallogr.* **1983**, *16*, 548–558. [CrossRef]
34. Clegg, W.; Scott, A.J. CCDC 1575175: Experimental Crystal Structure Determination. *CSD Commun.* **2017**. [CrossRef]
35. *Materials Studio, 6.0*; Acceryls Inc.: San Diego, CA, USA, 2012.
36. Sun, H. COMPASS: An ab Initio Force-Field Optimized for Condensed-Phase ApplicationsOverview with Details on Alkane and Benzene Compounds. *J. Phys. Chem. B* **1998**, *102*, 7338–7364. [CrossRef]
37. Lan, G.; Jin, S.; Li, J.; Wang, J.; Li, J.; Chen, S.; Li, L. The study of external growth environments on the crystal morphology of ε-HNIW by molecular dynamics simulation. *J. Mater. Sci.* **2018**, *53*, 12921–12936. [CrossRef]
38. Li, J.; Jin, S.; Lan, G.; Xu, Z.; Wu, N.; Chen, S.; Li, L. The effect of solution conditions on the crystal morphology of β-HMX by molecular dynamics simulations. *J. Cryst. Growth* **2019**, *507*, 38–45. [CrossRef]

39. Xia, X.; Jiang, D. Determination and correlation of solubilities of catechol. *CIESC J.* **2007**, *58*, 1082–1085. [CrossRef]
40. Parmar, M.M.; Khan, O.; Seton, L.; Ford, J.L. Polymorph selection with morphology control using solvents. *Cryst. Growth Des.* **2007**, *7*, 1635–1642. [CrossRef]
41. Kitamura, M.; Ishizu, T. Growth kinetics and morphological change of polymorphs of L-glutamic acid. *J. Cryst. Growth* **2000**, *209*, 138–145. [CrossRef]
42. Ter Horst, J.H.; Kramer, H.J.M.; van Rosmalen, G.M.; Jansens, P.J. Molecular modelling of the crystallization of polymorphs. Part I: The morphology of HMX polymorphs. *J. Cryst. Growth* **2002**, *237*, 2215–2220. [CrossRef]
43. Wunderlich, H.; Mootz, D. Die Kristallstruktur von Brenzcatechin: Eine Neubestimmung. *Acta Crystallogr. B* **1971**, *27*, 1684–1686. [CrossRef]
44. Zhao, Q.; Liu, N.; Wang, B.; Wang, W. A study of solvent selectivity on the crystal morphology of FOX-7 via a modified attachment energy model. *RSC Adv.* **2016**, *6*, 59784–59793. [CrossRef]
45. Liang, Z.; Chen, J.-F.; Ma, Y.; Wang, W.; Han, X.; Xue, C.; Zhao, H. Qualitative rationalization of the crystal growth morphology of benzoic acid controlled using solvents. *CrystEngComm* **2014**, *16*, 5997–6002. [CrossRef]
46. Einstein, A. Über einen die Erzeugung und Verwandlung des Lichtes betreffenden heuristischen Gesichtspunkt [AdP 17, 132 (1905)]. *Annalen der Physik* **2005**, *14*, 164–181. [CrossRef]
47. Frenkel, D.; Smit, B. *Understanding Molecular Simulation: From Algorithms to Applications*; Academic Press: Cornwall, UK, 2001; Volume 1, pp. 84–105.
48. Shi, W.; Xia, M.; Lei, W.; Wang, F. Solvent effect on the crystal morphology of 2,6-diamino-3,5-dinitropyridine-1-oxide: A molecular dynamics simulation study. *J. Mol. Graph. Model.* **2014**, *50*, 71–77. [CrossRef] [PubMed]
49. Gupta, K.M.; Yani, Y.; Poornachary, S.K.; Chow, P.S. Atomistic Simulation To Understand Anisotropic Growth Behavior of Naproxen Crystal in the Presence of Polymeric Additives. *Cryst. Growth Des.* **2019**, *19*, 3768–3776. [CrossRef]
50. Li, J.; Jin, S.; Lan, G.; Ma, X.; Ruan, J.; Zhang, B.; Chen, S.; Li, L. Morphology control of 3-nitro-1,2,4-triazole-5-one (NTO) by molecular dynamics simulation. *CrystEngComm* **2018**, *20*, 6252–6260. [CrossRef]
51. Sun, T.; Liu, Q.; Xiao, J.; Zhao, F.; Xiao, H. Molecular Dynamics Simulation of Interface Interactions and Mechanical Properties of CL-20/HMX Cocrystal and Its Based PBXs. *Acta Chim. Sinica* **2014**, *72*, 1036. [CrossRef]

© 2020 by the authors. Licensee MDPI, Basel, Switzerland. This article is an open access article distributed under the terms and conditions of the Creative Commons Attribution (CC BY) license (http://creativecommons.org/licenses/by/4.0/).

Article

Numerical Model Study of Multiple Dendrite Motion Behavior in Melt Based on LBM-CA Method

Yu Bai, Yingming Wang, Shijie Zhang, Qi Wang and Ri Li *

Simulation Laboratory, School of Materials Science and Engineering, Hebei University of Technology, Tianjin 300132, China; 201721803007@stu.hebut.edu.cn (Y.B.); 201731804057@stu.hebut.edu.cn (Y.W.); 201821803012@stu.hebut.edu.cn (S.Z.); 201831804007@stu.hebut.edu.cn (Q.W.)
* Correspondence: sdzllr@163.com

Received: 27 December 2019; Accepted: 25 January 2020; Published: 27 January 2020

Abstract: In this paper, a new method is proposed to solve the solute field of moving grains, and a Cellular automaton (CA)-Lattice Boltzmann method (LBM)-Semi rebound format(Ladd) coupling model which can accurately simulate the motion behavior of multiple dendrites is established. The growth process of microstructure in the solidification process of Al-4.7% Cu alloy ingot was calculated by Cellular automaton (CA) method, the momentum, heat, and mass transfer processes were calculated by Lattice Boltzmann method (LBM), and the melt-dendrite sharp interface interaction was treated by Ladd method. The reliability of the model is verified, and then the growth and movement of single dendrite and multiple dendrites under the action of gravity field are simulated. The simulation results show that the growth and movement mode of multiple dendrites are quite different from that of single dendrite, which is shown in two aspects: (1) the original motion state of dendrites is changed by the combination of flow field, which slows down the falling speed of dendrites to a certain extent; (2) the fusion of solute field between dendrites changed the original growth mode of boundary dendrites and increased their rotation speed.

Keywords: multi-dendrite motion; CA-LBM model; dendritic growth; natural convection; numerical simulation

---

1. Introduction

During the solidification process of the alloy, the fine grains formed on the surface of the mold will fall off, and the solidified dendrite arms will also be remelted and fractured, resulting in a large number of free equiaxed dendrites in the liquid phase area, which will move under the action of natural convection and gravity [1]. The movement and falling process of a large number of group grains not only has an important influence on the formation of positive segregation at the top of ingot, A-type and V-type segregation [2], but also is the main reason for the formation of triangular cone-shaped negative segregation at the bottom of large ingot [3,4].Therefore, it is of great significance to add the calculation of grain movement process into the numerical model of ingot macrosegregation to improve the prediction accuracy.

At present, the phase field method is mostly used to simulate grain movement in the world. In 2008, Do-Quang M et al. [5] used the phase field-virtual domain method to simulate the growth and movement of single dendrite under the action of gravity. In 2012, karagadde and Bhattacharya [6] used the enthalpy method (EF) to calculate the growth of dendrites, the volume of fluid method (VOF) to calculate the movement behavior of dendrites, and the immersion boundary method (IBM) to deal with the solid-liquid interface. The simulation results show that the multi dendrite growth pattern is significantly different from that of the single dendrite. In 2013, Medvedev et al. [7] used Phase Field-Lattice Boltzmann method (LBM) coupling model to calculate the dendrite growth and movement behavior of aluminum copper alloy under the action of shear flow and pipe flow. In 2015,

Rojas and Takaki [8] used a PF-LBM model to simulate the growth and movement of dendrites under shear flow, and analyzed the effect of solution flow on the growth and movement of dendrites. In the same year, Takaki et al. [9] added GPU technology to the program code, which greatly improved the simulation efficiency and scale, and used the technology to simulate the settlement behavior of single dendrite in the gravity field. In 2017, Qi et al. [10] proposed a new phase-field model incorporating dendrite-melt two-phase flow, and modified the boundary layer of growth kinetics equation, so that it can better reflect the relationship between the growth rate of dendrite tip and the flow direction of fluid. In 2018, Takaki et al. [11] established a new phase field model to simulate the growth and movement of multi-dendrite, and coupled this model with lattice Boltzmann to simulate the growth, movement, collision, and growth behavior after bonding of multi-dendrite.

However, the grid of phase field method is very small and the amount of calculation is huge, which greatly limits the number of equiaxed dendrites. It is impossible to simulate the solidification process of ingots with a large number of equiaxed and columnar dendrites [12]. The Cellular automaton (CA) method has a small amount of calculation and a fast calculation speed, and is undoubtedly more suitable for calculating multi-dendritic motion behavior. Currently, only the work of Liu et al. [13] used the CA method to calculate the moving dendrite. He can only simulate the settlement of a single dendrite, and the dendrite cannot rotate, obviously this is different from the actual situation. This paper improves the calculation accuracy of the concentration field and simulates the movement of multiple dendrites.

In addition, the temperature, flow, and solute fields need to be calculated when simulating dendrite motion in the melt. The LBM method developed in recent years can effectively calculate the passage process of dendrites in the melt, so it has been widely used. This paper uses the LBM method to calculate three fields. The dendrite in the melt must interact with the melt during the movement. This article uses the Ladd method to deal with this effect, because the Ladd method uniformly processes the solid phase and the liquid phase, so as to avoid the mass and momentum loss caused by the solid phase node covering the liquid phase node in the process of dendrite movement. However, the difficulty in dealing with the solid-liquid interaction lies in the calculation of the solute field. The solute diffusion coefficient of the two phases is quite different, therefore it cannot directly treat the two phases as a whole to deal with the boundary. Therefore, this paper presents a method to deal with the solute field of multi grain movement, which realizes the calculation of solute field in the real sense of dendrite movement.

2. Materials and Methods

In this paper, Al-4.7% Cu alloy is selected as the research object, and the physical parameters are shown in Table 1.

Table 1. Physical properties of Al-4.7% Cu alloy.

Physical Parameter	Symbol	Value
Melting temperature	T_m (K)	933.3
Liquidus temperature	T_L (K)	917
Solidus temperature	T_S (K)	821
Liquidus slope	M (m·K/%)	−3.44
Thermal diffusivity	A ($m^2 \cdot s^{-1}$)	2.7×10^{-7}
Fluid viscosity	N ($m^2 \cdot s^{-1}$)	1.2×10^{-6}
Diffusivity in liquid	D ($m^2 \cdot s^{-1}$)	3.0×10^{-9}
Partition coefficient	k	0.145
Liquid density	P (kg·m^{-3})	2606

2.1. CA Model

The solute equilibrium model (ZS model) proposed by Zhu [14] is used in this paper. The growth driving force is the difference between the equilibrium crystallization concentration and the actual liquid concentration at the interface. The actual liquid concentration C_L can be calculated by LBM, and the equilibrium crystallization concentration C_L^{eq} can be calculated by the following formula according to the equilibrium crystallization theory:

$$C_L^{eq} = C_0 + [T_L - T_L^{eq} + \Gamma K f(\phi, \theta_0)]/m \tag{1}$$

where C_0 is the initial concentration of the alloy; m is the slope of the liquidus; T_L is the actual temperature of the interface; T_L^{eq} is the liquidus temperature at the initial concentration of C_0, Γ is the Gibbs Thomson coefficient; K is the average curvature at the solid/liquid interface, $f(\phi, \theta_0)$ is the anisotropic function of the interface energy. K can be calculated from the spatial distribution of the interface solid phase ratio.

$$K = [(\frac{\partial f_S}{\partial x})^2 + (\frac{\partial f_S}{\partial y})^2]^{-3/2} \cdot [2 \frac{\partial f_S}{\partial x} \frac{\partial f_S}{\partial y} \frac{\partial^2 f_S}{\partial x \partial y} - (\frac{\partial f_S}{\partial x})^2 \frac{\partial^2 f_S}{\partial y^2} - (\frac{\partial f_S}{\partial y})^2 \frac{\partial^2 f_S}{\partial x^2}] \tag{2}$$

According to the Gibbs-Thomson formula, the interface energy anisotropy function in Equation (1) can be expressed as:

$$f(\phi, \theta_0) = \Psi(\phi, \theta_0) + \frac{\partial^2}{\partial \phi^2} \Psi(\phi, \theta_0) = 1 - \delta \cos[4(\phi - \theta_0)] \tag{3}$$

where $\delta = 15\varepsilon$ is the anisotropy coefficient (ε is the anisotropic strength of the interface energy), $\psi(\phi, \theta_0)$ is the anisotropy function of the interface energy, ϕ is the angle between the normal direction of the solid-liquid interface and the horizontal direction, and θ_0 is the preferred growth direction. The anisotropy function $\psi(\phi, \theta_0)$ and the growth angle ϕ can be calculated from Equations (4) and (5).

$$\psi(\phi, \theta_0) = 1 + \varepsilon \cos[4(\phi - \theta_0)] \tag{4}$$

$$\phi = \begin{cases} \cos^{-1}[\frac{\partial f_S}{\partial x}[(\frac{\partial f_S}{\partial x})^2 + (\frac{\partial f_S}{\partial y})^2]^{-1/2}] & \frac{\partial f_S}{\partial y} \geq 0 \\ 2\pi - \cos^{-1}[\frac{\partial f_S}{\partial x}[(\frac{\partial f_S}{\partial x})^2 + (\frac{\partial f_S}{\partial y})^2]^{-1/2}] & \frac{\partial f_S}{\partial y} < 0 \end{cases} \tag{5}$$

When $C_L < C_L^{eq}$, the solid fraction increment Δf_s in a time step is calculated by the following formula:

$$\Delta f_S = \frac{(C_L^{eq} - C_L)}{C_L^{eq}(1-k)} \tag{6}$$

In order to partially eliminate anisotropy, this paper uses an improved eight-neighbor capture method proposed by Zhu [15].

2.2. LBM Model

According to Boussinesq's approximation, the effect of latent heat and solutes on the density during the solidification process can be expressed by the following formula:

$$\rho = \rho_0[1 - \beta_T(T - T_0) - \beta_C(C - C_0)] \tag{7}$$

where ρ_0, T_0, and C_0 represent the initial density, temperature, and concentration of the liquid phase, respectively, and T and C are the temperature and concentration of the liquid phase at the current moment. β_T and β_C are the volume expansion coefficients of temperature and concentration changes, respectively. The resultant force of the fluid particles is:

$$\begin{aligned} F &= g\rho_0[1 - \beta_T(T - T_0) - \beta_C(C - C_0)] + (-\rho_0 g) \\ &= g\rho_0[-\beta_T(T - T_0) - \beta_C(C - C_0)] \end{aligned} \quad (8)$$

The distribution function of the flow field can be expressed as:

$$f_i(x + e_i \Delta t, t + \Delta t) = f_i(x, t) + \frac{1}{\tau_f}(f_i^{eq}(x, t) - f_i(x, t)) + F_i \quad (9)$$

where F_i is the component force of the particle under the external force field in the i direction, and its magnitude is expressed as:

$$F_i = (1 - \frac{1}{2\tau_f})\omega_i[3\frac{e_i - u}{c^2} + 9\frac{e_i \cdot u}{c^4}]\Delta t \cdot F \quad (10)$$

where ω_i is the weight coefficient in each direction, which represents the probability of particles moving in different directions, which can be expressed as:

$$\omega_i = \begin{cases} 4/9 & i = 0 \\ 1/9 & i = 1, 2, 3, 4 \\ 1/36 & i = 5, 6, 7, 8 \end{cases} \quad (11)$$

The equilibrium distribution function and relaxation time of the flow field in Equation (9) are respectively expressed as:

$$f_i^{eq}(x, t) = \omega_i \rho(1 + 3\frac{e_i \cdot u}{c^2} + \frac{9}{2}\frac{(e_i \cdot u)^2}{c^4} - \frac{3}{2}\frac{u \cdot u}{c^2}) \quad (12)$$

$$\tau_f = 3\nu/(c^2 \Delta t) + 0.5 \quad (13)$$

where ν is the dynamic viscosity of the fluid. The macroscopic density ρ and velocity u are obtained by adding the distribution function.

$$\rho = \sum_{i=0}^{8} f_i \quad (14)$$

$$u = (\sum_{i=0}^{8} e_i f_i + F \cdot \Delta t/2)/\rho \quad (15)$$

The distribution functions of temperature field and solute field are similar to those of flow field:

$$h_i(x + e_i \Delta t, t + \Delta t) = h_i(x, t) + \frac{1}{\tau_\alpha}(h_i^{eq}(x, t) - h_i(x, t)) + H_i \quad (16)$$

$$g_i(x + e_i \Delta t, t + \Delta t) = g_i(x, t) + \frac{1}{\tau_D}(g_i^{eq}(x, t) - g_i(x, t)) + G_i \quad (17)$$

where $h_i(x, t)$ and $g_i(x, t)$ are the distribution functions of temperature field and solute field at the position x at time t, respectively, and $h_i(x, t)$ and $g_i(x, t)$ represent the temperature field and the equilibrium distribution function of the solute field, respectively defined as:

$$h_i^{eq}(x, t) = \omega_i T(1 + 3\frac{e_i \cdot u}{c^2} + \frac{9}{2}\frac{(e_i \cdot u)^2}{c^4} - \frac{3}{2}\frac{u \cdot u}{c^2}) \quad (18)$$

$$g_i^{eq}(x,t) = \omega_i C(1 + 3\frac{e_i \cdot u}{c^2} + \frac{9}{2}\frac{(e_i \cdot u)^2}{c^4} - \frac{3}{2}\frac{u \cdot u}{c^2}) \tag{19}$$

The relaxation time τ_α in the temperature field and the relaxation time τ_D in the solute field can be obtained by using the corresponding diffusion coefficients:

$$\tau_\alpha = 3\alpha/(c^2 \Delta t) + 0.5 \tag{20}$$

$$\tau_D = 3D/(c^2 \Delta t) + 0.5 \tag{21}$$

where α is a temperature diffusion coefficient, and D is a concentration diffusion coefficient. Macro temperature and concentration are:

$$T = \sum_{i=0}^{8} h_i(x,t); C = \sum_{i=0}^{8} g_i(r,t) \tag{22}$$

The source term H_i of the temperature field and the source term G_i of the concentration field can be expressed as:

$$H_i = \omega_i \Delta T; G_i = \omega_i \Delta C \tag{23}$$

In the formula, ΔT and ΔH respectively represent the latent heat released by the solidification of the alloy and the excluded solutes.

2.3. Ladd Method to Calculate the Solid-liquid Interface Interaction Force

In Figure 1, x_b is the particle boundary point, x_l is the liquid phase lattice point of the boundary node along the c_{-i} direction, x_s is the solid phase lattice point of the boundary node along the c_i direction, and u_b is the particle velocity. The calculation formulas of x_s and x_b are:

$$x_s = x_l + \Delta t \cdot c_i \tag{24}$$

$$u_b = V_b + W_b(x_b - x_c) \tag{25}$$

where Δt is the time step, V_b and W_b are the translational and rotational speeds, respectively, and x_c is the center of mass of the solid particles. The solid-liquid distribution functions at time t are $f_i(x_l, t)$ and $f_{-i}(x_s, t)$. After $\Delta t/2$ time, the two particles move to the boundary and collide. The distribution function at this time is:

$$f_{-i}(x_b, t + \Delta t/2) = f_i(x_l, t) + 2\omega_{-i}\rho \frac{c_{-i}u_b}{c_s^2} \tag{26}$$

$$f_i(x_b, t + \Delta t/2) = f_{-i}(x_s, t) + 2\omega_i \rho \frac{c_i u_b}{c_s^2} \tag{27}$$

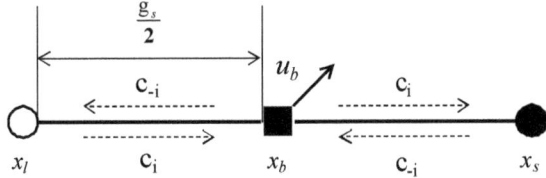

Figure 1. Sketch map of half step rebound format.

After the $\Delta t/2$ time, the fluid particles bounce to the corresponding lattice points respectively. At this time, the distribution functions of the liquid phase and solid phase lattice points are:

$$f_{-i}(x_l, t + \Delta t) = f_{-i}(x_b, t + \Delta t/2) \tag{28}$$

$$f_i(x_s, t+\Delta t) = f_i(x_b, t+\Delta t/2) \tag{29}$$

Force exerted by fluid particles on solid particles:

$$F_i = \frac{\Delta x^2}{\Delta t}\left[f_{-i}(x_f, t+\Delta t) + f_i(x_f, t) - f_i(x_s, t+\Delta t) - f_{-i}(x_s, t)\right]c_i \tag{30}$$

The total force F on the solid particles is:

$$F = \frac{\Delta x^2}{\Delta t}\sum_{x_b}\sum_i [f_{-i}(x_f, t+\Delta t) + f_i(x_f, t) - f_i(x_s, t+\Delta t) - f_{-i}(x_s, t)]c_i \tag{31}$$

The force moment on the solid particles is:

$$T_t = \sum_{x_b} F_{x_b}(x_b - x_c) \tag{32}$$

According to Newton's second law, the translation speed and rotation speed of the grains can be calculated respectively as:

$$V = \frac{F+G}{M_S}dt \quad W = \frac{T_t}{I_S}dt \tag{33}$$

where M_S and I_S are mass and moment of inertia, respectively, and G is the combined force of gravity and buoyancy.

2.4. Processing of Solute Fields at Moving Boundaries

In order to accurately calculate the solute field during movement, a solute extrapolation method for calculating the solute field at the moving boundary is proposed.

As shown in Figure 2, the white and black grid points are the liquid and solid grid points, the solid ellipse is the position of the dendrite at the previous moment, the dotted ellipse is the current position. The gray grid points are the covered liquid nodes because of dendrite movement. The concentration C_L of the covered liquid grid points will be distributed in a certain proportion, which is:

$$\Delta C_L^1 = A \cdot C_L \Delta C_L^2 = B \cdot C_L \Delta C_L^3 = C \cdot C_L \tag{34}$$

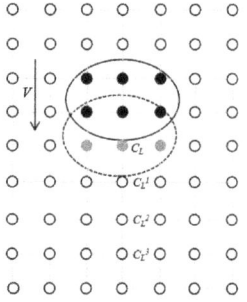

Figure 2. A schematic diagram of solid fall.

$\Delta C_L^1, \Delta C_L^2$, and ΔC_L^3 respectively represent the concentration assigned to three lattice points along the direction of dendrite movement. The distance between the lattice points is the grain movement distance. A, B, and C are the distribution coefficients and satisfy the following relations:

$$A + B + C = 1 \quad A > B > C \tag{35}$$

In this paper, the undetermined coefficient method is used, and the distribution coefficients of different proportions are used to calculate. Then the results are compared with the experimental results of Liu's single dendrite drop. The empirical values of A, B, and C are 0.7, 0.2, and 0.1, respectively. The assigned concentration will be added to the three lattice points along the direction of dendrite movement.

$$C_L^n = C_L^n + \Delta C_L^n \tag{36}$$

3. Verification

First, the accuracy of the Ladd method for processing moving boundaries is evaluated through the settlement process of circular particles in an infinitely long pipe. Then, the accuracy of the solute distribution model established in this paper is verified by calculating the solute field conservation of the single dendrite in the moving state. The LBM model, the dendrite growth of CA model, and the rationality of the single dendrite movement model have been confirmed in relevant researches [16,17], so this article will not be described here.

3.1. Settling of a Circular Particle in an Infinitely Long Tube

The circular particle accelerates to settle down in the infinite tube due to gravity. As the speed increases, the resistance of the ball increases. When the three forces of resistance, gravity, and buoyancy are balanced, the particle falls at a uniform speed. Glowinski pointed out in the literature [18] that, for a circular particle settling in an infinitely long pipe, when its physical parameters are determined, the settling speed is also determined at a steady state.

When the liquid is at a low Reynolds number, the resistance Ff to circular particles moving in an infinitely long pipeline is directly proportional to the settlement speed [19] and can be expressed as:

$$F_f = 4\pi K \eta v \tag{37}$$

where v is the falling velocity of circular particles, η is the dynamic viscosity of the fluid, K is the correction factor, and its value reflects the influence of the resistance of the pipe wall facing the particles. K is expressed as:

$$K = \frac{1}{\ln W_i - 0.9157 + 1.7244(W_i)^{-2} - 1.7302(W_i)^{-4} + 2.4056(W_i)^{-6} - 4.5913(W_i)^{-8}} \tag{38}$$

where $W_i = W/D$, W is the width of a long square tube and D is the diameter of the circular particle. When particles are in a state of three forces equilibrium, the resistance can be obtained as:

$$F_f = \frac{1}{4}\pi D^2 (\rho_S - \rho_L) g \tag{39}$$

ρ_S and ρ_L are solid density and liquid density, and g is gravity acceleration. According to Equations (37)–(39), the final falling speed of the circular particle can be calculated as:

$$v = \frac{D^2(\rho_S - \rho_f)\mathbf{g}}{16K\eta} \tag{40}$$

The size of the circular pipe in this paper is 4 cm × 8 cm. The side length of each grid is 0.01 cm, with a total of 400 × 800 grids. The particle diameter is taken as 0.48 cm, the fluid density is $\rho_L = 1.0$ g/cm^3 and the solid density is $\rho_S = 1.02$ g/cm^3, and the fluid viscosity is $\eta = 0.33$ g/(cm·s). The relaxation time τ is taken as 0.8, and the gravity acceleration is $g = 980.0$ cm/s^2. At the initial moment, the circular particle are placed at points (2 cm, 6 cm), and they are at rest before the calculation starts.

The simulation results are shown in Figure 3. The particle dropped due to gravity, and two symmetrical vortices formed on both sides of the particle, which is consistent with what Do-Quang [5] described in his literature. According to the simulation value of particle settlement speed, it can be seen that the simulation results in this paper agree well with the theoretical analytical solution. Therefore, the method used in this paper can be used to calculate the moving boundary problem.

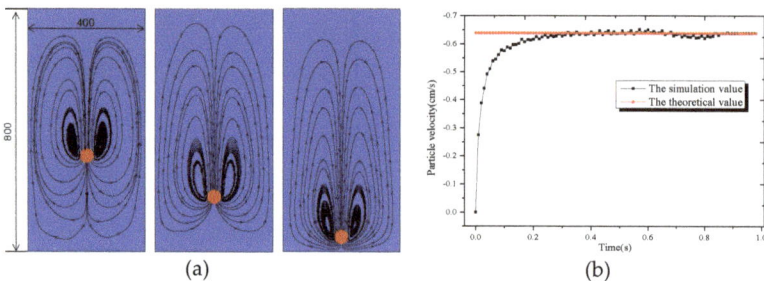

Figure 3. Diagram of circular particle settlement: (**a**) Flow field; (**b**) Settlement velocity.

3.2. Calculation of the Solute Field

Figure 4 shows the morphology of dendrite growth under different solute field treatment methods. Figure 4a is the solute distribution method and Figure 4b is the solute extrapolation method.

In the first method, the lower end of the dendrite grew faster, because the dendrite covered the original high-concentration solute domain during the settlement of the dendrite. Through the uniform distribution of solutes, the solute gradient at the upstream end of the dendrite is reduced, so it has a faster growth rate. In contrast, dendrite growth using the solute extrapolation method was more uniform. This article compares the two methods to calculate the average regional concentration at the time of solidification time of 0.3 s, 0.5 s, 0.7 s. It is found that the solute field of Method 1 changes greatly, and the average solute concentrations at the corresponding moments are 4.682%, 4.673%, 4.658% (standard is 4.7%), solute field conservation is always 4.7% in Method 2. It can be seen that the solute field treatment in the second method is more accurate.

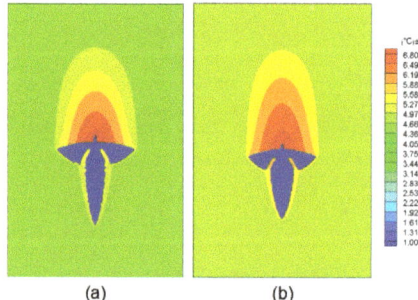

Figure 4. Dendritic morphology evolution diagram under different methods. (**a**) Method 1; (**b**) Method 2.

3.3. Multiple Dendrites Rotation

At the initial time, 5 crystal nuclei with preferential growth angles of 0 and a solid phase rate of 0.2 were placed in the middle of the simulation region. When the number of solidification grids is 3000, the dendrites stop growing. At this time, a uniform rotating flow is applied in the simulation region. It can be seen from the Figure 5 that the dendrite can maintain its original shape even after being rotated for multiple turns.

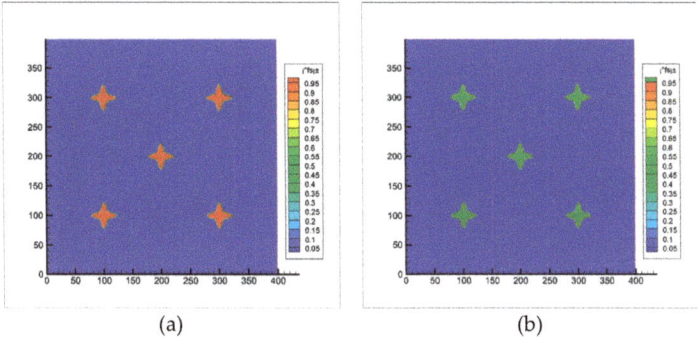

Figure 5. Dendritic morphology before and after rotation (**a**) Before rotation; (**b**) After rotation.

This article gives the formula for calculating the solid phase rate error, as follows:

$$\Delta = (f_{Sin} - f_{S0})/f_{S0} \times 100 \tag{41}$$

where f_{S0} is the initial solid phase rate of the dendrite, f_{Sin} is the solid phase rate after dendrite rotation. According to Equation (41), it is calculated that the change of the solid phase rate of the equiaxed crystal is maintained within 0.2% after 10 revolutions. In the calculations of this paper, the amplitude of dendrite rotation is small (rad <1, rad represents the radian of dendrite rotation), so it can be considered that the numerical model for calculating dendrite rotation established in this paper has no great influence on the dendrite morphology.

4. Discussion

This article is based on the following assumptions. The melt is an incompressible Newtonian fluid. The solid and liquid phases have the same thermal conductivity, and the heat transfer boundary conditions are adiabatic boundary conditions. Solute diffusion only occurs in the liquid phase, there is no solute diffusion in the solid phase, and the boundary condition for mass transfer is the non-diffusion boundary condition [13].

4.1. Single Dendrite Movement

As shown in Figure 6, the dendrite arm at the front end of the dendrite grows faster in the moving state, and a secondary dendrite arm is generated. The upper dendrite arm has a lower speed and almost stops growing. The reason is that the dendrite movement compresses the lower solute boundary layer, the concentration gradient becomes larger, and the temperature gradient becomes lower; the upper solute boundary layer is stretched, the concentration gradient becomes smaller, and the temperature gradient increases. Therefore, the growth of the dendrite arms at the upstream side will be further promoted. As the falling speed increases, this asymmetric growth phenomenon becomes more and more obvious. In actual solidification, when the dendrite grows asymmetrically, the dendrite will rotate. From Figure 6c$_2$, it can be seen that compared with pure falling dendrites (Figure 6b$_2$), the former dendrite arms have prominent tips, asymmetric growth is more obvious, and the latter dendrites are more uniform, which is consistent with the literature [6].

The solute distribution is also closely related to the flow of the solution around the dendrites. It can be seen from Figure 7 that the flow direction of the solution is different in the three movement states. The solution near the stationary dendrite flows upward, and the solution far away from the dendrite flows downward. For a moving dendrite, due to the high viscosity of the solution, a downward pulling force will be generated on the surrounding solution during the drop of the dendrite. This causes the solution near the dendrite to flow downwards and the solution away from the dendrites to flow upwards. In addition, for moving dendrites, two vortices will be generated behind it.

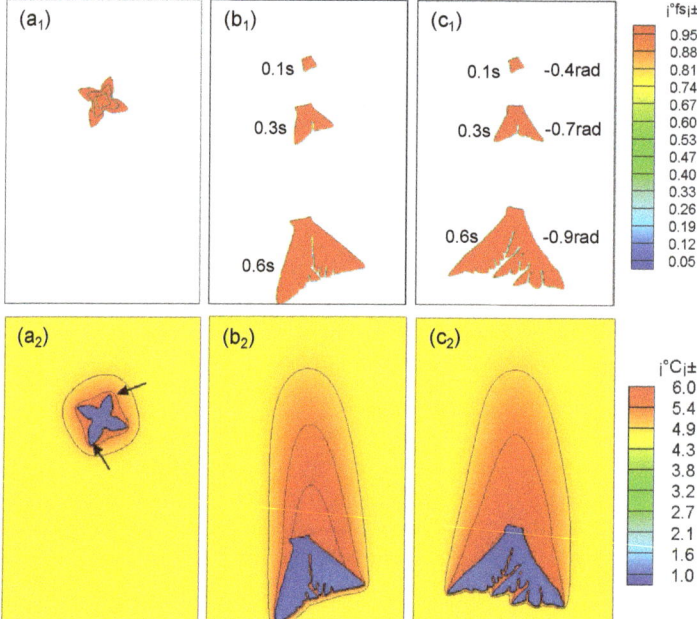

Figure 6. Effects of different motion states on dendrite growth. (**a₁**), (**b₁**), (**c₁**) the growth morphology diagram of dendrite under the condition of stationary, translational falling without rotation, and rotational falling; (**a₂**), (**b₂**), (**c₂**) the solute distribution diagram of dendrite under the condition of stationary, translational falling without rotation, and rotational falling.

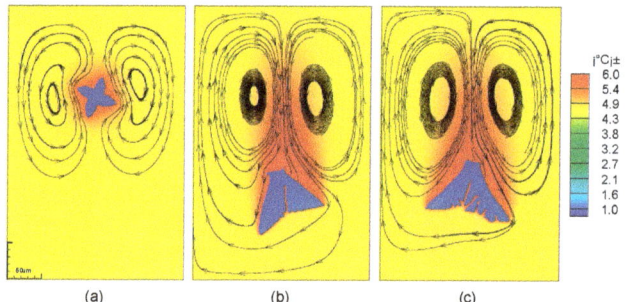

Figure 7. Flow field distribution in the process of dendrite falling; (**a**) stationary dendrites; (**b**) falling dendrites; (**c**) falling and rotating dendrites.

It can be seen from Figure 8 that the solution flows faster around the moving dendrites. When the solidification time is 0.5 s, the flow velocity of the solution around the dendrite in the stationary, pure falling, and rotating falling states is 0.008 mm/s, −0.01 mm/s, −0.009 mm/s. This is because the rotation of the dendrite makes the contact surface larger between the lower end and the solution, resulting in an increase in the resistance of dendrites, which reduces the falling speed of the rotating dendrite to a certain extent.

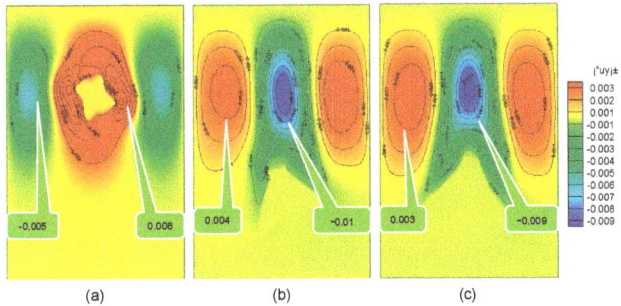

Figure 8. Flow velocity in the process of dendrite falling. (**a**) Stationary dendrites; (**b**) falling dendrites; (**c**) falling and rotating dendrites.

4.2. Multi-Dendrite Movement

As the dendrites grow, the solute boundary layers around the dendrites come into contact and fuse with each other to form a high-concentration solute domain. The growth of dendrite arms in this domain is inhibited, and the growth of dendrite arms away from this domain is promoted. The asymmetry of the dendrite is aggravated, and the rotation speed of the dendrite is increased.

The rotation of dendrites is different from that of single dendrites during solidification. See Table 2 for the change of preferential dendrite growth angle. The initial preferred growth angle of No. 4 dendrite is 0.1 rad. The solutes between the dendrites are close to each other and form a high concentration solute domain, therefore the growth of the right dendrite arm of No. 4 dendrite is inhibited, and the growth of the left dendrite arm is promoted. This causes the mass on the left side of the dendrite to be greater than the mass on the right side, so the dendrite is subjected to a counterclockwise torque. When the solidification time is 0.4 s, the dendrite rotates 0.3 rad in a counterclockwise direction. For the No. 1 dendrite, the lower part of the simulated domain is less affected by the solute field. It starts to rotate at 0.2 s and falls to the bottom at 0.5 s. At this time, the preferred growth direction of the dendrite is −0.5 rad. For the No. 9 dendrite, the dendrite arm at the lower end of the dendrite extends to the high-concentration domain, and the growth is inhibited. Therefore, the dendrite is more symmetrical on the left and right, and the torque is less. During its movement, the dendrite only rotated 0.2 rad. The No. 6 dendrite is located in the center of the simulation area. The uncertainty of its growth behavior leads to the complexity of its movement behavior. It can be seen that the dendrite rotates 0.3 rad clockwise when the solidification time is 0.4 s. At 0.5 s, the dendrite rotates counterclockwise by 0.1 rad, and the dendrite rotates left and right.

Table 2. Preferred growth angle of dendrites at different times.

	Dendrite Rotation Angle (rad)			
Time (s)	No. 1	No. 4	No. 4	No. 9
0	−0.2	0.1	−0.3	−0.6
0.1	−0.2	0.1	−0.3	−0.6
0.2	−0.3	0	−0.4	−0.7
0.3	−0.4	0	−0.6	−0.8
0.4	−0.5	0.3	−0.6	−0.8
0.5	−0.5	0.5	−0.5	−0.8

Multi-dendritic effects are also manifested in interactions between fluids. As shown in Figure 9a, as the dendrites grow, the fluid vortices begin to merge, forming a strong convection between the dendrites. This strong convection hinders the vertical drop of the dendrite and has a lateral force on the dendrite. The dendrite starts to move laterally, and the grains on both sides have a tendency of centrifugal movement. And the movement of the grains in the central domain shows a trend of swinging left and right.

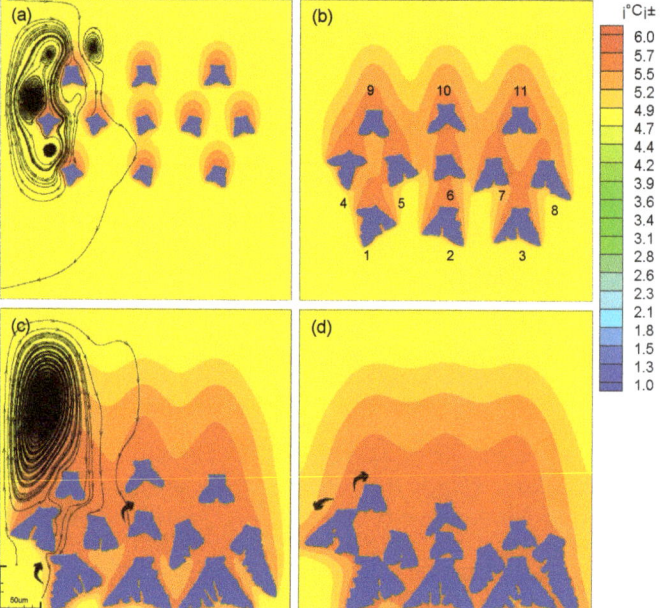

Figure 9. The evolution and concentration distribution of multi-falling-dendrites. (**a**) t_1=200; (**b**) t_2=500; (**c**) t_3=800; (**d**) t_4=1200.

Figure 10 shows the falling speed of the four dendrites. Each dendrite undergoes a process of acceleration and deceleration, and their absolute speeds are lower than those of the single dendrites. It is shown that during the growth of multi-dendritic, due to the interaction between the dendrites, a part of the gravity is offset and the dendrite's moving speed is reduced to a certain extent.

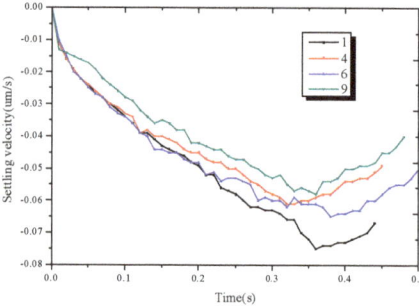

Figure 10. The falling velocity of dendrite 1, dendrite 4, dendrite 6, dendrite 9.

Compared with the growth of multiple dendrites in the moving state, the growth mode of the dendrites in the stationary state is much simpler. The growth of the dendrites in the middle of the simulated domain is suppressed, and the growth of dendrites near the boundary is promoted. (As shown in Figure 11)

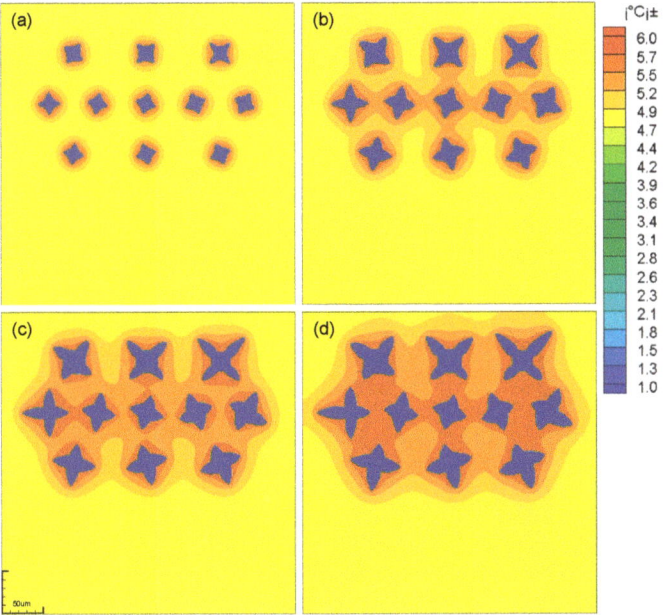

Figure 11. The evolution and concentration distribution of dendrites without motion. (**a**) $t_1 = 200$; (**b**) $t_2 = 500$; (**c**) $t_3 = 800$; (**d**) $t_4 = 1000$.

From the above analysis, the movement behavior of multi-dendrites is related to the melt convection and solute overlap between the dendrites. In response to this phenomenon, scholars proposed to apply an external force field (such as an electromagnetic field) to the dendrite to offset gravity, thereby changing the movement state of the dendrite, and then controlling the solute segregation of the casting [20,21].

5. Conclusions

In this work, a CA-LBM-Ladd coupling model for calculating multi-dendritic motion was established, and an extrapolated distribution method for calculating the solute distribution around the dendrite under the state of dendritic motion in the melt was proposed, which realizes the calculation of the solute field in the real state of dendrite movement. The CA-LBM-Ladd coupling model was verified, and then the motion of single and multiple dendrites was simulated using this model. The simulation results show that: 1) the superposition of the flow fields between the multiple dendrites causes the movement state to change; 2) the superposition of solute field results in the change of concentration gradient, changes the growth mode of dendrite, and then changes the movement state. This is quite different from the growth mode of single dendrite.

Author Contributions: Conceptualization, Y.B.; data curation, Y.B.; formal analysis, Y.B.; investigation, Y.B.; methodology, S.Z.; software, Y.W.; validation, Y.B.; visualization, Q.W. project administration, R.L.; Funding acquisition, R.L. All authors have read and agreed to the published version of the manuscript.

Funding: This research was funded by National Natural Science Foundation of China, (No. 51475138 and 51975182)

Acknowledgments: This work was supported by a grant from the National Natural Science Foundation of China (No. 51475138). Y.B. would like to thank Ri Li (Hebei University of Technology) for providing academic guidance.

Conflicts of Interest: The authors declare no conflict of interest.

Abbreviations

A detailed list of symbols abbreviations used in this paper is listed and a brief explanation is given.

Symbols	Unit	Meaning
f_s	mass%	Solid fraction
C_L	mass%	Actual solute concentration at interface
C_L^{eq}	mass%	Liquid-phase equilibrium crystallization concentration at the interface
Δf_s	mass%	Solid fraction increment
k	—	Equilibrium partition coefficient of solute
C_0	mass%	Initial concentration of the alloy
T_L	K	Actual temperature of the interface
T_L^{eq}	K	Equilibrium liquidus temperature
m	k/mass%	Liquidus slope
Γ	m·K	Gibbs-Thomson coefficient
K	1/m	Average curvature at the solid / liquid interface
ψ	—	Anisotropic function
ε	—	Anisotropic strength of interface energy
δ	—	Anisotropic coefficient
θ_0	deg	Preferred growth direction
ϕ	deg	Growth angle
τ_f	—	Relaxation time of flow field
τ_a	—	Relaxation time of temperature field
τ_D	—	Relaxation time of concentration field
ν	m²/s	Fluid viscosity
α	m²/s	Thermal diffusivity
D	m²/s	Concentration diffusion coefficient
ρ_0	Kg/m³	Initial density of fluid
T_0	K	Initial temperature
β_T	K^{-1}	Volume expansion coefficient of temperature change
β_C	Mass%$^{-1}$	Volume expansion coefficient of concentration change
ω_i	—	Weight coefficient
e_i	m/s	Discrete velocity
c	m/s	Lattice velocity
M_S	Kg	Mass
I_S	Kg·m²	Moment of inertia
g	m/s²	Gravitational acceleration
F	N	The force of fluid on dendrite
T_t	N·m	Force moment
F_i	N	The component force of the particle in the i direction
G_i	mass%	Source term of concentration field
H_i	K	Source term of temperature field
u	m/s	Macroscopic velocity
ΔT	K	Undercooling

References

1. Flemings, M.C. Our Understanding of Macrosegregation: Past and Present. *Trans. Iron Steel Inst. Jpn.* **2000**, *40*, 838–841. [CrossRef]
2. Wu, M.; Ludwig, A.; Bührig-Polaczek, A.; Fehlbier, M.; Sahm, P.R. Influence of convection and grain movement on globular equiaxed solidification. *Int. J. Heat Mass Transf.* **2003**, *46*, 2819–2832. [CrossRef]
3. Wu, M.; Ludwig, A. Modeling equiaxed solidification with melt convection and grain sedimentation—II. Model verification. *Acta Mater.* **2009**, *57*, 5632–5644. [CrossRef]
4. Liu, B.; Xu, Q.; Jing, T.; Shen, H.; Han, Z. Advances in multi-scale modeling of solidification and casting processes. *JOM* **2011**, *63*, 19–25. [CrossRef]

5. Do-Quang, M.; Amberg, G. Simulation of free dendritic crystal growth in a gravity environment. *J. Comput. Phys.* **2008**, *227*, 1772–1789. [CrossRef]
6. Karagadd, S.; Bhattacharya, A.; Tomar, G.; Dutta, P. A coupled VOF-IBM-enthalpy approach for modeling motion and growth of equiaxed dendrites in a solidifying melt. *J. Comput. Phys.* **2012**, *231*, 3987–4000. [CrossRef]
7. Medvedev, D.; Varnik, F.; Steinbach, I. Simulating mobile dendrites in a flow. *Procedia Comput. Sci.* **2013**, *18*, 2512–2520. [CrossRef]
8. Rojas, R.; Takaki, T.; Ohno, M. A phase-field-lattice Boltzmann method for modeling motion and growth of a dendrite for binary alloy solidification in the presence of melt convection. *J. Comput. Phys.* **2015**, *298*, 29–40. [CrossRef]
9. Takaki, T.; Rojas, R.; Ohno, M.; Shimokawabe, T.; Aoki, T. GPU phase-field lattice Boltzmann simulations of growth and motion of a binary alloy dendrite. *IOP Conf. Ser. Mater. Sci. Eng.* **2015**, *84*, 012006. [CrossRef]
10. Qi, X.; Chen, Y.; Kang, X.; Li, D.; Gong, T. Modeling of coupled motion and growth interaction of equiaxed dendritic crystals in a binary alloy during solidification. *Sci. Rep.* **2017**, *7*, 45770. [CrossRef]
11. Takaki, T.; Sato, R.; Rojas, R.; Ohno, M.; Shibuta, Y. Phase-field lattice Boltzmann simulations of multiple dendrite growth with motion, collision, and coalescence and subsequent grain growth. *Comput. Mater. Sci.* **2018**, *147*, 124–131. [CrossRef]
12. Conti, M. Solidification of binary alloys: Thermal effects studied with the phase-field model. *Phys. Rev. E* **1997**, *55*, 765–771. [CrossRef]
13. Liu, L.; Pian, S.; Zhang, Z.; Bao, R.; Li, H.; Chen, A. cellular automaton-lattice Boltzmann method for modeling growth and settlement of the dendrites for Al-4.7%Cu solidification. *Comput. Mater. Sci.* **2018**, *146*, 9–17. [CrossRef]
14. Zhu, M.; Stefanescu, D.M. Virtual front tracking model for the quantitative modeling of dendritic growth in solidification of alloys. *Acta Mater.* **2007**, *55*, 1741–1755. [CrossRef]
15. Zhu, M.; Hong, C. A modified cellular automaton model for the simulation of dendritic growth in solidification of alloys. *ISIJ Int.* **2001**, *41*, 436–445. [CrossRef]
16. Yin, H.; Felicelli, S.D.; Wang, L. Simulation of a dendritic microstructure with the lattice Boltzmann and cellular automaton methods. *Acta Mater.* **2011**, *59*, 3124–3136. [CrossRef]
17. Mohsen, A.Z.; Hebi, Y. Comparison of cellular automaton and phase field models to simulate dendrite growth in hexagonal crystals. *J. Mater. Sci. Technol.* **2012**, *28*, 137–146.
18. Glowinski, R.; Pan, T.W.; Hesla, T.I.; Joseph, D.D.; Périaux, J. A Fictitious Domain Approach to the Direct Numerical Simulation of Incompressible Viscous Flow past Moving Rigid Bodies: Application to Particulate Flow. *J. Comput. Phys.* **2001**, *169*, 363–426. [CrossRef]
19. Happel, J.; Brenner, H. *Low Reynolds Number Hydrodynamics*; Martinus Nijhoff Publishers: Boston, MA, USA, 1983; pp. 113–119.
20. Bi, C.; Guo, Z.; Liotti, E.; Xiong, S. Quantification study on dendrite fragmentation in solidification process of alluminum alloys. *Acta Metall. Sin.* **2015**, *51*, 677–684.
21. Kaldre, I.; Fautrelle, Y.; Etay, J.; Bojarevics, A.; Buligins, L. Investigation of Liquid Phase Motion Generated by the Thermoelectric Current and Magnetic Field Interaction. *Magnetohydrodynamics* **2010**, *46*, 371–380. [CrossRef]

© 2020 by the authors. Licensee MDPI, Basel, Switzerland. This article is an open access article distributed under the terms and conditions of the Creative Commons Attribution (CC BY) license (http://creativecommons.org/licenses/by/4.0/).

Article

A Novel Shadowgraphic Inline Measurement Technique for Image-Based Crystal Size Distribution Analysis

Dominic Wirz [1], Marc Hofmann [1], Heike Lorenz [2], Hans-Jörg Bart [1], Andreas Seidel-Morgenstern [2] and Erik Temmel [2,3,*]

[1] Chair of Separation Science and Technology, TU Kaiserslautern, Gottlieb-Daimler-Straße, 67663 Kaiserslautern, Germany; Dominic.Wirz@mv.uni-kl.de (D.W.); m_hofman@rhrk.uni-kl.de (M.H.); bart@mv.uni-kl.de (H.-J.B.)
[2] Max Planck Institute for Dynamics of Complex Technical Systems, Sandtorstraße 1, 39106 Magdeburg, Germany; lorenz@mpi-magdeburg.mpg.de (H.L.); seidel@mpi-magdeburg.mpg.de (A.S.-M.)
[3] Sulzer Chemtech Ltd., Gewerbestraße 28, 4123 Allschwil, Switzerland
* Correspondence: erik.temmel@sulzer.com or temmel@mpi-magdeburg.mpg.de

Received: 30 July 2020; Accepted: 19 August 2020; Published: 21 August 2020

Abstract: A novel shadowgraphic inline probe to measure crystal size distributions (CSD), based on acquired greyscale images, is evaluated in terms of elevated temperatures and fragile crystals, and compared to well-established, alternative online and offline measurement techniques, i.e., sieving analysis and online microscopy. Additionally, the operation limits, with respect to temperature, supersaturation, suspension, and optical density, are investigated. Two different substance systems, potassium dihydrogen phosphate (prisms) and thiamine hydrochloride (needles), are crystallized for this purpose at 25 L scale. Crystal phases of the well-known KH_2PO_4/H_2O system are measured continuously by the inline probe and in a bypass by the online microscope during cooling crystallizations. Both measurement techniques show similar results with respect to the crystal size distribution, except for higher temperatures, where the bypass variant tends to fail due to blockage. Thiamine hydrochloride, a substance forming long and fragile needles in aqueous solutions, is solidified with an anti-solvent crystallization with ethanol. The novel inline probe could identify a new field of application for image-based crystal size distribution measurements, with respect to difficult particle shapes (needles) and elevated temperatures, which cannot be evaluated with common techniques.

Keywords: optical measurement techniques; crystal size measurement; inline probe; crystal needles

1. Introduction

Crystallization is widely applied in agricultural, pharmaceutical, or chemical industry, with an enormous variety of duties and products. Besides purification and concentration of substances, crystallization is mainly applied to produce a particulate phase exhibiting defined properties. The particular requirements can be manifold. While for fine chemicals or active pharmaceutical ingredients (API) the solid-state form is commonly of major interest, a certain crystal size distribution (CSD) and crystal shape is usually demanded. Any of these particular properties affect the product functionality [1], as well as other processes in the downstream procedure (filtration, drying, etc.) [2,3]. Hence, control of the crystal size and solid-state form (e.g., polymorphism) is one important challenge that spreads over all fields of industrial crystallization [4].

Different options like cooling, adding an antisolvent, or evaporation of the solvent can be applied to create the corresponding driving force for the crystallization process. In order to compensate the supersaturation, a solid phase is formed with a specific shape through nucleation or growth, depending on the thermodynamic state of the system. However, crystallization kinetics are complex, nonlinearly connected to the driving force, and highly affected by initial conditions, process disturbances and hydrodynamic effects, which appear during operation.

Beside these fundamental aspects of the solid phase formation, the operation mode of a crystallization process plays a major role. In pharmaceutical industry the vast majority of crystallizations are still operated batch-wise since the process design is straightforward and can be based on the experience of many decades. Usually, a defined seeding strategy, combined with a controlled cooling policy, are utilized to meet the rigorous product specifications. This is the most common approach for process development and commercial manufacturing, but it also has various disadvantages. Batch processes suffer from batch-to-batch variations of the product and potentially high manufacturing costs, due to dead times for charging and cleaning [5–7]. To overcome these drawbacks, the crystallization can be carried out continuously, which increases the productivity and especially the space time yield. Hence, continuous operation has received increasing attention in recent years as a key element for improving crystallization based production [7–9]. Simple design methods for continuous crystallizations, like for batch-processes, are still missing. The continuous operation relies essentially on a precise control of nucleation and crystal growth. Thus, the crystal size distribution (and shape) must be well known, together with the liquid phase conditions, to reach a certain steady state, which yields the desired product.

Hence, in order to optimize existing processes, to develop new ones, or to transfer a batch-wise operation into a continuous mode, it is essential to monitor the solid phase formation together with the actual state of the liquid phase. The measurement of integral parameters, like temperature or concentration, is state of the art while determining the solid phase state, e.g., the crystal size is still quite challenging. Hence, this contribution is concerned with the evaluation of a novel inline probe for the measurement of CSD. In the first part, the technology is explained and compared to commonly applied analysis methods. In the following, the experimental approach is introduced, which is used to investigate the quality and application range of the new probe. Altogether, the results of three different techniques for the measurement of CSD are compared with the example of two substance systems in a wide range of operation conditions.

1.1. Particulate Measurement Techniques

The measurement of CSD is a challenging task due to the limited variety of monitoring techniques. However, the choice of the right measurement principle, with sufficient temporal and spatial resolution, can be decisive for process control and performance. With respect to this, a comprehensive summary of available techniques is given in the following paragraphs.

An overview for particulate measurement systems, with multiphase flow, in general is given elsewhere [10,11], and focusses on the determination of particulate properties like particle size and shape, flow field visualization, or concentration measurements. In contrast to the broad range of applications discussed in the cited literature, this article focusses on crystallization. One can distinguish between offline/atline, online, and inline measurement techniques. Offline measurement systems are used to analyze samples of the process in a laboratory, resulting in a relatively long deadtime between sampling and analysis results. Atline measurements are basically offline methods, that are placed close to the sampling point to minimize transportation distances. Hence, the time of the whole analytical processes is reduced. Nevertheless, both kinds of measurement operation mode are usually difficult to automate, often need manual adjustment, and are too slow for an efficient process control. For fast analysis and direct process control, online and inline measurements are indispensable.

Online measurements are commonly used together with a bypass. A representative sample is continuously withdrawn from the process into the measurement system and then returned to the apparatus. Inline techniques, often designed as probes, collect the information at the point of interest.

The acquired data can be exploited in both cases, either to adjust the process conditions or to serve as an input for actuators on a model-predictive control [11].

For the evaluation of the transient particulate phase during crystallization processes, offline/atline approaches are slow and tedious (sampling, washing, etc.), with the drawback of influence on the particle size distribution (PSD) during the involved procedure. The only alternative is inline or online determination, to directly control the product properties, which only can lead to high quality products [3,7,12,13].

The most commonly used quantitative particle measurement methods applied in crystallization are either laser diffractometer (LD) or focused beam reflectance measurement (FBRM). The LD is a classical offline technique and measures the refraction of light to determine the diameter of an equivalent sphere. Therefore, this technique is mostly used for spherical and compact particles, because it is well known that this technique struggles with non-spherical crystals, especially if these have high aspect ratios [14,15]. FBRM is the state of the art technique in crystallization, because it is commercially available and easy to use [4]. A probe that is inserted in the apparatus and measures the chord lengths of the crystals from backscattered light originating from a fast-rotating laser beam [7,16–24], yielding a one-dimensional chord length distribution (CLD) instead of a real crystal PSD. The conversion of a CLD to a PSD is based on models and assumptions, but is lacking for complex and elongated particle shapes [25–28]. Another limitation of the technique is the one-dimensional distribution, which proves insufficient at higher dimensions, like with the width and length of a needle-shaped crystal [3,22,29,30].

So far, only optical imaging measurement techniques, together with sophisticated image processing algorithms, manage to determine efficiently the particle shape of a particulate phase [13,30]. Other noninvasive niche methods rely, for instance, on the use of supersonic wave probes for the characterization of the dispersed phase [31–33].

In the simplest optical image processing approach, a camera is placed in front of a transparent reactor [34–36], e.g., a stirred glass vessel, for continuous monitoring. This concept often suffers from image distortion due to the curved reactor wall, poor contrast ratios, and limitations with respect to the solid content or suspension density. In addition, the acquired images are mostly evaluated manually, because their image quality varies in illumination and contrast. In order to benefit from online and inline techniques, it is important to have an automated, or at least fast, image processing algorithm. Therefore, it is of major importance to utilize a camera setup that acquires images with high contrast, sharp edges, and constant image quality.

Online approaches via bypass variants are costly at industrial scale and therefore commonly applied at laboratory scale. In this case, the suspension is isokinetically withdrawn from the process, passing a cuvette or flow-through cell for analysis. Hence, high contrast images with constantly sharp particles can be acquired if the focus is in the middle of the cell. Various publications have demonstrated the use of this technique for the determination of the PSD for resilient crystals that do not tend to break [37–39]. Another variant is with a stereoscopic imaging system for reconstruction of the 3D-shape of crystals [29,30,40].

Other variants are image-giving inline probes, which have been developed and established in the last twenty years [12,41–45]. These common incident-light probes are inserted into the reactor, acquiring the information at the point of interest. Available systems use entocentric lenses and are therefore quite compact, but unfortunately suffer from a small focal plane. As a result, most of the particles in the measurement area may appear blurry, which can lead to inaccurately imaged particles and an erroneous PSD. Thus, these are mostly used for qualitative analyses, like monitoring secondary nucleation or phase transformation, in combination with an FBRM that measures the quantitative particle chord length [17,18,20,23,46,47].

The ultimate approach is with tomographic methods that have excellent temporal and spatial resolution, and where applicability is possible even with high solid content and in opaque media. This measurement principle was developed in human medicine and is completely noninvasive. Today, this technique is also applied in various fields of process engineering technology.

Its disadvantages are a high space requirement and high costs, thus only at laboratory scale can applications be found. The techniques used in this article, beside sieving, are image-based techniques: an established bypass online microscope and the telecentric shadowgraphic probe for the evaluation of this new measurement technique. Both imaging techniques are described in detail below.

1.2. Shadowgraphic Optical Probe and Online Microscope

The in situ data of the experimental investigations in this study were recorded with two measurement systems: a commercial QICPIC online microscope (Sympatec, Germany) installed in a bypass and a shadowgraphic inline probe. The online microscope was using the transmitted light technique, where a pulsed light source is vis-à-vis of a camera. The measurement volume was formed by a flow-through cuvette that was placed between the optics and the light source. The parts were adjusted, so that the focus plane was in the middle of the cuvette. Thus, particles that were transported through the cell were captured sharply, with constant image quality.

A peristaltic pump fed the suspension via a temperature-controlled bypass to the cuvette and back to the crystallizer. The measurement volume had a fixed width of 2 mm given by the cuvette geometry. The camera and lens provided a square field of view of 5 mm in height and width at a resolution of 1024 pixel × 1024 pixel. The microscope software supports an autofocus function to adjust the focus plane of the camera in the middle of the flow cell, alternatively it could be adjusted manually.

The shadowgraphic probe is a further development of the so-called optical multimode online probe (OMOP) [48,49]. The primary design of the probe is based on a transmitted light technique and consists of two opposite protection tubes in a measurement flange.

One tube contains the illumination unit and the other one the camera and a telecentric lens. The light source consists of a LED placed in the focus point of a plano-convex lens, which emits a parallel light beam. The light beam is passed through the measurement volume between the two tubes, which are sealed with inspection windows in the front. Through the parallel light, high contrast images of the particles within the measurement volume can be acquired, even if the particles are nearly transparent.

In contrast, commercial image-based probes apply an incident light technique with an endocentric optics. In comparison, these probes have lower contrast ratios and natural image distortion is easily caused by the illumination and the optics [50,51]. Therefore, image analysis is quite challenging and requires significant effort to achieve quantitative results [52]. Furthermore, the depicted particles captured by the camera appear smaller, the larger the distance from the entocentric lens, and need sophisticated correction. Due to a comparably small focus areas, only a limited number of particles can be evaluated.

Alternatively, telecentric lenses provide a distant independent image of the particles when using parallel light. An aperture in the image-sided focus point of the lens filters out non-parallel light beams and, thus, shadowgraphic pictures are generated by parallel light only. Hence, no prior calibration is necessary. In addition, these lenses have higher depth of field compared to entocentric lenses, therefore having a large measurement volume instead of a focus plane. Hence, various versions of the OMOP for different applications, like capturing droplets [53,54], bubbles [55–57], or sprays [58,59] were developed.

In order to achieve easier access to an apparatus and to promote industrial applications, two probes have been designed recently; first, a robust DN 80 variant for industrial applications, with the full functionality of the two-sided OMOP principle [60]. Second, as a further development, a single sided endoscopic probe in DN 50 version was designed for laboratory scale [56,61], and was used for the experiments of this article. The probe has an adjustment mechanism where the position inside the apparatus and width of the measurement volume can be changed, even during the crystallization process, to adapt it to the increasing particle concentration (see Figure 1).

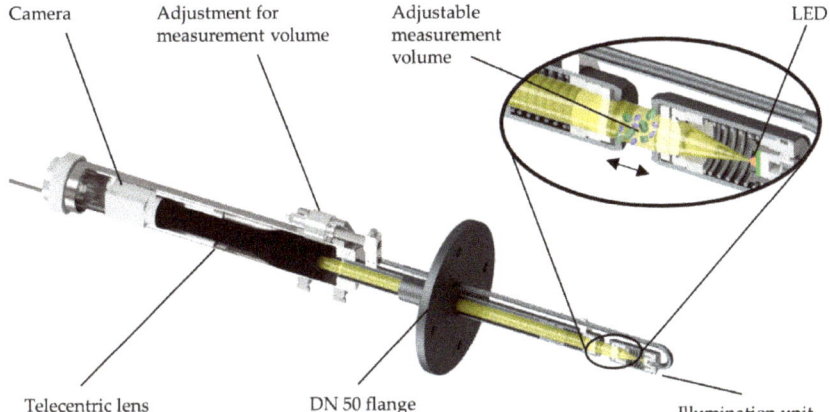

Figure 1. One-sided DN 50 telecentric shadowgraphic probe for laboratory scale published in [56,61].

2. Materials and Methods

2.1. Experimental Setup

For the experiments in this study, the following setup was used for all experiments (Figure 2):

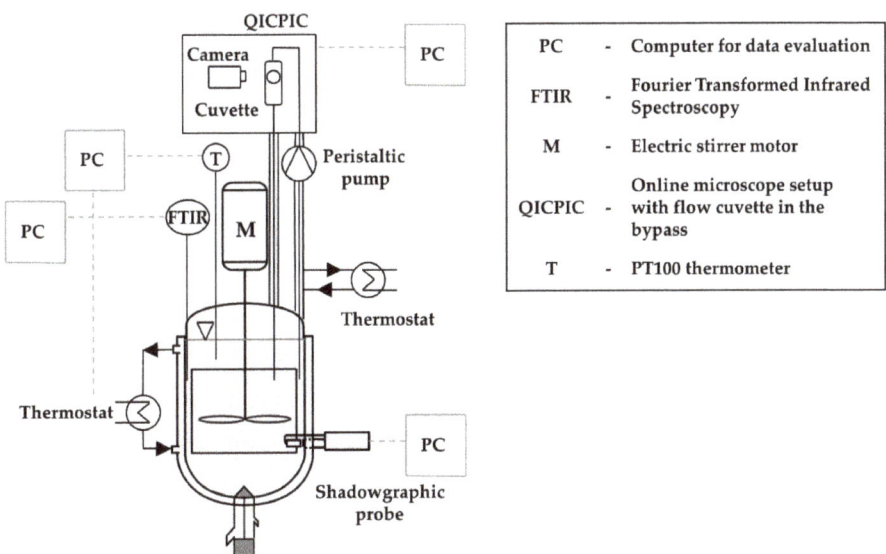

Figure 2. Scheme of the 25 L utilized double jacket draft tube crystallizer setup [62].

The temperature-controlled 25 L double jacket draft tube crystallizer was equipped with a propeller-type stirrer (diameter 150 mm, BASF, Ludwigshafen, Germany) and a PT100 was used to monitor the temperature (T in Figure 2). In order to measure the liquid phase composition, an attenuated total reflection Fourier-transform infrared spectroscope (ATR-FTIR, Thermo Fisher Scientific, Waltham, Massachusetts, USA) was used (FTIR in Figure 2).

The measurement depth of the shadowgraphic probe was set to 2 mm to be consistent with the flow cuvette of the online microscope. The probe was fitted with a Basler 1440-73 gm camera and a

1× telecentric lens, and the measurement volume was in the middle between the draft tube and the reactor wall. The bypass for the online microscope was inserted at the top of the reactor, directly above the shadowgraphic probe to ensure that the withdrawn suspension was similar. A high bypass flow rate (≈40 L/h) was chosen to ensure unclassified sampling of the suspension. All bypass tubes were double jacketed and temperature-controlled by a thermostat, which were set to 1–1.5 K above the reactor temperature. Details of the QICPIC bypass setup are reported elsewhere [37,62].

Both measurement techniques were capturing 750 images at 25 fps for each measurement point to monitor the executed experiments. Additionally, solid-free offline samples of the liquid phase were taken every 5 min, which were analyzed gravimetrically to verify the inline data, and finally a representative suspension sample was taken from the bottom valve at the end of each experiment to allow for sieve analyses.

2.2. Substances

In order to analyze different crystal shapes, potassium dihydrogen phosphate (KH_2PO_4) and thiamine hydrochloride were used.

KH_2PO_4 grew bipyramidal, prismatic shaped crystals, depicted in Figure 3a, and thiamine hydrochloride monohydrate grew needle-like shape, as depicted in Figure 3b.

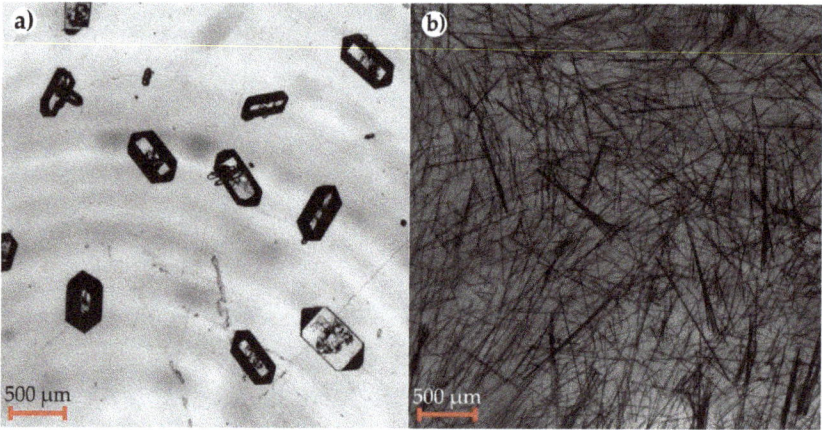

Figure 3. (a) KH_2PO_4 crystals; (b) Thiamine hydrochloride monohydrate crystals captured by the shadowgraphic probe during the experiments.

KH_2PO_4 grows resilient crystals, that are very suitable for validation purposes, because they do not tend to break in a bypass, during filtration, or sieving. In order to design the crystallization process, the solubility and the kinetics must be known, especially because KH_2PO_4 tends to form longer or shorter body prisms, depending on the operation conditions. The crystallization kinetics in aqueous solutions and the solubility are well known and reported [62,63]. The latter can be described by the following polynomial expression:

$$\begin{aligned} C_{sat}(T)[\text{wt.}-\%] &= 15.24 \text{ wt.}-\% + 2.06 \times 10^{-1} \tfrac{\text{wt.}-\%}{°C} T + 1.01 \\ &\quad \times 10^{-2} \tfrac{\text{wt.}-\%}{°C^2} T^2 - 1.45 \times 10^{-4} \tfrac{\text{wt.}-\%}{°C^3} T^3 + 1.23 \\ &\quad \times 10^{-6} \tfrac{\text{wt.}-\%}{°C^4} T^4 \end{aligned} \quad (1)$$

Thiamine hydrochloride exists in five solid-state forms, according to the literature [64,65]. The desired form in industrial applications is a pseudo-monohydrate that crystallizes in contact with water, and contains 0.5 to 1 mole water per mole thiamine in the needle-like shaped crystals (see Figure 3b).

However, this monohydrate is metastable at ambient conditions and converts fast into the thermodynamic stable thiamin hydrochloride hemihydrate. Hence, the solid-liquid equilibrium of the pseudo-monohydrate is difficult to measure, and therefore barely investigated [65,66]. Fortunately, data of the less-soluble thiamine hydrochloride hemihydrate in some binary solvents are reported, and therefore the experimental design was based on data of a binary water/ethanol mixture [67]. Some physiochemical properties of the thiamin hydrochloride and KH_2PO_4 are shown in Table 1.

Table 1. Properties of the utilized substances KH_2PO_4 and thiamin hydrochloride.

Substance Properties	Symbol	KH_2PO_4	Thiamine HCl Anhydrate	Unit
Solid density	ρ_{solid}	2340	1.4	[kg/m^3]
Molar mass	M	136.09	337.27	[g/mol]
Purity	Pr	≥99	≥98	[%]
Vendor		Applichem	Sigma Aldrich	

2.3. Experimental Procedure

Table 2 depicts the conditions of the executed experiments, one with thiamine hydrochloride (Exp. 5) and four with KH_2PO_4 (Exp. 1–4). For Exp. 1–3 the saturation temperature was 35 °C and the cooling rate, the final temperature, and the mass of the seed loading was varied for each experiment to evaluate the limits, with respect to the suspension density and crystal size. The seed fraction was sieved and had a normal distributed initial size of 212–300 µm for all experiments, except for Exp. 3. Smaller seeds with a range of 150–212 µm were used in this case to alter the initial suspension and optical density, to evaluate the impact of these parameters on the crystal size measurements. The fourth experiment was carried out with a larger initial concentration (according to a saturation temperature of 56.5 °C), since it is well known that the bypass of online microscopes often tends to block under these conditions. In addition, an anti-solvent crystallization, Exp. 5, of thiamine hydrochloride was performed to evaluate the applicability of the shadowgraphic probe to fragile crystal systems.

Table 2. Process conditions of the performed experiments with KH_2PO_4 and thiamine hydrochloride.

Exp.	$\Delta T/\Delta t$ [°C/h]	T_{Sat} [°C]	T_{Seed} [°C]	T_{end} [°C]	m_{H2O} [°C]	m_{EtOH} [kg]	m_{solute} [kg]	m_{seeds} [kg]
Exp. 1—KH_2PO_4	−7.5	35	34	27	21	-	6.5	0.05
Exp. 2—KH_2PO_4	−10	35	34	15.5	21	-	6.5	0.1
Exp. 3—KH_2PO_4	−10	35	34	22	21	-	6.5	0.2
Exp. 4—KH_2PO_4	−12	56.5	56.4	43	21	-	9.5	0.1
Exp. 5—thiamine hydrochl.	-	≈30	-	25	5.6	12	4	-

KH_2PO_4 was added to the reactor according to Equation (1) (see Table 2), the impeller speed was set to 250 rpm and the crystallizer was heated a priori for 0.5 h to 5–10 K above the respective saturation temperature to ensure complete dissolution and equal starting conditions for all experiments. Afterwards, the clear solutions were slightly subcooled (0.1–1 K) and the seeds were added at temperature T_{seed} (see Table 2) and time t = 0 h. Subsequently, cooling was executed as a simple linear cooling ramp after the seed's addition, with a certain slope between −7.5 °C/h and −12 °C/h (see Figure 4a for an exemplary temperature curve of Exp. 3—KH_2PO_4). Immediately after seeding (t = 0 h), the particle size was measured simultaneously every 5 min by the online microscope and the shadowgraphic probe. The experiments ended at the final temperature, T_{end}, if either the suspension density was too high, resulting in too much overlapping of the single crystals, or excessive nucleation was observed. During the last measurement of the crystal size by the optical methods an unclassified suspension sample was taken from the bottom valve of the reactor, and was immediately filtered using a strainer and a filter paper. Then, the filter cake was washed with an adjusted ethanol/water-mixture to prevent nucleation or dissolution of the crystals through the residual mother liquor. Afterwards, the crystals were dried and sieved to determine the mass-based size distribution. During the experiments with KH_2PO_4, the state of the liquid phase was monitored by a calibrated ATR-FTIR and by solid-free liquid samples, which were taken every 5 min, simultaneous to the particle size measurements (see Figure 4b).

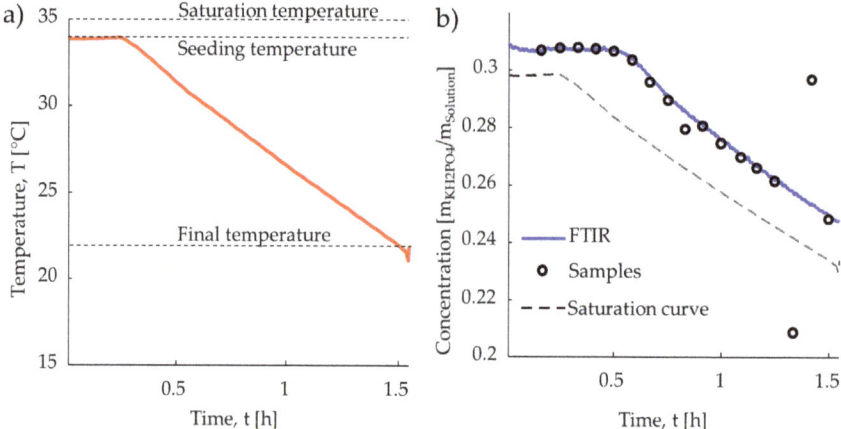

Figure 4. (a) Temperature profile of Exp. 3—KH_2PO_4: saturation, seeding, linear cooling ramp, and final temperature. (b) Concentration profile of Exp. 3—KH_2PO_4: FTIR, offline samples, and saturation curve.

From ATR-FTIR spectra the mass fraction ($m_{KDP}/m_{solution}$) was calculated by an existing calibration, successfully applied in the past [68] to evaluate the suspension density, where c_{FTIR} is the concentration of the FTIR and c_0 is the initial concentration:

$$\rho_{Susp}(t) = (c_0 - c_{FTIR}(t)) + \frac{m_{seed}}{m_{solution}} \quad (2)$$

$$S(t) = \frac{c_{FTIR}(t)}{c_{sat}(T(t))} \quad (3)$$

The supersaturation, S, is the driving force in crystallization, and was calculated according to Equation (3), with the concentration at saturation, $c_{sat}(T(t))$.

The experiments started at small supersaturations, which gradually increased due to cooling. After about 0.6 h sufficient solid surface was present in the crystallizer to counterbalance the supersaturation generation and the driving force started to decrease, exemplarily shown in Figure 5a for Exp. 3—KH_2PO_4.

The concentration of the liquid phase versus temperature in the binary phase diagram for all KH_2PO_4 experiments is given in Figure 5b. Exp.1–3 have almost identical conditions with a saturation temperature of 35 °C and only Exp. 4 was saturated at an elevated temperature. A significant influence of the seed load or the cooling ramp on the concentration profile is not clearly visible due to the fast crystallization kinetics of KH_2PO_4.

In addition, a fifth experiment with thiamin hydrochloride was carried out as an anti-solvent crystallization via primary nucleation. The thiamine hydrochloride was dissolved in water and added to the reactor. At t_{start} = 0 h the ethanol was added as the antisolvent. The initial masses were calculated based on literature data [67]. The experiment was carried out at a constant temperature of 25 °C, and the impeller speed was set to 250 rpm, similar to the experiments with KH_2PO_4.

The shadowgraphic probe took pictures of the suspension every minute from the beginning, and after a significant number of crystals were visible (t = 1 h), the online microscope in the bypass was put into operation.

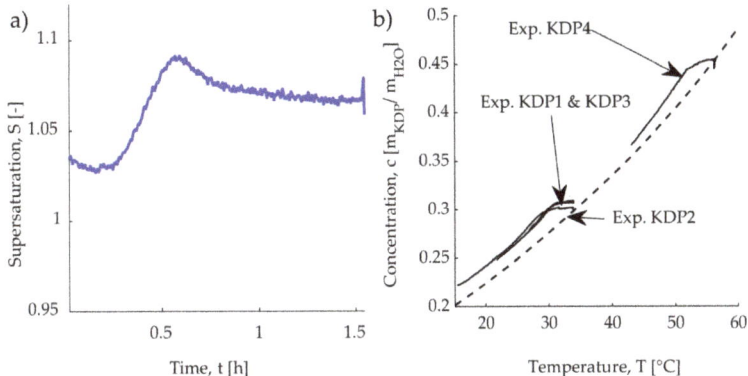

Figure 5. (a) Supersaturation profile calculated for Exp. 3—KH$_2$PO$_4$ with the FTIR data, (b) Liquid phase concentration for Exp. 1–4 KH$_2$PO$_4$, measured with the FTIR depicted in a part of the binary phase diagram of KH$_2$PO$_4$/H$_2$O. The dashed line is the saturation curve according to Equation (1).

The experiment was carried out until the concentration was too high to identify single crystals. Taking a suspension sample similar to the experiments with KH$_2$PO$_4$, was not possible since the thin, needle-like crystals of the thiamine hydrochloride monohydrate broke during filtration and further handling. A representative sieving analysis was, therefore, not possible and the crystal length and width were only determined via image processing.

2.4. Image Processing

The pictures of the probe and the microscope were evaluated with an existing MATLAB routine [69]. Based on the difference in contrast between the crystals and the background, the crystals could be isolated through contrast enhancement and binarization. In addition, a dynamic background subtraction out of a picture series was carried out to eliminate scratches or immobile adherent particles. The KH$_2$PO$_4$ crystals, especially, have large bright areas in the crystal center (see Figure 3a) that would lead to erroneous object identification. Therefore, morphological closing and region filling was utilized to fill the empty areas within the crystals.

The evaluation focuses on single crystals only, assuming agglomerates are of negligible number. Therefore, two shape descriptors, the numerical eccentricity, ε, (Equation (4)) of an ellipse, and solidity, s, for the description of the convexity (Equation (5)), were used to exclude agglomerates from further evaluation.

$$\varepsilon = \frac{\sqrt{a^2 + b^2}}{a} \tag{4}$$

$$s = \frac{A}{A_{convex}} \tag{5}$$

The use of these parameters was based on experience and was successfully applied for single crystals of KH$_2$PO$_4$ in the past. Particles are classified as single crystals if the eccentricity was within 0.4 to 1, and the solidity from 0.95 to 1, respectively. As a result, gas bubbles and overlapping crystals were excluded in the data evaluation. This was manually crosschecked by comparing the crystal detection results with the original images. Furthermore, crystals touching the image border were not considered in the evaluation, since incomplete objects lead to erroneous size calculations. More details about the algorithm are reported in the literature [69].

In order to evaluate the geometry of the observed objects and to calculate meaningful distributions, it is necessary to choose a reasonable characteristic length. For slightly elongated bipyramidal crystals with a square prism body, the width of the square cross-sectional area can be used as a characteristic length, L, to describe its size. This length can be obtained by the minimal Feret's diameter, L_{Fmin} of the projected area of the crystal, but the orientation of the crystal to the picture must be considered. The relevant orientation for the representation of the minimal Feret's diameter is given by the rotation around an axis, passing the two pyramid tips along the elongated direction of the crystal. Imaging a rotation around this axis, while constantly measuring the minimal Feret's diameter, gives values between a minimum and a maximum value, L and $\sqrt{2}L$ for L_{Fmin}. Assuming that the orientation of the particles is normally distributed, for the probability of all rotations between the two extreme values, an arithmetic mean L_1 can be described according to:

$$L \approx L_1 = \frac{2 L_{Fmin}}{1 + \sqrt{2}} \qquad (6)$$

Based on this averaged crystal width of the square prism body, the distributions can be compared with the sieving analysis, where L_1 is measured. There are advanced methods for the correction of the crystal orientation reported in the literature [36], but for the sake of simplicity and computational effort, this approach was chosen.

The single crystals detected and measured by the image analysis were sorted in 400 size classes between 1 and 2000 µm, based on their corrected crystal size, and normalized by the class width, giving the number distribution, $f(L)$. Normalization by the integral $\int f_j(L)dL$ yields the density distribution q_j of dimension j:

$$q_j = \frac{fj(L)}{\int f_j(L)dL} \qquad (7)$$

Additionally, the percentiles, p, were utilized for the comparison of the online microscope and the shadowgraphic probe's results according to:

$$Q_j(L_p) = \int_0^{L_p} q_j(L)dL = p \qquad (8)$$

The results were depicted and evaluated in terms of number distribution ($j = 0$) and mass distribution ($j = 3$), mainly, but other characteristic values of distributions can be used, as well [70,71].

Thiamin hydrochloride crystals were characterized utilizing the same algorithm, but without any correction of the Feret's diameter for the orientation. The minimal and maximal Feret's diameter were interpreted as the width and the length of the needles.

Further, for both substance systems an optical density was calculated based on the acquired and binarized images of the shadowgraphic probe and the QICPIC. For this purpose, the number of all black pixels in an image was divided by its resolution. This ratio was used in the following, called optical suspension density, and helps to interpret the results.

3. Results and Discussion

In the following, the results of Exp. 3 with the highest initial suspension density will be discussed in detail, because the Exp. 1 and Exp. 2 show similar results and have the same starting saturation temperature. Furthermore, Exp. 4 will be shown, as it has a higher starting saturation temperature. Afterwards the results of the thiamine hydrochloride crystals will be discussed.

3.1. Comparison of the Crystal Size Measurement Techniques with KH_2PO_4 Crystals

Figure 6 gives an example of captured KH_2PO_4 crystals, with the online microscope in comparison with the shadowgraphic probe. Both pictures show good contrast ratios, which simplifies the subsequent image processing. The imaged crystals have clear edges and can be accurately detected and measured by the algorithm. Thus, the number, q_0, and mass, q_3, density functions of the distributions could be calculated as shown in Figure 7. In the number density functions (see Figure 7a,b), the evolution of the fines content can be visualized, while the mass density function (see Figure 7c,d) serves to illustrate the evolution of the larger crystal fractions.

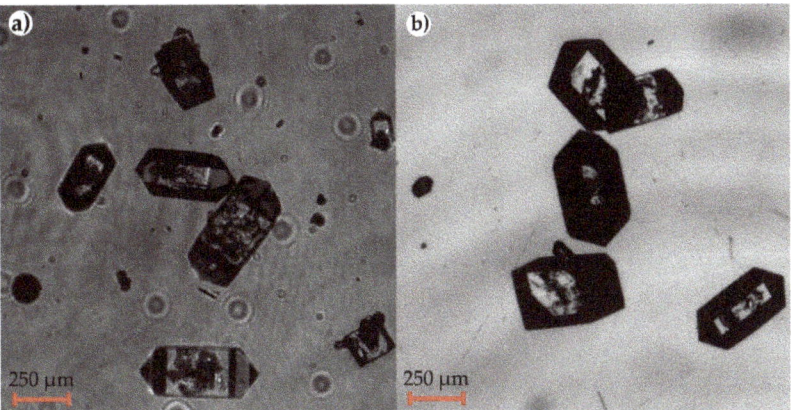

Figure 6. KH_2PO_4 crystals captured with (**a**) QICPIC (online microscope with bypass); (**b**) shadowgraphic probe (OMOP, inline probe). The images are enlarged for a better view of the edges.

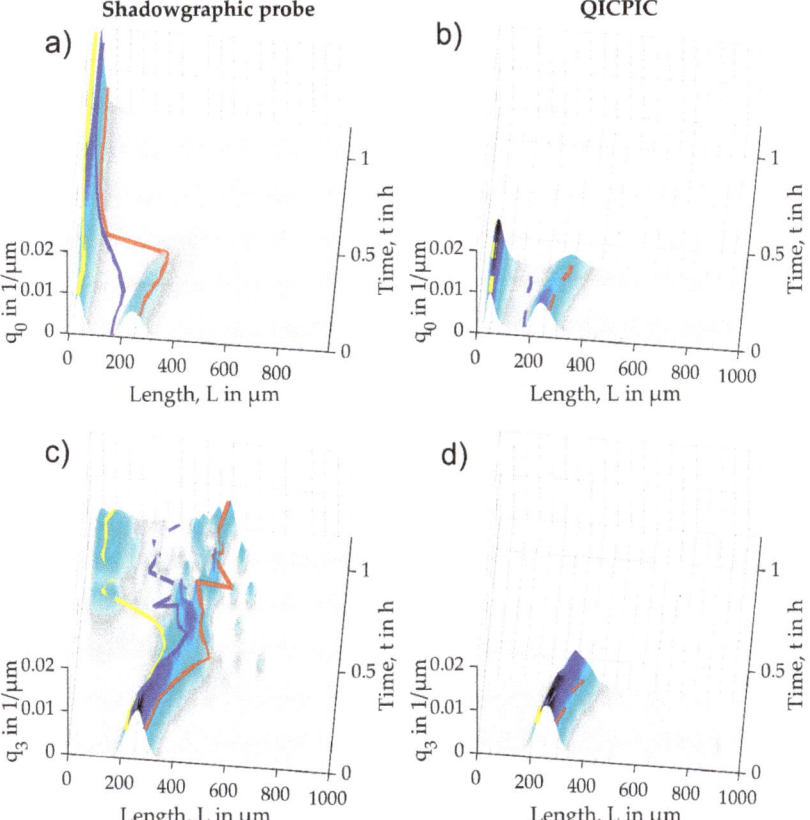

Figure 7. KH_2PO_4—Exp. 3 distributions (**a**) q_0-distribution shadowgraphic probe; (**b**) q_0-distribution QICPIC; (**c**) q_3-distribution shadowgraphic probe; (**d**) q_3-distribution QICPIC. *Solid lines*—percentiles of the shadowgraphic probe; *Dashed lines*—percentiles of the QICPIC; *yellow and red* −0.15 and 0.85 percentile distribution; *blue*—transient mean sizes of the distribution.

In general, both techniques give similar distributions. The number distributions show, for t = 0 h two major crystal fractions, one with about 50 µm and another one with 184 µm. As seen in the mass distributions, the small fraction is not visible and was probably caused by fine grain KH_2PO_4 particles in the seeds and dust. After a time of about t = 0.6 h, where the largest supersaturation was present (see. Figure 5a) a shift in crystal size towards bigger crystals due to growth can be observed (see Figure 7). Obviously, a certain threshold driving force must be present for the seeds to become active. Several reasons are known for this behavior, e.g., the crystal surfaces need to heal before macroscopic growth can take place or impurities block growth centers. However, a detailed study of the mechanism is not the focus of this article. The q_0-distributions (Figure 7a,b) also show that the number of smaller particles increases at the same time, caused by nucleation. After crystal growth can be observed, a significant broadening of the seed fraction is visible (see Figure 7c,d for t > 0.6 h), which can be attributed to growth rate dispersion. This influence can also be seen with a look at the percentiles, therefore they are shown as the top view in Figure 8, on the 3D diagram in Figure 7. For a better overview the surface plot is not shown in Figure 8.

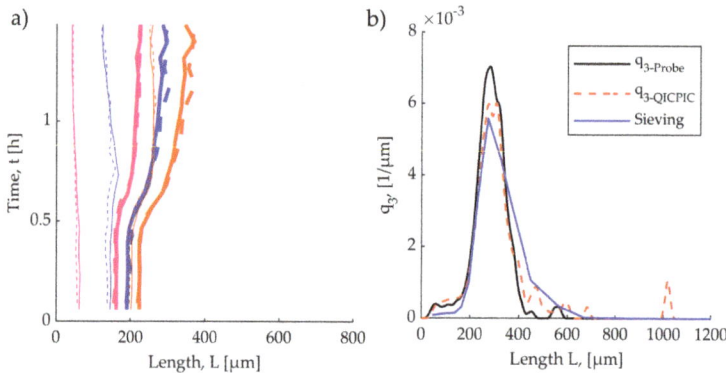

Figure 8. (**a**) Percentiles and mean values of the crystal size distributions from Figure 7 *thin lines*—percentiles of the q_0-distribution; *bold lines*—percentiles of the q_3-distributions; *Solid lines*—percentiles and mean values shadowgraphic probe; *dashed lines*—percentiles and mean values QICPIC; *brown; magenta* −0.15 and 0.85 percentile of the corresponding distribution; *blue*—mean size of the corresponding distribution. (**b**) Comparison of the last mass distribution: shadowgraphic probe, QICPIC, sieve analysis.

The percentiles of the number, and mass density distribution, match well for both optical measurement techniques (Figure 8a), and show almost identical curves. Hence, an explicit classification effect of one measurement technique, either caused by the sample withdrawal to the bypass or by the measurement gap of the inline probe, can be excluded. For t = 0 h, the percentiles match with the initial size range of the seeds (see Table 3), and confirm a reasonable measurement of the crystal size. The change of the crystal size can be tracked properly over the whole experimental time. This can be confirmed with respect to the q_3-distributions obtained by the sieve analysis of the suspension sample at the end of the experiment (see Figure 8b).

Table 3. Mass-averaged crystal widths evaluated with the optical measurement techniques at the start and the end on Exp. 1–4 KH_2PO_4.

			Mass-Averaged Crystal Width L [µm] at the Start		Mass-Averaged Crystal Width L [µm] at the End		
Exp. no.	Seed Size [µm]	Experimental Time [h]	QICPIC	Probe	QICPIC	Probe	Sieving
Exp. 1	212–300	1.15	257	280	590	584	596
Exp. 2	212–300	2.01	248	257	645	582	660
Exp. 3	150–212	1.42	189	188	303	284	281
Exp. 4	212–300	1.18	258	259	-	347	526

The comparison shows that the distributions measured by the optical measurement techniques fit well with the sieve analyses. The fraction of 0–200 µm is underestimated by the sieve analysis compared to the optical techniques. Probably, a part of the fines is lost during solid/liquid separation, washing, and sieving. In the range of 450–600 µm a slightly higher density for larger particles in the sieve analysis is visible. Since the image analysis focuses on single crystals, the agglomerates are not considered in the imaging techniques. In contrast, the sieve analysis also has agglomerates in the distribution, therefore this shift can be addressed to a small amount of agglomerates present in the sample. In summary, both optical measurement techniques are suitable to evaluate crystal size distributions. The deviations are in the typical error range, except for the final crystal size of Exp. 2. However, the main sources of deviation in image-based size determination, in general, is the image conversion and the binarization. About two pixels on the edges was the common deviation during the capturing by the camera, and an additional two pixel uncertainty occurred during the thresholding for the binarization. This sums up to four pixel in total, which equals 20 µm with a pixel size of 5 µm (depending on the camera and the lens used) for both techniques, and is the typical error range for image-based size evaluation in general.

Several parameters were changed during experiments Exp. 1–3, initial seed loading, seed size, final process temperature, and the final crystal size, as well as the optical and suspension density. The latter two cannot be controlled directly but are a result of various process parameters. None of the changes led to a significant impact on the deviation between both optical measurement techniques, since a good agreement was found for density functions of all experiments (see Appendix A for the other detailed results of Exp. 1 and 2). A comparison of the initial seed sizes of all experiments (see Table 3) confirms a suitable determination of the crystal sizes between the QICPIC and the shadowgraphic probe. At the end of the experiments where larger crystals occurred, the probe measured slightly smaller crystal sizes than the QICPIC and the sieve analysis, but the deviations were still in the deviation of 20 µm mentioned above. Nevertheless, an effect of the measurement window of the shadowgraphic probe can be assumed. Larger particles tend to touch the image border, especially if the measurement window is smaller. Because the QICPIC has a larger measurement window (5 mm × 5 mm) than the probe (5 mm × 3.5 mm), this effect is maybe noticeable. Only Exp. 2 shows significant deviations for the measured final crystal sizes. For this experiment the percentiles (see Appendix A Figure A7) are almost identical for both techniques, except at the last two measurement points. For these measurements the percentiles show a significant drop, and the amount of measured crystals increases drastically. The supersaturation curve (see Appendix A Figure A8) shows an increase of the concentration and secondary nucleation occurs, which causes the decrease in the mean crystal size. Additionally, the probe has more agglomerates in the images, leading to fewer single particles being detected. A classifying effect may occur within the QICPIC bypass, where these agglomerates are seen less often, and therefore more single crystals are detected. Although the final crystal sizes show deviations in Exp. 2, the percentiles confirm a suitable transient crystal size determination up to the last 10 min. It is not clear if secondary nucleation will affect the measurement in general and this must be clarified in further investigations. The usage of larger measurement windows with a sophisticated algorithm for agglomerates can maybe solve this issue.

It is important to consider the number of particles measured in order to have a statistically verified PSD. Therefore, the total amount of measured crystals for each optical technique is shown, with their corresponding optical density, in Figure 9a. For the online microscope and the shadowgraphic probe it is clearly visible that the number of measured particles decreases over time, which is caused by the increasing number of larger crystals. This is a key issue of image analysis in general as there are particles in the system that overlap with smaller particles or other particles. Hence, these overlapping clusters of particles cannot be evaluated by the algorithm, which then leads to erroneous PSD's. The other reason is, that comparable larger particles, with respect to the image size, have a higher probability of being cut off by the measurement window. Therefore, these particles are likewise not detected and lead to a smaller number of measured particles. Nevertheless, both optical techniques measure a few thousand particles for each distribution, guaranteeing a statistically sufficient amount for a representative distribution. The results also show that the online microscope detects more particles than the shadowgraphic probe. This is an expectable phenomenon, since the measurement window of the probe is smaller, due to a smaller camera sensor size, in comparison to the microscope.

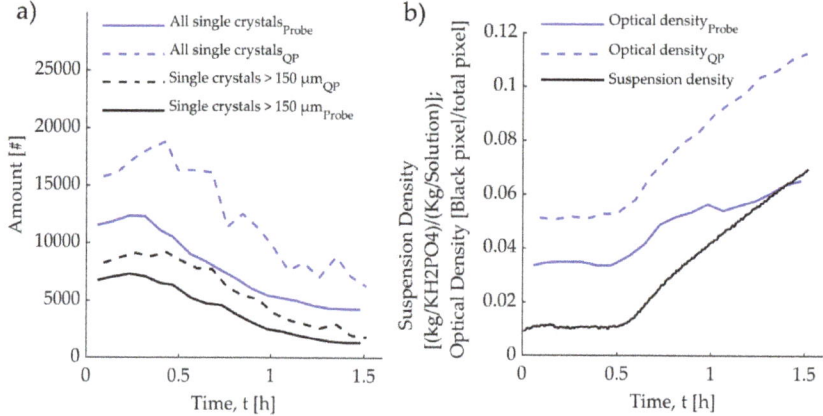

Figure 9. (**a**) Quantity of single crystals analyzed (**b**) the optical and suspension densities during experiment Exp. 3—KH$_2$PO$_4$.

The suspension density, according to Equation (2), derived from ATR-FTIR data, confirms the measured increase in crystal size at t = 0.6 h. The optical density based on pixel ratios shows a similar trend (see Figure 9b), although the optical densities do not match with the mass-based suspension density. Effects such as the overshadowing of smaller particles caused by larger ones, and overlapping, affect these values measured with the optical techniques. Therefore, the optical density is additionally connected to the dispersity of the particulate phase. Furthermore, the suspension density, determined by the concentration measurement, is a global value, while the optical measurement techniques provide local information. This means that the optical methods can recognize overall trends in the suspension density but are not suitable for its representation. Nevertheless, it could be shown that crystal size evaluation is possible and not affected by the suspension density, at least up to 6% in Exp. 3, and up to 8% in Exp. 2.

The suspension density can either be over- or underestimated with optical methods in comparison with the suspension density calculated with the FTIR data of Exp. 2, as given in Figure 10. At the start of the experiment, where a narrow distribution of one crystal size was present, the optical suspension density was less that the mass-based suspension density. This changed during the experiment, because the optical density was additionally connected to the dispersity of the system. Multimodal distributions increase the optical density, especially the fine particle content increases the particulate content in the pictures, and lead to higher optical densities. As a result, the optical density cannot be used to determine the suspension density directly, because the particulate state must be taken into account as well.

3.2. Investigation of KH$_2$PO$_4$ at Elevated Temperature

The experiment EXP. 4—KH$_2$PO$_4$ started at an elevated seeding temperature of 56.4 °C. After seeding, both optical measurement techniques were able to measure the initial crystal size distribution. After t = 0.5 h the reactor reached a temperature of around 50 °C and crystallization occurred in the bypass of the online microscope, which led to a blockage. Therefore, the bypass was closed down and only the shadowgraphic probe was used to evaluate the state of the particulate phase (see Figure 11).

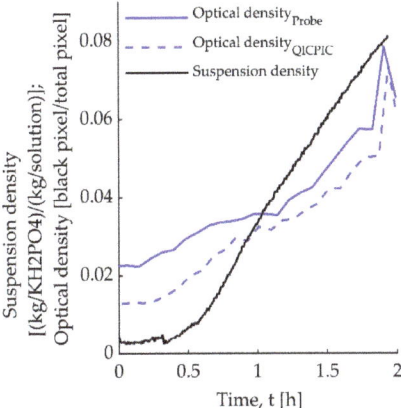

Figure 10. ATR-FTIR suspension density in comparison to the optical densities (microscope and probe) based on pixel ratios for Exp. 2—KH_2PO_4.

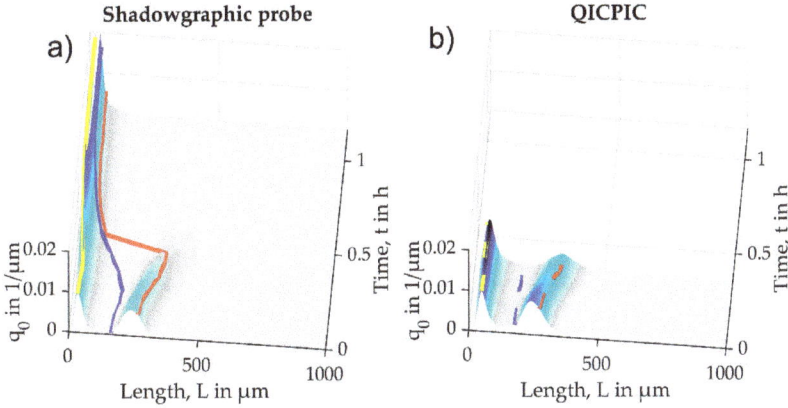

Figure 11. q_0-distributions of Exp. 4—KH_2PO_4 (**a**) shadowgraphic probe; (**b**) QICPIC; *Solid lines*—percentiles shadowgraphic probe; *Dashed lines*—percentiles QICPIC; *yellow and red* −0.15 and 0.85 percentile distribution; *blue*—transient mean sizes of the distribution.

Both principles have a good match with the distributions of the seeds, and similar growth is visible for both techniques at starting conditions. After the bypass was blocked, the shadowgraphic probe recognized a broadening of the mass-based distribution, therefore the larger fractions are no longer visible in the number distribution. This is also caused by nucleation of smaller crystals that are dominant in number compared to the larger grown seeds. This smaller fraction increased rapidly in number and therefore, accounts for about 15% of solid fraction in the mass-based distribution. To sum up, the inline probe can be utilized at conditions where massive nucleation and fast crystal growth leads to blocking of a bypass-based measurement technique, which requires a precise temperature control when withdrawing samples. This shows clearly that the probe opens a new field of applications, where other measurement systems fail.

3.3. Crystallization of Thiamin Hydrochloride Monohydrate

The crystallization of thiamine hydrochloride monohydrate was performed as nucleation from aqueous solution with ethanol as antisolvent. Nucleation was observed by the shadowgraphic probe after approximately t = 0.3 h after adding the antisolvent. These crystals were only a few pixels in width, and a certain time was necessary to overcome the lower detection limit of the probe, therefore an exact time cannot be referred. After t = 0.6 h, a representative amount of crystals was visible within the images. The amount increased significantly up to the time of t = 1 h, when the bypass was put into operation. As the suspension passed the flow cuvette, the online microscope was not capable of setting an autofocus automatically, therefore the focus had to be adjusted manually. An example of the images taken by the optical methods is given in Figure 12.

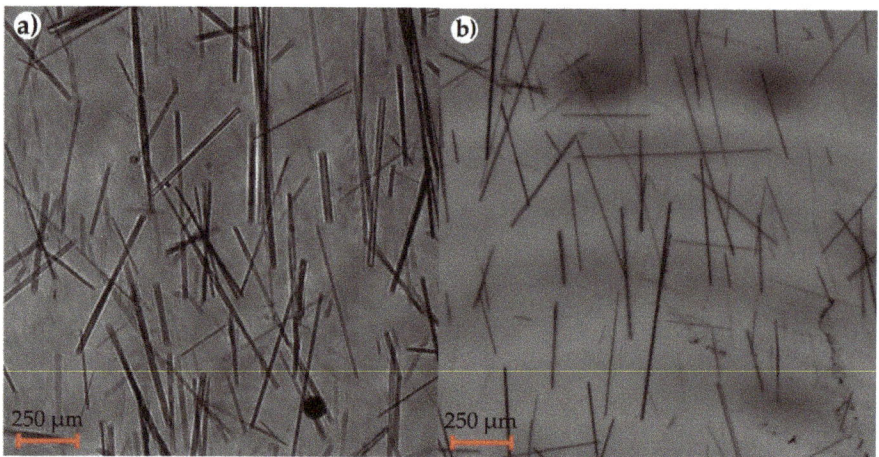

Figure 12. Thiamine hydrochloride (t = 1 h) monohydrate crystals (**a**) online microscope (**b**) shadowgraphic probe. The images are enlarged for a better view of the edges.

The crystals captured with the online microscope appear blurry, without clear edges, and a proper image evaluation is not possible. In addition, the bypass could only be utilized for a few minutes before the suspension flow blocked. In contrast, the shadowgraphic probe was still capable of capturing images with sharp edges and suitable image quality for a crystal size evaluation. For the present case, the telecentric lens in the shadowgrphic probe shows a clear advantage, because no focus has to be adjusted, and the depth of field covers the whole measured volume. After t = 1.22 h, the experiment was ended due to a significant increase in suspension density. At this point even a qualitative evaluation of the captured images was not possible, either with the online microscope nor with the shadowgraphic probe. Figure 13 shows images captured by the shadowgraphic probe of the suspension in different states.

At about t = 1 h the crystals grew as thin needles, as described above. They varied in length and width but were mostly isolated single crystals. Already a few minutes later (t = 1.16 h), the suspension density increased significantly, which may have been supported by natural breakage and secondary nucleation. Hence, the broken crystal pieces increase the total particle number, additionally. Due to the increased number of crystals, single particles have an increased probability of colliding with each other and forming agglomerates, which can be seen at the time mark for t = 1 h in Figure 13c,d.

For this state of the system an image evaluation of the crystals is quite challenging and may not be solved with a conventionally image analysis based on binary object identification, because the thin needles overlap and single crystals cannot be identified [72]. Interestingly, the needles tend to align with the flow direction in the measured volume of the shadowgraphic probe, especially at a higher solid content. Because the gap is comparably small to the vessel, the flow inside the gap is hindered and mostly laminar, even if the flow around it is turbulent. A group of researchers reported an algorithm which was utilized to determine the size of high-aspect-ratio crystals. They found that an irregular alignment hinders clear object detection at a higher solid content [72]. Therefore, the alignment of the needle-like crystals in the shadowgraphic probe could be an opportunity to investigate the crystallization in such difficult systems, with high aspect ratios of the particles.

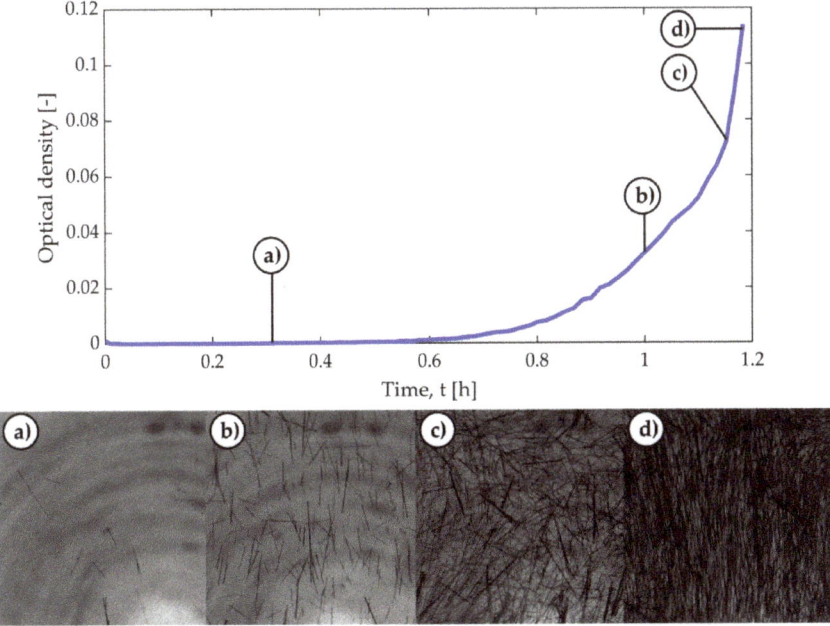

Figure 13. Optical density for the thiamine anti solvent crystallization based on pixel ratios of the shadowgraphic probe images. The images at the bottom were captured with the shadowgraphic probe at different experiment times; (**a**) t = 0.3 h first single crystals, (**b**) t = 1 h recognizable number of crystals, (**c**) t = 1.16 h first agglomerates and increased overlapping of the crystals and (**d**) t = 1.18 h last possible measurement point, afterwards the suspension density was too high to capture further images.

Although a crystal size determination with the presented methods was not possible after t = 1 h, a measurement of the optical suspension density was still possible (see Figure 13).

The optical density shows a rapidly increasing amount of crystals that started at around t = 0.3 h. From that point, the optical density increased with exponential progression. The width and length distributions at different experiment times are shown in Figure 14.

The diagram for t = 0 h was made as a reference, where mostly dust was detected. At t = 0.63 h the first reliable distribution shows that the single needle-like crystals have between 500–1000 µm in length and 20–40 µm width. At t = 0.97 h the crystal sizes are about the same, while their amount has significantly increased. The last diagram at t = 1.18 h depicts only a detection of small particles, which clearly shows that the simple image algorithm that was applied failed to isolate the crystals, and is the limit in PSD evaluation, at least with the methods that were used.

The comparison between the thiamine hydrochloride and KH_2PO_4 experiments shows clearly that the limiting optical density for image-based measurement systems depends on several properties of the particulate phase, e.g., size, size distribution, and shape. KH_2PO_4 could be measured up to an optical density of 8% for the shadowgraphic probe and 11% for the online microscope, with a suspension density (FTIR) of 7%. The measurement had already failed for thiamine at a value of 3% to 4% optical density of the probe (see Figure 13, t = 1 h). Hence, a clear suspension, or optical density limit, of application for both utilized techniques cannot be given here. It has to be determined for each substance system individually, and the methods for the image evaluation must be adjusted for the specific case in order to maximize the applicability.

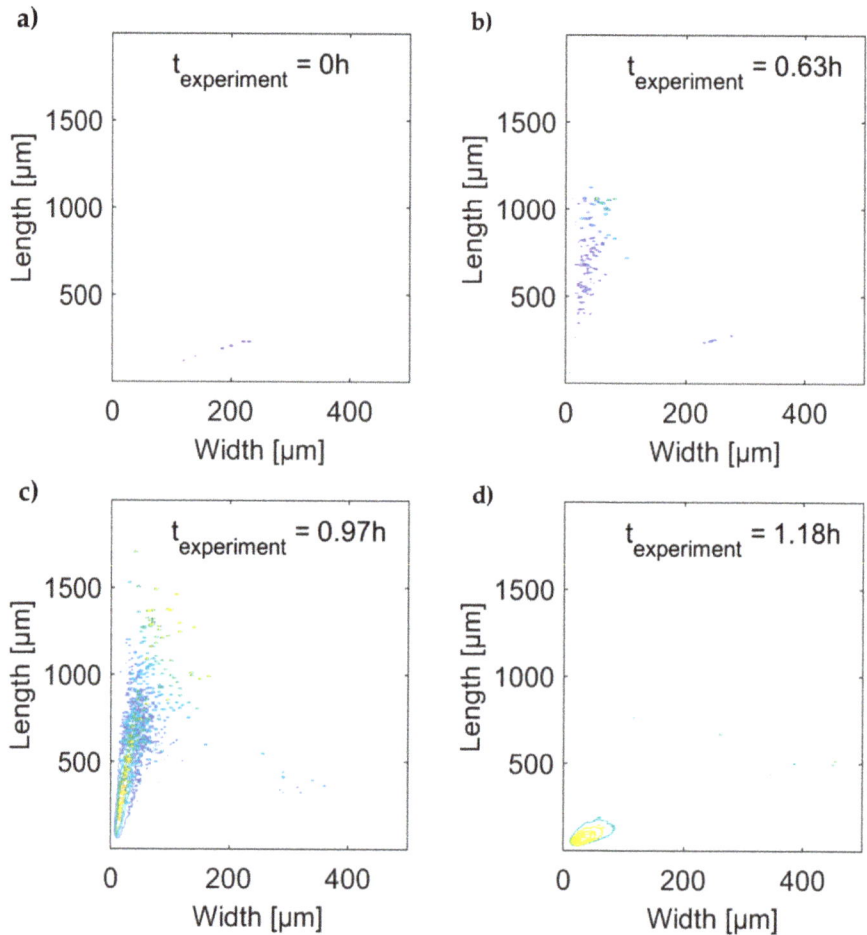

Figure 14. Length and width distributions of the needle-like crystals of thiamine hydrochloride monohydrate at different time points of Exp. 5, measured with the shadowgraphic probe. (**a**) measured crystals in the clear solution, (**b**) measurement of the first single needles, (**c**) significant number of crystals, (**d**) failed measurement as result of massive overlapping.

4. Conclusions

A novel inline shadowgraphic probe was utilized to determine the transient crystal size distribution in different crystallization processes, based on acquired greyscale images. For validation, three experiments in a well-known seeded KH_2PO_4/H_2O cooling crystallization were carried out, and the crystal size distributions between the new shadowgraphic probe were compared with an established bypass online microscope and sieve analysis. The measured number and mass distributions showed a good agreement for both image-based techniques in all three experiments. The percentiles support the results, as they exhibit similar trends and values with an average deviation of 10–20 µm. Classifying effects, such as shifts in the crystal size distributions could not be observed up to a characteristic crystal size of 600 µm. This is confirmed by sieve analyses of suspension samples that were compared to the final mass-based distributions of the optical techniques. The experiments were performed for different seed loadings, and hence, different suspension densities up to 8%, and an optical density up to 11%, without any influence with respect to the measured distributions.

In addition, a fourth experiment, with the same substances at a higher starting saturation, temperature was executed. It was shown that the probe can be applied to these elevated temperatures and conditions, while a measurement with the bypass variant failed due to a blockage within the bypass tubes.

Needle-like thiamine hydrochloride monohydrate was crystallized from a clear thiamine/water solution when adding ethanol as an anti-solvent. The process was only investigated using the shadowgraphic probe, since measurements with the bypass online microscope failed due to blocking. In addition, the images captured by the online microscope had poor image quality, due to blurry edges of the imaged objects. The needle-like crystals could be measured in length and width up to a suspension density of three percent, until the image algorithm based on binarization failed, due to missing segmentation methods and massive particle overlapping. It was shown that the shadowgraphic probe can be applied to systems that form fragile crystals, where other techniques fail. It was found that the needle-like crystals align with the flow direction in the measurement gap, which offers a great potential for different image processing routines at a higher solid content.

A suspension density limit for the applied techniques cannot be generally determined. The measured optical density on pixel ratios does not necessarily match with the mass-based suspension density, but it can identify trends. The optical suspension density is mutually connected to the particulate state, such as size, distribution, and particle shape, and it must be evaluated for each system individually.

It was shown that the shadowgraphic probe is capable of monitoring the transient evolution of the PSD in a crystallization processes, with an extended range of operation conditions, and was compared to an established online bypass variant and sieve analysis. While bypass variants mainly suffer from blockage at high temperatures and supersaturations, the shadowgraphic probe can be applied under these conditions. In view of industrial application, it is desirable to extend the range of operation up to industrial conditions, i.e., suspension density, temperature, pressure, and chemical resistance. It is well known that image analysis fails at a high solid content, but mathematical algorithms (e.g., neuronal networks) have significantly developed in the recent years to overcome this gap. Hence, it is desirable to enhance the analysis range to industrially relevant conditions (e.g., larger suspension densities). With endoscopic probes, in combination with appropriate image analysis software, processes can be designed and scaled to industrially relevant size, as the development is less based on experience than on intrinsic data, including the particulate state.

In the state-of-the-art crystallization processes, the particulate phase is mostly not monitored and therefore largely unknown, which is one of the key problems in developing continuous crystallization processes.

Author Contributions: Conceptualization, E.T. and D.W.; methodology, E.T. and D.W.; investigation, formal analysis, validation, data curation and visualization, E.T. and D.W.; experiments, E.T., M.H. and D.W. writing—original draft preparation, E.T., D.W., M.H., H.L., A.S.-M., and H.-J.B.; writing—review and editing, D.W. and E.T., H.L., H.-J.B. and A.S.-M.; supervision, H.L., H.-J.B. and A.S.-M.; project administration and funding acquisition, H.-J.B. and A.S.-M. All authors have read and agreed to the published version of the manuscript.

Funding: This research received no external funding.

Acknowledgments: The support of Holger Eisenschmidt is gratefully acknowledged.

Conflicts of Interest: The authors declare no conflict of interest. The funders had no role in the design of the study; in the collection, analyses, or interpretation of data; in the writing of the manuscript, or in the decision to publish the results.

Appendix A

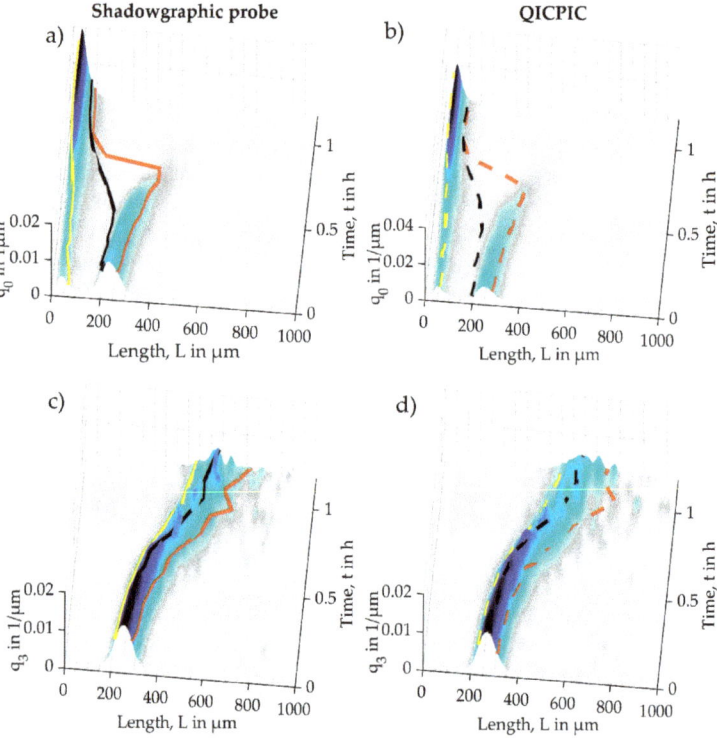

Figure A1. Exp. 1—KH$_2$PO$_4$ distributions (**a**) q_0-distribution shadowgraphic probe; (**b**) q_0-distribution QICPIC; (**c**) q_3-distribution shadowgraphic probe; (**d**) q_3-distribution QICPIC. *Solid lines*—percentiles of the shadowgraphic probe; *Dashed lines*—percentiles of the QICPIC; *yellow and red* −0.15 and 0.85 percentile distribution; *blue*—transient mean sizes of the distribution.

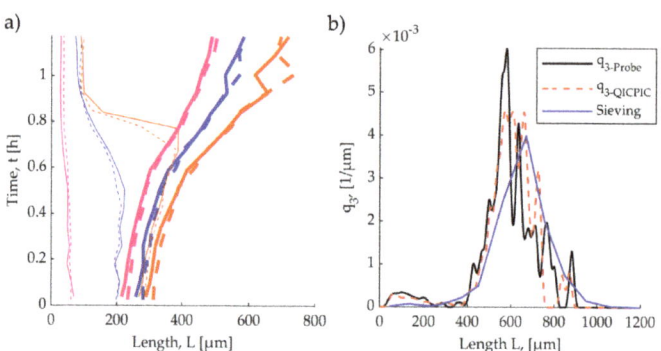

Figure A2. (**a**) Percentiles and mean values of the crystal size distributions for Exp. 1—KH$_2$PO$_4$ *thin lines*—percentiles of the q_0-distribution; *bold lines*—percentiles of the q_3-distributions; *Solid lines*—percentiles and mean values shadowgraphic probe; *Dashed* lines—percentiles and mean values QICPIC; *brown; magenta* −0.15 and 0.85 percentile of the corresponding distribution; *blue*—mean size of the corresponding distribution. (**b**) Comparison of the mass distribution: shadowgraphic probe, QICPIC, sieve analysis.

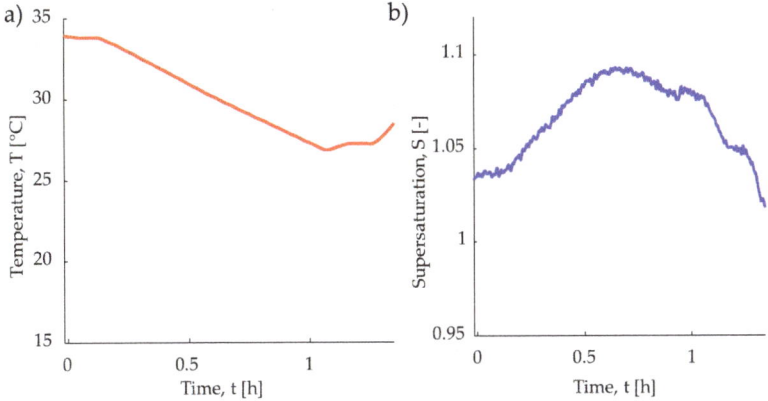

Figure A3. (a) Temperature profile of the Exp. 1—KH_2PO_4: saturation, seeding, linear cooling ramp, and final temperature. (b) Concentration profile of Exp. 3—KH_2PO_4: FTIR, offline samples and saturation curve.

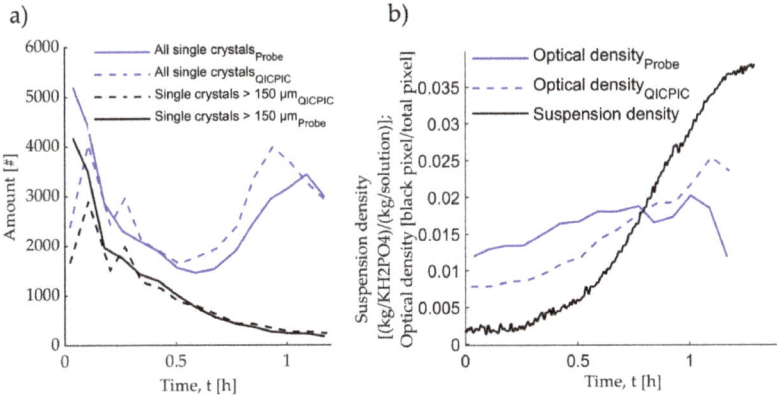

Figure A4. (a) Quantity of single crystals analyzed (b) the optical and suspension densities during experiment Exp. 1—KH_2PO_4.

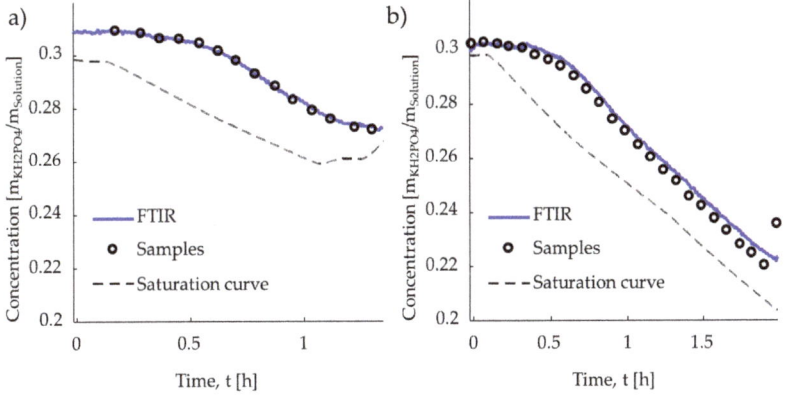

Figure A5. Concentration profile of (a) Exp. 1—KH_2PO_4 and (b) Exp. 2—KH_2PO_4; FTIR, offline samples and saturation Equation (1).

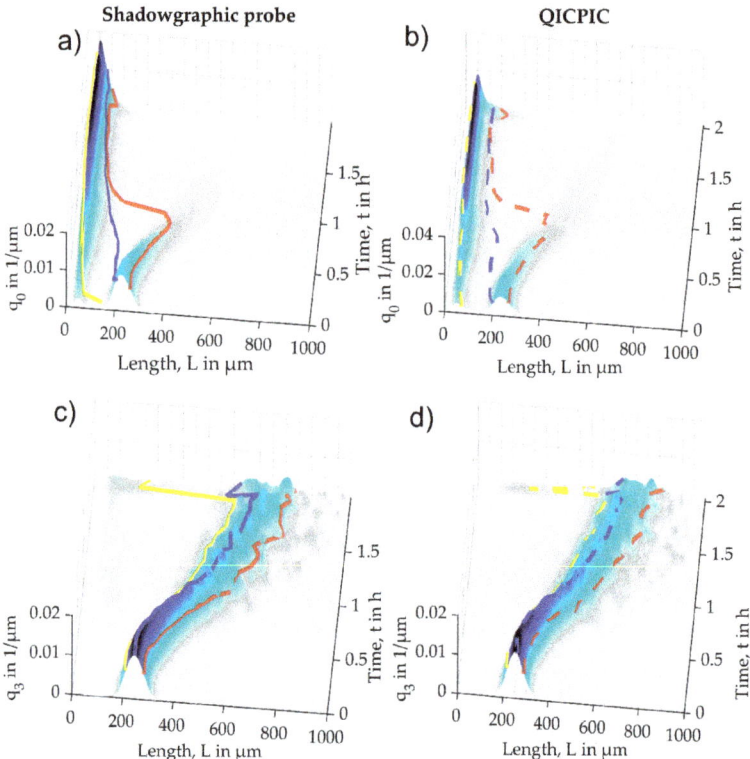

Figure A6. Exp. 2—KH$_2$PO$_4$ distributions (**a**) q_0-distribution shadowgraphic probe; (**b**) q_0-distribution QICPIC; (**c**) q_3-distribution shadowgraphic probe; (**d**) q_3-distribution QICPIC. *Solid lines*—percentiles of the shadowgraphic probe; *Dashed lines*—percentiles of the QICPIC; *yellow and red* −0.15 and 0.85 percentile distribution; *blue*—transient mean sizes of the distribution.

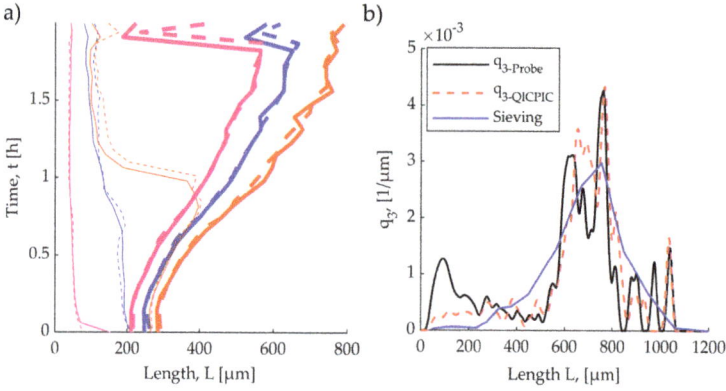

Figure A7. (**a**) Percentiles and mean values of the crystal size distributions for Exp. 2—KH$_2$PO$_4$ *thin lines*—percentiles of the q_0-distribution; *bold lines*—percentiles of the q_3-distributions; *Solid lines*—percentiles and mean values shadowgraphic probe; *Dashed lines*—percentiles and mean values QICPIC; *brown; magenta* −0.15 and 0.85 percentile of the corresponding distribution; *blue*—mean size of the corresponding distribution. (**b**) Comparison of the mass distribution: shadowgraphic probe, QICPIC, sieve analysis.

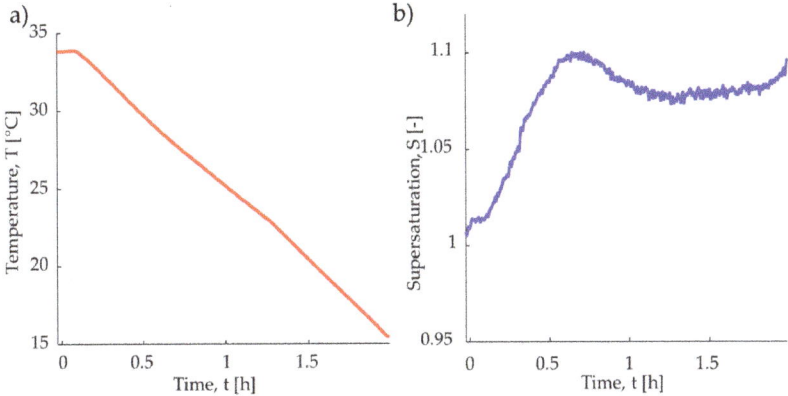

Figure A8. (**a**) Temperature profile of the Exp. 2—KH_2PO_4: saturation, seeding, linear cooling ramp and final temperature. (**b**) Concentration profile of Exp. 3—KH_2PO_4: FTIR, offline samples and saturation curve.

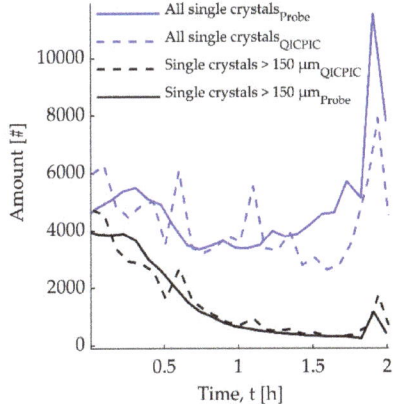

Figure A9. Quantity of single crystals analyzed during experiment Exp. 2—KH_2PO_4.

References

1. Chow, K.; Tong, H.H.Y.; Lum, S.; Chow, A.H.L. Engineering of Pharmaceutical materials: An industrial perspective. *J. Pharm. Sci.* **2008**, *97*, 2855–2877. [CrossRef] [PubMed]
2. Wibowo, C.; Chang, W.-C.; Ng, K.M. Design of integrated crystallization systems. *AIChE J.* **2001**, *47*, 2474–2492. [CrossRef]
3. Braatz, R.D. Advanced control of crystallization processes. *Annu. Rev. Control* **2002**, *26*, 87–99. [CrossRef]
4. Ulrich, J.; Frohberg, P. Problems, potentials and future of industrial crystallization. *Front. Chem. Sci. Eng.* **2013**, *7*, 1–8. [CrossRef]
5. Vetter, T.; Burcham, C.L.; Doherty, M.F. Regions of attainable particle sizes in continuous and batch crystallization processes. *Chem. Eng. Sci.* **2014**, *106*, 167–180. [CrossRef]
6. Chen, J.; Sarma, B.; Evans, J.M.B.; Myerson, A.S. Pharmaceutical Crystallization. *Cryst. Growth Des.* **2011**, *11*, 887–895. [CrossRef]
7. Tahir, F.; Krzemieniewska-Nandwani, K.; Mack, J.; Lovett, D.; Siddique, H.; Mabbott, F.; Raval, V.; Houson, I.; Florence, A. Advanced control of a continuous oscillatory flow crystalliser. *Control Eng. Pract.* **2017**, *67*, 64–75. [CrossRef]
8. Su, Q.; Nagy, Z.K.; Rielly, C.D. Pharmaceutical crystallisation processes from batch to continuous operation using MSMPR stages: Modelling, design, and control. *Chem. Eng. Process.* **2015**, *89*, 41–53. [CrossRef]

9. Lawton, S.; Steele, G.; Shering, P.; Zhao, L.; Laird, I.; Ni, X.-W. Continuous Crystallization of Pharmaceuticals Using a Continuous Oscillatory Baffled Crystallizer. *Org. Process. Res. Dev.* **2009**, *13*, 1357–1363. [CrossRef]
10. Schlüter, M. Lokale Messverfahren für Mehrphasenströmungen. *Chem. Ing. Tech.* **2011**, *83*, 992–1004. [CrossRef]
11. Lichti, M.; Bart, H.-J. Particle Measurement Techniques in Fluid Process Engineering. *Chembioeng Rev.* **2018**, *5*, 79–89. [CrossRef]
12. Nagy, Z.K.; Fevotte, G.; Kramer, H.; Simon, L.L. Recent advances in the monitoring, modelling and control of crystallization systems. *Chem. Eng. Res. Des.* **2013**, *91*, 1903–1922. [CrossRef]
13. Yu, L.X.; Lionberger, R.A.; Raw, A.S.; D'Costa, R.; Wu, H.; Hussain, A.S. Applications of process analytical technology to crystallization processes. *Adv. Drug Deliv. Rev.* **2004**, *56*, 349–369. [CrossRef] [PubMed]
14. Ma, Z.; Merkus, H.G.; De Smet, J.G.A.E.; Heffels, C.; Scarlett, B. New developments in particle characterization by laser diffraction: Size and shape. *Powder Technol.* **2000**, *111*, 66–78. [CrossRef]
15. Ma, Z.; Merkus, H.G.; Scarlett, B. Extending laser diffraction for particle shape characterization: Technical aspects and application. *Powder Technol.* **2001**, *118*, 180–187. [CrossRef]
16. Cogoni, G.; De Souza, B.P.; Frawley, P.J. Particle Size Distribution and yield control in continuous Plug Flow Crystallizers with recycle. *Chem. Eng. Sci.* **2015**, *138*, 592–599. [CrossRef]
17. Jia, C.-Y.; Yin, Q.-X.; Zhang, M.-J.; Wang, J.-K.; Shen, Z.-H. Polymorphic Transformation of Pravastatin Sodium Monitored Using Combined Online FBRM and PVM. *Org. Process. Res. Dev.* **2008**, *12*, 1223–1228. [CrossRef]
18. Kail, N.; Marquardt, W.; Briesen, H. Process Analysis by Means of Focused Beam Reflectance Measurements. *Ind. Eng. Chem. Res.* **2009**, *48*, 2936–2946. [CrossRef]
19. Fujiwara, M.; Chow, P.S.; Ma, D.L.; Braatz, R.D. Paracetamol Crystallization Using Laser Backscattering and ATR-FTIR Spectroscopy: Metastability, Agglomeration, and Control. *Cryst. Growth Des.* **2002**, *2*, 363–370. [CrossRef]
20. Kee, N.C.S.; Woo, X.Y.; Goh, L.M.; Rusli, E.; He, G.; Bhamidi, V.; Tan, R.B.H.; Kenis, P.J.A.; Zukoski, C.F.; Braatz, R.D. Design of Crystallization Processes from Laboratory Research and Development to the Manufacturing Scale. *Am. Pharm. Rev.* **2008**, *11*, 110–125.
21. Kougoulos, E.; Jones, A.G.; Wood-Kaczmar, M.W. Process Modelling Tools for Continuous and Batch Organic Crystallization Processes Including Application to Scale-Up. *Org. Process Res. Dev.* **2006**, *10*, 739–750. [CrossRef]
22. Barrett, P.; Glennon, B. In-line FBRM Monitoring of Particle Size in Dilute Agitated Suspensions. *Part. Part. Syst. Charact.* **1999**, *16*, 207–211. [CrossRef]
23. Leyssens, T.; Baudry, C.; Hernandez, M.L. Optimization of a Crystallization by Online FBRM Analysis of Needle-Shaped Crystals. *Org. Process Res. Dev.* **2011**, *15*, 413–426. [CrossRef]
24. Schöell, J.; Bonalumi, D.; Vicum, L.; Mazzotti, M.; Müller, M. In Situ Monitoring and Modeling of the Solvent-Mediated Polymorphic Transformation ofl-Glutamic Acid. *Cryst. Growth Des.* **2006**, *6*, 881–891. [CrossRef]
25. Ruf, A.; Worlitschek, J.; Mazzotti, M. Modeling and Experimental Analysis of PSD Measurements through FBRM. *Part. Part. Syst. Charact.* **2000**, *17*, 167–179. [CrossRef]
26. Hobbel, E.F.; Davies, R.; Rennie, F.W.; Allen, T.; Butler, L.E.; Waters, E.R.; Smith, J.T.; Sylvester, R.W. Modern Methods of On-Line Size Analysis for Particulate Process Streams. *Part. Part. Syst. Charact.* **1991**, *8*, 29–34. [CrossRef]
27. Clark, N.N.; Turton, R. Chord length distributions related to bubble size distributions in multiphase flows. *Int. J. Multiph. Flow* **1988**, *14*, 413–424. [CrossRef]
28. Simmons, M.J.H.; Langston, P.A.; Burbidge, A.S. Particle and droplet size analysis from chord distributions. *Powder Technol.* **1999**, *102*, 75–83. [CrossRef]
29. Schorsch, S.; Ochsenbein, D.R.; Vetter, T.; Morari, M.; Mazzotti, M. High accuracy online measurement of multidimensional particle size distributions during crystallization. *Chem. Eng. Sci.* **2014**, *105*, 155–168. [CrossRef]
30. Schorsch, S.; Vetter, T.; Mazzotti, M. Measuring multidimensional particle size distributions during crystallization. *Chem. Eng. Sci.* **2012**, *77*, 130–142. [CrossRef]
31. Mougin, P.M.J. In Situ and On-Line Ultrasonic Attenuation Spectroscopy for Particle Sizing during the Crystallisation of Organic Fine Chemicals. Ph.D. Thesis, Heriot-Watt University, Edinburgh, UK, 2008.

32. Pertig, D.; Buchfink, R.; Petersen, S.; Stelzer, T.; Ulrich, J. Inline Analyzing of Industrial Crystallization Processes by an Innovative Ultrasonic Probe Technique. *Chem. Eng. Technol.* **2011**, *34*, 639–646. [CrossRef]
33. Pertig, D.; Fardmostafavi, M.; Stelzer, T.; Ulrich, J. Monitoring concept of single-frequency ultrasound and its application in dynamic crystallization processes. *Adv. Powder Technol.* **2015**, *26*, 874–881. [CrossRef]
34. Calderon De Anda, J.; Wang, X.Z.; Roberts, K.J. Multi-scale segmentation image analysis for the in-process monitoring of particle shape with batch crystallisers. *Chem. Eng. Sci.* **2005**, *60*, 1053–1065. [CrossRef]
35. Dharmayat, S.; Calderon De Anda, J.; Hammond, R.B.; Lai, X.; Roberts, K.J.; Wang, X.Z. Polymorphic transformation of l-glutamic acid monitored using combined on-line video microscopy and X-ray diffraction. *J. Cryst. Growth* **2006**, *294*, 35–40. [CrossRef]
36. Wang, X.Z.; Calderon De Anda, J.; Roberts, K.J. Real-Time Measurement of the Growth Rates of Individual Crystal Facets Using Imaging and Image Analysis. *Chem. Eng. Res. Des.* **2007**, *85*, 921–927. [CrossRef]
37. Borchert, C.; Temmel, E.; Eisenschmidt, H.; Lorenz, H.; Seidel-Morgenstern, A.; Sundmacher, K. Image-Based in Situ Identification of Face Specific Crystal Growth Rates from Crystal Populations. *Cryst. Growth Des.* **2014**, *14*, 952–971. [CrossRef]
38. Eggers, J.; Kempkes, M.; Mazzotti, M. Measurement of size and shape distributions of particles through image analysis. *Chem. Eng. Sci.* **2008**, *63*, 5513–5521. [CrossRef]
39. Patience, D.B.; Rawlings, J.B. Particle-shape monitoring and control in crystallization processes. *AIChE J.* **2001**, *47*, 2125–2130. [CrossRef]
40. Kempkes, M. Monitoring of Particle Size and Shape in Crystallization Processes. Ph.D. Thesis, ETH-Zürich, Zürich, Switzerland, 2009.
41. Scott, D.M.; Sunshine, G.; Rosen, L.; Jochen, E. Industrial applications of process imaging and image processing. In *Process Imaging for Automatic Control*; Intelligent Systems and Smart Manufacturing; McCann, H., Scott, D.M., Eds.; SPIE: Boston, MA, USA, 2001.
42. Kempkes, M.; Darakis, E.; Khanam, T.; Rajendran, A.; Kariwala, V.; Mazzotti, M.; Naughton, T.J.; Asundi, A.K. Three dimensional digital holographic profiling of micro-fibers. *Opt. Express* **2009**, *17*, 2938–2943. [CrossRef]
43. Bluma, A.; Höpfner, T.; Rudolph, G.; Lindner, P.; Beutel, S.; Hitzmann, B.; Scheper, T. Adaptation of in-situ microscopy for crystallization processes. *J. Cryst. Growth* **2009**, *311*, 4193–4198. [CrossRef]
44. Ojaniemi, U.; Puranen, J.; Manninen, M.; Gorshkova, E.; Louhi-Kultanen, M. Hydrodynamics and kinetics in semi-batch stirred tank precipitation of l-glutamic acid based on pH shift with mineral acids. *Chem. Eng. Sci.* **2018**, *178*, 167–182. [CrossRef]
45. Kacker, R.; Maaß, S.; Emmerich, J.; Kramer, H. Application of inline imaging for monitoring crystallization process in a continuous oscillatory baffled crystallizer. *AIChE J.* **2018**, *64*, 2450–2461. [CrossRef]
46. Wang, X.Z.; Roberts, K.J.; Ma, C. Crystal growth measurement using 2D and 3D imaging and the perspectives for shape control. *Chem. Eng. Sci.* **2008**, *63*, 1173–1184. [CrossRef]
47. Barrett, P.; Glennon, B. Characterizing the Metastable Zone Width and Solubility Curve Using Lasentec FBRM and PVM. *Chem. Eng. Res. Des.* **2002**, *80*, 799–805. [CrossRef]
48. Mickler, M.; Bart, H.-J. Optical Multimode Online Probe: Erfassung und Analyse von Partikelkollektiven. *Chem. Ing. Tech.* **2013**, *85*, 901–906. [CrossRef]
49. Mickler, M.; Boecker, B.; Bart, H.-J. Drop swarm analysis in dispersions with incident-light and transmitted-light illumination. *Flow Meas. Instrum.* **2013**, *30*, 81–89. [CrossRef]
50. Amokrane, A.; Maaß, S.; Lamadie, F.; Puel, F.; Charton, S. On droplets size distribution in a pulsed column. Part I: In-situ measurements and corresponding CFD–PBE simulations. *Int. J. Chem. Eng.* **2016**, *296*, 366–376. [CrossRef]
51. Maaß, S.; Rojahn, J.; Hänsch, R.; Kraume, M. Automated drop detection using image analysis for online particle size monitoring in multiphase systems. *Comput. Chem. Eng.* **2012**, *45*, 27–37. [CrossRef]
52. Bart, H.-J.; Hlawitschka, M.W.; Mickler, M.; Jaradat, M.; Didas, S.; Chen, F.; Hagen, H. Tropfencluster—Analytik, Simulation und Visualisierung. *Chem. Ing. Tech.* **2011**, *83*, 965–978. [CrossRef]
53. Steinhoff, J.; Bart, H.-J. Settling Behavior and CFD Simulation of a Gravity Separator. In *Extraction 2018*; Davis, B.R., Moats, M.S., Wang, S., Gregurek, D., Kapusta, J., Battle, T.P., Schlesinger, M.E., Alvear Flores, G.R., Jak, E., Goodall, G., et al., Eds.; Springer International Publishing: Cham, Switzerland, 2018; pp. 1997–2007.
54. Schmitt, P.; Hlawitschka, M.W.; Bart, H.-J. Centrifugal Pumps as Extractors. *Chem. Ing. Tech.* **2020**, *262*, 12215. [CrossRef]
55. Lichti, M.; Bart, H.-J. Bubble size distributions with a shadowgraphic optical probe. *Flow Meas. Instrum.* **2018**, *60*, 164–170. [CrossRef]

56. Lichti, M.; Cheng, X.; Stephani, H.; Bart, H.-J. Online Detection of Ellipsoidal Bubbles by an Innovative Optical Approach. *Chem. Eng. Technol.* **2019**, *42*, 506–511. [CrossRef]
57. Wirz, D.; Lichti, M.; Bart, H.-J. Erfassung partikulärer Prozessgrößen im Dreiphasensystem. *Chem. Ing. Tech.* **2018**, *90*, 1318. [CrossRef]
58. Schulz, J.; Bart, H.-J. Analysis of entrained liquid by use of optical measurement technology. *Chem. Eng. Res. Des.* **2019**, *147*, 624–633. [CrossRef]
59. Schulz, J.; Schäfer, K.; Bart, H.-J. Entrainment Control Using a Newly Developed Telecentric Inline Probe. *Chem. Ing. Tech.* **2020**, *92*, 256–265. [CrossRef]
60. Lichti, M.; Bart, H.-J.; Roth, C. Vorrichtung für Bildaufnahmen eines Messvolumens in einem Behälter. European Patent 3.067.685.A1; Europäisches Patentamt München, Germany, 3 June 2020.
61. Lichti, M. Optische Erfassung von Partikelmerkmalen: Entwicklung einer Durchlichtmesstechnik für Apparate der Fluidverfahrenstechnik. Ph.D. Thesis, TU Kaiserslautern, Kaiserslautern, Germany, 2018.
62. Temmel, E. Design of Continuous Crystallization Processes. Ph.D. Thesis, Fakultät für Verfahrens- und Systemtechnik, Otto von Guericke University, Magdeburg, Germany, 2016.
63. Temmel, E.; Eicke, M.; Lorenz, H.; Seidel-Morgenstern, A. A Short-Cut Method for the Quantification of Crystallization Kinetics. 2. Experimental Application. *Cryst. Growth Des.* **2016**, *16*, 6756–6768. [CrossRef]
64. Watanabe, A.; Tasaki, S.; Wada, Y.; Nakamachi, H. Polymorphism of thiamine hydrochloride. II. Crystal structure of thiamine hydrochloride hemihydrate and its stability. *Chem. Pharm. Bull.* **1979**, *27*, 2751–2759. [CrossRef]
65. Chakravarty, P.; Berendt, R.T.; Munson, E.J.; Young, V.G.; Govindarajan, R.; Suryanarayanan, R. Insights into the dehydration behavior of thiamine hydrochloride (vitamin B1) hydrates: Part II. *J. Pharm. Sci.* **2010**, *99*, 1882–1895. [CrossRef]
66. Chakravarty, P.; Suryanarayanan, R. Characterization and Structure Analysis of Thiamine Hydrochloride Methanol Solvate. *Cryst. Growth Des.* **2010**, *10*, 4414–4420. [CrossRef]
67. Li, X.; Han, D.; Wang, Y.; Du, S.; Liu, Y.; Zhang, J.; Yu, B.; Hou, B.; Gong, J. Measurement of Solubility of Thiamine Hydrochloride Hemihydrate in Three Binary Solvents and Mixing Properties of Solutions. *J. Chem. Eng. Data* **2016**, *61*, 3665–3678. [CrossRef]
68. Eisenschmidt, H.; Bajcinca, N.; Sundmacher, K. Optimal Control of Crystal Shapes in Batch Crystallization Experiments by Growth-Dissolution Cycles. *Cryst. Growth Des.* **2016**, *16*, 3297–3306. [CrossRef]
69. Borchert, C.; Sundmacher, K. Crystal Aggregation in a Flow Tube: Image-Based Observation. *Chem. Eng. Technol.* **2011**, *34*, 545–556. [CrossRef]
70. Stieß, M. *Mechanische Verfahrenstechnik—Partikeltechnologie 1*, 3rd ed.; Springer: Berlin/Heidelberg, Germany, 2009.
71. Randolph, A.D.; Larson, M.A. *Theory of Particulate Processes: Analysis and Techniques of Continuous Crystallization*, 2nd ed.; Academic Press: San Diego, CA, USA, 1988.
72. Larsen, P.A.; Rawlings, J.B.; Ferrier, N.J. An algorithm for analyzing noisy, in situ images of high-aspect-ratio crystals to monitor particle size distribution. *Chem. Eng. Sci.* **2006**, *61*, 5236–5248. [CrossRef]

© 2020 by the authors. Licensee MDPI, Basel, Switzerland. This article is an open access article distributed under the terms and conditions of the Creative Commons Attribution (CC BY) license (http://creativecommons.org/licenses/by/4.0/).

Article

Microplates for Crystal Growth and in situ Data Collection at a Synchrotron Beamline

Miao Liang [1,2], Zhijun Wang [1,3], Hai Wu [1,2], Li Yu [1,2], Bo Sun [1,3], Huan Zhou [1,3], Feng Yu [1,3], Qisheng Wang [1,3,*] and Jianhua He [1,4,*]

1. Shanghai Institute of Applied Physics, Chinese Academy of Sciences, Shanghai 201800, China; liangmiao@sinap.ac.cn (M.L.); wangzj@sari.ac.cn (Z.W.); wuhai@sinap.ac.cn (H.W.); yuli@sinap.ac.cn (L.Y.); sunbo@zjlab.org.cn (B.S.); zhouhuan@zjlab.org.cn (H.Z.); yufeng@zjlab.org.cn (F.Y.)
2. University of Chinese Academy of Sciences, Beijing 100049, China
3. Shanghai Synchrotron Radiation Facility, Shanghai Advanced Research Institute, Chinese Academy of Sciences, Shanghai 201204, China
4. The Institute for Advanced Studies, Wuhan University, Wuhan 430072, China
* Correspondence: wangqisheng@sinap.ac.cn (Q.W.); hejianhua@whu.edu.cn (J.H.)

Received: 30 July 2020; Accepted: 6 September 2020; Published: 9 September 2020

Abstract: An efficient data collection method is important for microcrystals, because microcrystals are sensitive to radiation damage. Moreover, microcrystals are difficult to harvest and locate owing to refraction effects from the surface of the liquid drop or optically invisible, owing to their small size. Collecting X-ray diffraction data directly from the crystallization devices to completely eliminate the crystal harvesting step is of particular interest. To address these needs, novel microplates combining crystal growth and data collection have been designed for efficient in situ data collection and fully tested at Shanghai Synchrotron Radiation Facility (SSRF) crystallography beamlines. The design of the novel microplates fully adapts the advantage of in situ technology. Thin Kapton membranes were selected to seal the microplate for crystal growth, the crystallization plates can support hanging drop and setting drop vapor diffusion crystallization experiments. Then, the microplate was fixed on a magnetic base and mounted on the goniometer head for in situ data collection. Automatic grid scanning was applied for crystal location with a Blu-Ice data collection system and then in situ data collection was performed. The microcrystals of lysozyme were selected as the testing samples for diffraction data collection using the novel microplates. The results show that this method can achieve comparable data quality to that of the traditional method using the nylon loop. In addition, our method can efficiently and diversely perform data acquisition experiments, and be especially suitable for solving structures of multiple crystals at room temperature or cryogenic temperature.

Keywords: microcrystals; microplate; grid scanning; in situ data collection

1. Introduction

In recent years, in order to efficiently obtain the structure of protein, various processing steps of the protein crystallography have been improved and optimized, especially with the development of in situ X-ray crystallography [1]. The method of in situ diffraction has been developed to directly collect datasets from the location where crystals were grown using X-ray diffraction, which eliminates the process of transferring the crystal sample to nylon loop and avoids the influence of human factors on the quality of the crystals. This method was originally used to screen crystals and verify the quality of crystals, but now this method is mainly used to collect multiple data sets to obtain high-resolution protein structures, because in situ diffraction can greatly improve the efficiency of data collection, and at the same time, is very suitable for some special crystals. For example, some proteins can only produce microcrystals, and microcrystals are fragile to transfer with a normal nylon loop. In addition to the development of in situ data collection methods, microcrystals are sensitive to radiation damage, so it is necessary to collect data from multiple crystals in a small wedge and integrate the diffraction images of multiple crystals into a full dataset. Since the determination of radiation-sensitive crystal structures requires a large number of crystals, high-throughput data collection methods are also needed to improve efficiency. A number of successful devices have already been designed for in situ data collection at different light sources. So far, microfluidic devices [2–4], chips [5], and regular 96-well crystallization plates [6] are the main methods reported for in situ data collection.

Microfluidic devices including capillaries and nano-droplets are usually used for protein crystallization and then directly for in situ data collection at room temperature using such devices. Microfluidic devices provide sufficient convectionless space for high-throughput crystal growth and screening. For example, Pinker et al. reported a microfluidic device ChipX [7] that can, not only obtain high-quality crystal and diffraction data, but also perform in situ characterization without directly processing crystals. Furthermore, the microfluidic device ChipX is good for in situ data collection for fragile crystals. However, ChipX is not suitable for flash-cryocooling of crystals, which will cause the liquid to freeze. In addition, the microfluidic device based on capillary is also used for in situ data collection of protein crystals [8]. It has been reported that two sets of devices containing aqueous solution and oil respectively are designed to cooperate with the capillary [9]. The aqueous solution including the protein solution and precipitant is used for protein crystallization, and the oil is used to separate the aqueous solution to form a separate crystallization environment, and then the microcapillary containing the crystal is directly installed on the beamline station for data collection after the crystallization experiment. The in situ method based on nano-droplets is also generally implemented by the microfluidic device [10]. The crystals in the capillary gradually move to the surface of the nano-droplets and are fixed near the droplet interface. During the whole process, there is no special operation to fix the crystals, crystals are fixed by the high surface tension of the droplet and used for further data collection. Furthermore, the in situ method based on nano-droplets can realize free interface diffusion crystallization and large-scale preparation of monodisperse crystals, which avoids crystal accumulation. However, the data is collected in the liquid phase at room temperature so that the crystal is susceptible to radiation damage [11].

Chip and film were reported to transfer or grow crystals, and then mount them on the goniometer head for data collection. These methods do not limit the size of crystals. It has a high hitting rate of X-ray during data collection. The amount of protein used is small and only a few hundred crystals are needed to obtain a full dataset of protein structures [12–15]. The chip material generally uses quartz with high light transmittance and low background scattering [16]. X-rays have a high hitting rate to the crystal using a chip-based sample delivery method. Using micro-nano processing technology, holes or grooves are etched on the chip, and then the crystal is fixed in the holes and grooves on the chip for diffraction data collection. Zarrine-Afsar et al. first proposed the application of the chip-based sample-delivery method for protein crystal data collection [17]. They grew the crystals in situ on a chip with a polyimide film attached to the lower end. The chip has a grooved array (silicon mesh). By adding glass beads in the grooved area of the chip, the random orientation of the crystals is induced by increasing the roughness. The chip can be used for the in situ growth of macromolecular crystals and serial data collection. When data is collected, it is allowed to rotate at a certain angle, and collect multiple sets of diffraction data for one crystal. The sample consumption is small and it is increased by adding glass beads. The surface roughness makes the crystals oriented randomly. Some structures have been obtained by this method. However, there are chip grids, polyimide films, and crystalline solutions that will cause high background scattering. A thin film, transferred with crystals, fixed to the goniometer head for diffraction data collection was reported by Li et al. [18]. Li et al. designed a bracket, and then attached the polyimide membrane to the resin bracket with SuperGlue. The microcrystals were transferred from the coverslip to the surface of the polyimide membrane by a micromesh loop, a glass capillary with a tip diameter less than 100 µm was used to remove excess liquid, and then crystals were sealed with another polyimide film to protect it from dehydration, but human factors are added during the sample transfer. Baxter et al. designed a high-density grid [19], covered with a polymer film or sleeve, for efficient data collection for multiple crystal samples, incubation chambers have been developed to support crystallization experiments on grids, but the grid cannot screen the crystallization conditions because there is only one type of desiccant in the incubation chamber.

Currently, a particular research interest is the possibility of data collection directly from crystallization plates, normal size 96-well plates are reported for crystal screening and in situ data collection [20]. With the development of various systems for plate setup and handling, the crystallization plates have been standardized to achieve compatibility with some beamline platforms, for example, the Structural Biology Center (SBC) beamline 19-ID, located at the Advanced Photon Source, USA. Significant progress has been made to provide plates with a low X-ray absorption profile and scattering properties, such as the MiTeGen In situ–1 Plate (MiTeGen, Inc.). However, with these kinds of methods, it is difficult to focus and align the crystals to the beam position once the plates rotate to a small wedge angle.

Here, a set of simple and inexpensive microplates (plate A and B) is designed for screening crystallization conditions and in situ data collection at room temperature or cryogenic temperature. Plate A is used for setting drop vapor diffusion crystallization. The assembly of plate B and a specially designed crystallization plate is applied for the hanging-drop vapor-diffusion method. The 12.5 μm thick Kapton membranes are used for crystal growth and sealing up the microplates. Lysozyme is used as the model protein for crystal growth, and lysozyme crystals are used to verify the practicability of the new device and method. Structure obtained by in situ method is compared with that obtained by the traditional method. The grid scanning, implemented from Blu-Ice [21] was used at the Shanghai Synchrotron Radiation Facility (SSRF) BL18U1 beamline for sample location and data collection. Results show that the devices can be used for screening crystallization conditions and in situ data collection from multiple crystals.

2. Materials and Methods

2.1. Design for Microplates

As shown in Figure 1, plate A and B are manufactured using a three-dimensional printer. The size of the microplate depends on the limitation of the goniometer and the spray area of the liquid nitrogen nozzle. The diameter of the spraying area of the liquid nitrogen nozzle is 6 mm at BL18U1, and the Y-axis limitation of the goniometer is ±3 cm. In order to enable the crystal samples in the microplate to be moved to the X-ray optical path, and the crystal samples to be covered by liquid nitrogen cooling gas, the size of the microplate is designed to be less than or equal to 20 mm. Plate A has five crystallization chambers, and each crystallization chamber is composed of two protein wells and a reservoir well, and the common space is designed in the crystallization chamber to support sitting vapor-diffusion crystallization experiments. Each protein well and reservoir well are hollowed out (Figure 1a), and the position of the bottom column is on the side of the protein wells to ensure that the bottom center of plate A and the bottom center of the protein wells are on a straight line (defined as the centerline). It is worth noting that the distance between any point of the bottom of protein wells and the centerline does not exceed 1 mm. The reason for this is that the deflection of plate A does not cause the crystal to deviate from the optical path during data collection. The size of the bottom column is designed to match the magnetic base to be stably fixed on it, and the size of the reservoir wells can be changed according to different crystallization conditions. Plate A is sealed by two Kapton membranes for sitting-drop vapor-diffusion crystallization experiments after loading the protein samples.

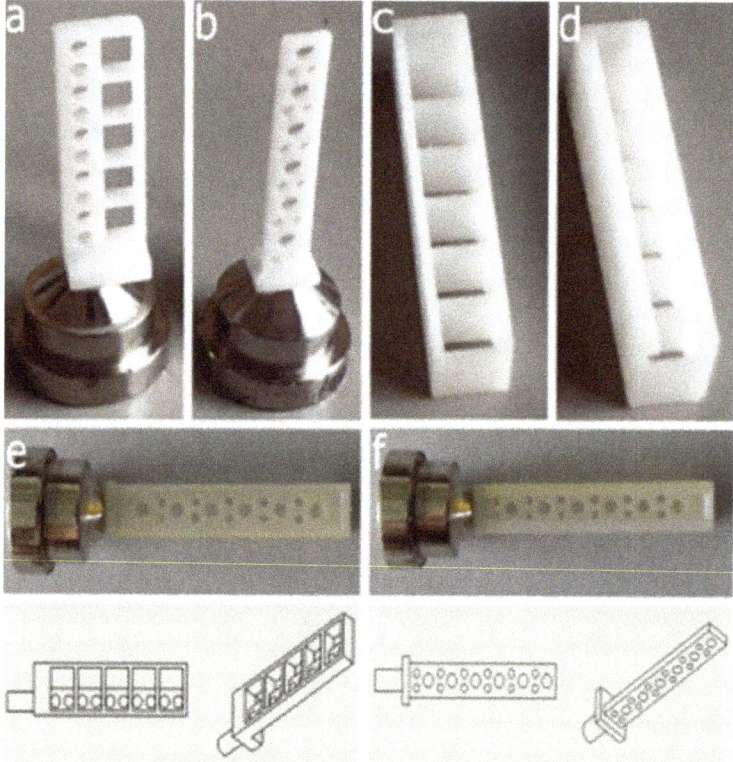

Figure 1. (a) Crystallization plate (Plate A) used for sitting-drop vapor-diffusion crystallization experiments; (b) Crystallization plate (Plate B) used for hanging-drop vapor-diffusion crystallization experiments; (c) Incubation chamber for Plate B by covering, crystallization buffers loaded into each chamber for screening of crystallization conditions for Plate B; (d) Incubation chamber for Plate B by insertion. (e) Assembly image for Plate B with incubation chamber in Figure 1c; (f) Assembly image for Plate B with incubation chamber in Figure 1d; (g) three-dimensional device structure of plate A; (h) three-dimensional device structure of plate B.

Plate B has six large protein wells and twelve small protein wells (Figure 1b). Incubation chambers (Figure 1c,d) are hollowed out. The hollowing of the reservoir well not only provides a preliminary observation of the crystal through the microscope but also facilitates the cleaning and secondary use of the plate. The size of the protein wells (Figure 1b) can be selected to fit different experimental settings, the bottom center of plate B and the bottom center of large protein wells are on a straight line (defined as the centerline), and the distance between any point of the bottom of the protein wells and the centerline does not exceed 1 mm. Unlike plate A, plate B utilizes a specially designed crystallization plate in the crystallization experiment. The Kapton membrane only needs to seal the side where the centerline is. After loading the protein samples onto plate B, then plate B was assembled with the incubation chambers (Figure 1c,d) for hanging-drop vapor-diffusion crystallization experiments (Figure 1e,f). There are six reservoir wells in the incubation chambers (Figure 1c,d), which is sealed by the Kapton membrane at the bottom side before loading the crystallization buffers into each well.

2.2. Crystal Growth

Chicken egg white lysozyme microcrystals smaller than 20 μm were used as testing samples. We used the controlled variable method to explore the crystallization conditions to grow microcrystals. For plate A (sitting-drop vapor-diffusion crystallization), the side of the centerline of plate A needs to be sealed with a Kapton membrane before the crystallization experiment, and then 0.8 μL of crystallization drops were added to each protein well and 8 μL of reservoir solution was added to each reservoir well. The lysozyme crystallization drops are prepared by dissolving 10 mg/mL of lysozyme protein in a reservoir solution. The reservoir solution is a solution with a concentration of 0.2 M citric acid and 0.2 M sodium acetate at a 1:1 ratio, 12% (w/v) NaCl, and then a mix 0.4 μL of a protein sample and 0.4 μL of a reservoir solution. Finally, the other side of plate A was sealed, lysozyme microcrystals were grown with the sitting-drop vapor-diffusion method, and microcrystals appeared after 6 h of incubation at 291 K. For plate B (hanging-drop vapor-diffusion method), the crystallization drops and reservoir solution used were the same as plate A. The side of the centerline of plate B also needs to be sealed with a Kapton membrane before the crystallization experiment, and then 0.8 μL of crystallization drops comprised of 0.4 μL of a protein sample and 0.4 μL of a reservoir solution were added to each large protein well and 0.4 μL of crystallization drops comprised 0.2 μL of a protein sample and 0.2 μL of a reservoir solution were added to each small protein well. The bottom of the crystallization plate was sealed with a Kapton membrane and 40 μL of reservoir solution was added to each reservoir well. Finally, plate B was assembled with the incubation chambers (Figure 1c,d) by covering or inserting, and lysozyme microcrystals were grown with the hanging-drop vapor-diffusion method. Microcrystals appeared after 6 h of incubation at 291 K.

2.3. Sample Loading

The bottom of the microplates were coated with glue and tightly fixed into the magnetic base (Figure 1a,b). For the sitting-drop vapor-diffusion experiment with plate A, the reservoir solution in the reservoir wells frost at cryogenic temperature and affect the protein wells, therefore, in our experiment, plate A was directly mounted on the goniometer head for data collection at room temperature. During the rotation of plate A, it is difficult for the reservoir solution and the crystallization drops to flow out due to atmospheric pressure and liquid tension in the actual test. For the hanging-drop vapor-diffusion experiment with plate B, we removed plate B from the incubation chambers (Figure 1c,d), and then directly added 0.2 μL of glycerol to each large protein well and added 0.1 μL of glycerol to each small protein well. Finally, the side away from the centerline was sealed with a Kapton membrane, and then plate B was mounted on the goniometer head for data collection at cryogenic temperature.

2.4. Data Collection

The experiment was carried out at the SSRF beamline BL18U1. The energy used for data collection was 12.662 keV, the photon flux at the sample point was 6×10^{11} phs/s, and the size of focused beam used for experiment was 20×20 μm^2.

Crystals in the microplates were aligned to beam position before collecting data at room temperature or cryogenic temperature. Firstly, the edge of the membrane was aligned to beam position at low magnification and then the microplate was rotated by 90° to locate the side of the centerline. Further steps were performed to bring the protein well to the beam and make the crystals align to beam position at high magnification. At room temperature, we used the optical method to locate lysozyme microcrystals due to high radiation damage, grid scanning was used to locate lysozyme microcrystals due to the protection of cryoprotectant at cryogenic temperature. The scanning area and the type of grid scanning were selected, BluIce system automatically scanned each grid point in the area with low-dose X-rays after setting the parameters, and then automatically calculated the initial diffraction value of each grid point and draw the diffraction pattern. Datasets were collected with a small wedge angle for each crystal.

Radiation damage is a factor that must be considered for data collection at room temperature or low temperature. X-rays will affect the life of the crystal. The life of the crystal is considered as the radiation dose that the crystal can receive without structural changes. Radiation damage is divided into overall radiation damage and local radiation damage. The overall radiation damage is not for a specific atom, it is mainly manifested in the diffraction pattern, which makes the resolution of protein crystals decrease, especially the high-resolution shell. Local radiation damage is the direct inelastic scattering of X-rays with the sample, through light absorption or Polly Compton scattering. Local radiation damage breaks some covalent bonds of protein molecules, such as disulfide bonds, which appear as a small group of atoms in the electron density map. Local radiation damage can cause researchers to misunderstand the crystal structure. The lifetime of the crystal depends on many factors, such as luminous flux density, wavelength, and protein sample composition (including molecular weight, heavy atom content). The absorption of X-ray by heavy atoms is higher than that of other atoms. Regarding the overall radiation damage, Owen et al. [22] proposed that the decrease in the diffraction quality of protein crystals produced by synchrotron radiation and the local changes in the protein structure can be measured by radiation dose. Local radiation damage depends on many factors, such as the folding of protein structures. This kind of radiation damage is difficult to judge from the diffraction quality, and can only be found by analyzing the structure. According to the radiation dose limit, we collected five diffraction images at room temperature and 40 diffraction images at cryogenic temperature. The exposure time for both was set to 0.5 s with a degree increment of 1°.

2.5. Data Processing and Analysis

Data collection at room temperature or cryogenic temperature required the use of multiple crystals to obtain a complete set of data. The crystals were screened on plates A and B, 20 crystals were selected at room temperature and 10 crystals were selected at cryogenic temperature for data collection. Each dataset was indexed by XDS [23], and then imported into the BLEND program [24] in CCP4 [25], and then we combined each dataset for integration and homogenization. We selected the best combination of data from several sets of data according to the quality of the data. The method of structure determination was adopted from the molecular replacement program CCP4, the search model of lysozyme was from the PDB1rcm of the protein database, and then Phenix [26] was used to refine the structure. The final model was modified by Coot [27].

3. Results and Discussion

3.1. Crystal Growth in Microplates

A typical result of crystal growth in microplate A with sitting-drop vapor-diffusion crystallization is shown in Figure 2a. Crystals can be clearly viewed from our beamline on-axis-video (OAV) system in Figure 2b. Conversely, crystal growth in microplate B with the hanging-drop vapor-diffusion method can also be achieved (Figure 2c,d). Using different desiccant in the reservoir well, we can grow different crystals in the microplates, which allows us to directly screen the crystallization conditions by the beamline video system. A typical result of the crystal growth under different crystallization conditions is shown in Figure 2e,f, which shows that crystallization conditions can be initially screened with microplate A and B under the microscope system (Figure 2e,f). Moreover, the suspension points or wells would be screened with grid scanning for further confirmation.

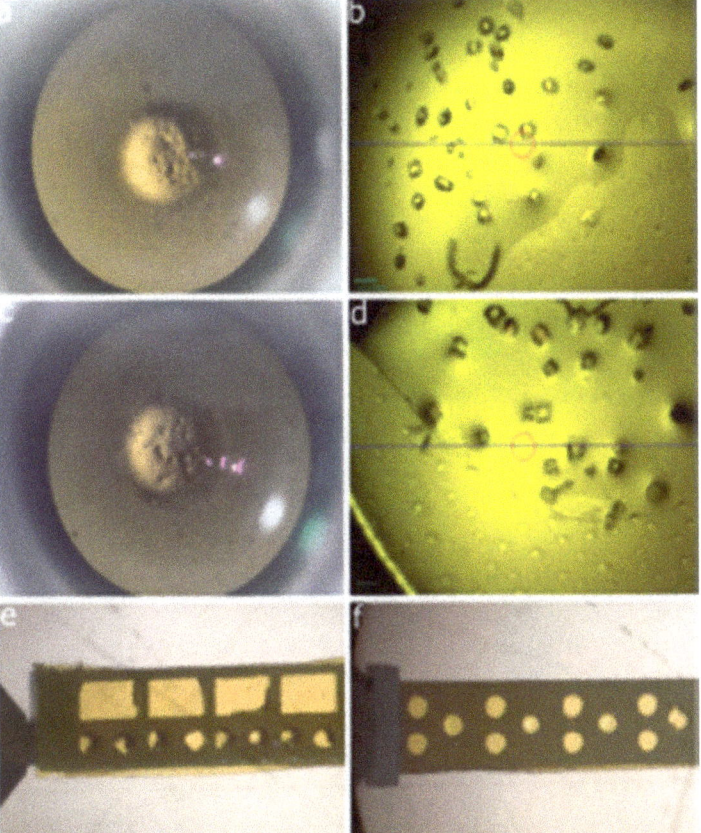

Figure 2. (**a**) Crystallization experiments, viewed under low magnification of the microscope, with plate A, the Kapton membrane is fixed on both sides of plate A to support crystallization experiments. (**b**) The growth of lysozyme microcrystals, viewed under high magnification of an on-axis-video system in a protein well of plate A, the length of microcrystal is less than 20 µm. (**c**) Crystallization experiments, viewed under low magnification of the microscope, with plate B, the Kapton membrane is fixed on the side where the centerline is of plate B to support crystallization experiments. (**d**) The growth of lysozyme microcrystals, viewed under high magnification of the on-axis-video system in a protein well of plate B. (**e**) The image of crystal growth under different crystallization conditions with plate A. (**f**) The image of crystal growth under different crystallization conditions with plate B.

3.2. In Situ Crystal Location

Once crystals were grown in microplate A or B, microplates were mounted on the goniometer head of the MD2 diffractometer for data collection, and it was important to quickly achieve precise alignment between the microcrystals and the incident X-rays. Unlike many other in situ data collection with traditional plates systems, our microplates can be moved into the beam position without adding any other motor systems. Moreover, crystals can be clearly focused to the beam position with the beamline centering system. Figure 3a,b shows the typical working position of microplates A and B at the beamline. However, it is extremely difficult to align the crystal to beam position with normal 96-well crystal plates, because normal sized crystal plates are difficult to rotate 90° to align the crystal to the beam position.

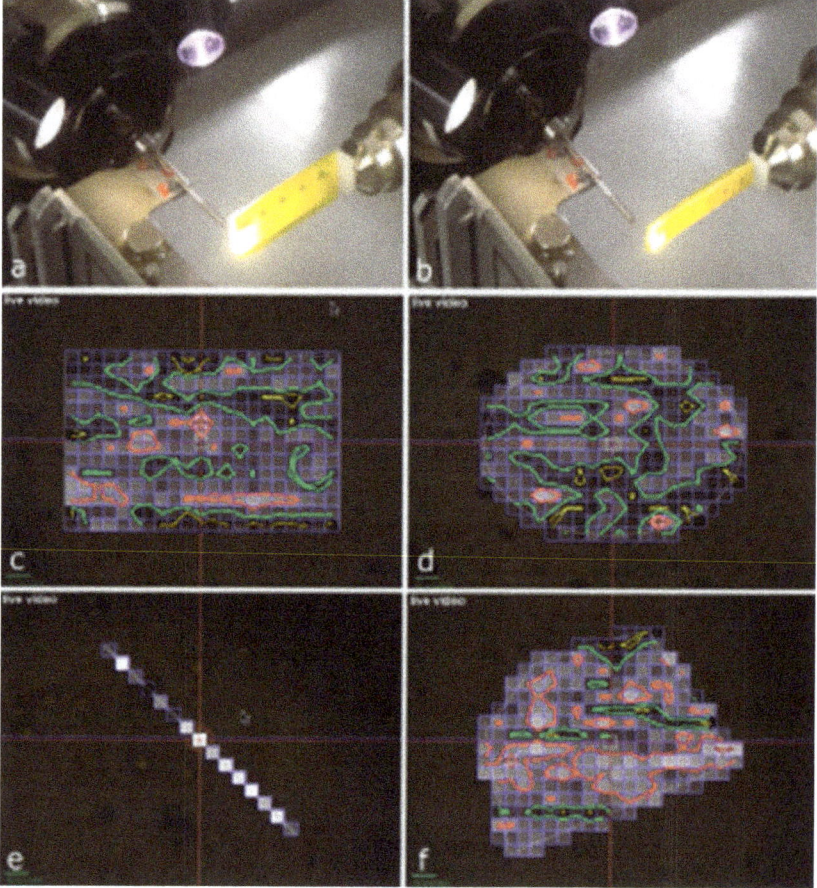

Figure 3. Microplates mounted on the goniometer head for data collection. (**a**) Plate A with 10 protein wells was installed on the goniometer head. (**b**) Plate B with 18 protein wells was installed on the goniometer head. Crystal location using different grid scanning types: (**c**) rectangle scanning area; (**d**) oval scanning area; (**e**) line scanning area; (**f**) polygon scanning area.

Crystal location results using grid scanning with plate B under cryogenic temperature are shown in Figure 3. Our results show that the crystal position, located on the Kapton membrane, can be easily identified by grid scanning. Four grid scanning types (rectangle, oval, line, polygon) can be used for plate B to promote the crystal location, the grid size, rotation wedge, and exposure time that can be defined by the user. The Kapton membrane does not affect the crystal location by grid scanning, moreover, to confirm that there is no effect of Kapton membrane on the crystal location, the effect of Kapton membrane is investigated in Section 3.3.

The combined use of the in situ diffraction plate and grid scan significantly improves the efficiency of diffraction data screening and collection. The common data collection of crystals requires each crystal to be separately installed on the goniometer, centered under the X-ray irradiation, and then removed and the sample changed. For this experiment, there are hundreds of lysozyme microcrystals in a protein well, and a set of plates can be loaded with thousands of crystals at the same time. Grid scanning can scan the entire protein well at once, and then it can quickly determine the position of all crystals in a protein well and the preliminary diffraction value of that position will be displayed.

Click the grid point with high diffraction value, and the system will automatically align it with the optical path, which can greatly reduce the time of sample exchange and positioning.

3.3. Background Scattering

Resolution is used as an important parameter for evaluating the quality of diffraction data and analyzing the effect of background scattering. The highest-resolution shells are determined using the following criteria: signal-to-noise ratio $[I/\sigma(I)] > 2$, redundancy > 2, completeness > 85%, and Rmerge < 1.

The background scattering of the crystal support material will affect the quality of the diffraction data of the protein crystal. When the background scattering of the support material is large, it will increase the noise points of the diffraction data, thereby masking the signal intensity of the diffraction points of the protein crystal. The selection of the material of the crystal support film requires special consideration: (1) High temperature resistance; high temperature will be generated after X-ray irradiation of the film material, which is likely to cause the film of some materials to be burnt and deformed. (2) High transparency; thin film materials with low light transparency are not conducive to the positioning of microcrystals. Thin films with high transparency can make the crystals visible on the microscope on the beamline station, which can efficiently locate the crystal position and facilitate data collection. (3) Chemical corrosion resistance; the crystal solution may contain corrosive components, or may easily chemically react with some thin film materials, resulting in the thin film materials being unusable. (4) Radiation resistance; synchrotron radiation X-ray has a high luminous flux, so the radiation dose is large, and the selected thin film material should be radiation resistant. At present, the commonly used films for serial crystallographic experiments based on fixed targets include Maylar film, polyimide film, synthetic cyclic olefin copolymer (COC), polycarbonate plastic, and other materials. Mylar film is a kind of polyester film with good light transmittance. The COC film is resistant to high temperature and chemical corrosion. Polyester film has a high melting point and good light transmission performance. The polyimide membrane, for example Kapton membrane, has many advantages and is best suited to all the requirements of our experiment, it is a kind of membrane with high temperature resistance, good light transmission performance, and chemical corrosion resistance. When the energy is 9–15 keV, the X-ray transmittance of 12.5 µm thick polyimide film can reach 99%, and it produces low background scattering under X-ray irradiation. Polyimide film has very good light transmittance, and microcrystals of approximately 10 µm are visible under the microscope. Therefore, it is very easy to locate the position of the crystal at the beamline station. Here, we only use Kapton membrane for our experiment.

In order to explore the influence of the background scattering of the in situ device including the crystallization drops and Kapton membranes on the quality of the protein crystal diffraction data, the air diffraction image (Figure 4a) and the in situ device diffraction image (Figure 4b) were collected respectively. Comparing the background scattering of air with the in situ device, it can be found that there is little difference between the two experiments (Figure 4a,b). In order to further prove the influence of the in situ device on the analysis of diffraction data, we collected the diffraction data of microcrystals by using the in situ device and a nylon loop. From the results of the resolution signal-to-noise ratio data (Figure 4c,d). It can be seen that there is little difference between the in situ device and the nylon loop. The in situ diffraction device has a major influence on the signal-to-noise ratio of the crystal diffraction data at approximately 4 Å, but the difference is basically negligible. The results show that the method of loading a microcrystal with the in situ device can also obtain high-quality diffraction data. The in situ device has lower background scattering and will not affect the quality of the crystal data.

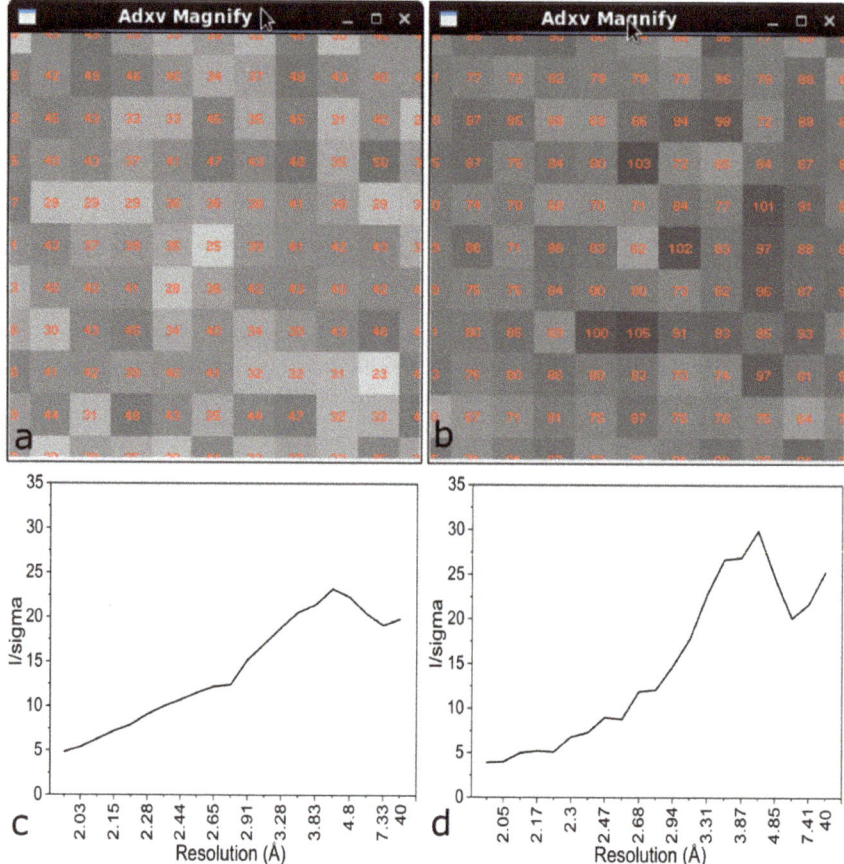

Figure 4. Background noise of air (**a**) and the in situ device (**b**) derived from the same coordinate point of the two diffraction patterns. (**c**) Signal-to-noise ratio in every resolution shell of the diffraction data obtained from a single lysozyme crystal in the nylon loop. (**d**) Signal-to-noise ratio in every resolution shell of the diffraction data obtained from multiple lysozyme crystals in the microplate.

3.4. Structure Determination with Microcrystals

We used lysozyme microcrystals (less than 20 μm) to verify the in situ microplates, and the data was collected at the BL18U beamline station of SSRF. When performing the experiments at room temperature, we used the visual method (Figure 2) to select 20 crystals from plate A for diffraction data collection. According to the radiation dose limit, five diffraction images were collected for each crystal (typical pattern in Figure 5b), and the relatively poor diffraction images were deleted, then each set of data was indexed with XDS. In the end, we selected five data sets with a total of 25 images, and these images were integrated through the blend program and then used for further structural analysis. According to the above-mentioned highest resolution criteria, we finally got the highest resolution of 2.15 Å. When performing the experiments at cryogenic temperatures, we used grid scanning (Figure 3) to select 10 crystals from plate B for diffraction data collection. A total of 40 diffraction images were collected from each crystal (typical pattern in Figure 5c). The data processing is the same as above. Finally, four data sets with 40 images were integrated through the blend program and then we got the highest resolution of 1.98 Å.

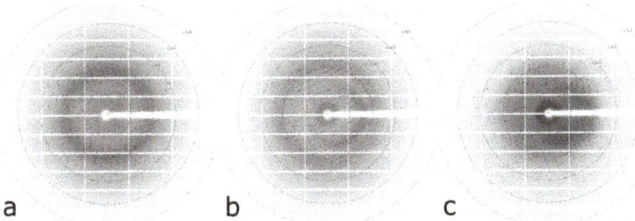

Figure 5. Typical diffraction pattern collected from crystal mounted by Nylon loop at 100 K (**a**) and the microplate A at room temperature (**b**) and the microplate B at 100 K (**c**).

In order to compare with the common method, the nylon loop was used for diffraction data collection. We used the same method to cultivate lysozyme microcrystals, and then the same data acquisition strategy was used to collect data from lysozyme at cryogenic temperature. According to the aforementioned data processing method, we finally integrated 60 diffraction images (typical pattern in Figure 5a) and obtained the highest resolution of 1.96 Å. The comparison of data collection through the in situ microplates and the data collection through the nylon loop is shown in Table 1.

Table 1. Statistical analysis of lysozyme using a nylon loop and the in situ plate for data collection. The data collection for multiple crystals was performed at a low temperature (100 K) and room temperature (RT), respectively.

	Nylon Loop (100 K)	Microplates (RT)	Microplates (100 K)
	Data collection		
Dominant size in sample (μm)	<20	<20	<20
Number of data sets	1	5	4
Number of images	60	25	40
Space group	P 4 2 2	P 4 2 2	P 4 2 2
	Unit cell		
a, b, c(Å)	79.33 79.33 36.94	79.21 79.21 37.83	79.80 79.80 36.96
a, b, c (°)	90.00 90.00 90.00	90.00 90.00 90.00	90.00 90.00 90.00
Energy (keV)	12	12	12
Resolution range (Å)	39.66–1.96	37.83–2.15	39.90–1.98
Number of unique reflections	8169	6796	7439
Completeness (%)	94.9(86.9)	97.5(95.9)	87.7(84.6)
Rmerge (%)	4.5(46.4)	10.8(88.0)	9.6(52.0)
$<I/\sigma(I)>$	19.8(4.8)	8.8(2.0)	25.3(3.9)
Redundancy	4.5(3.9)	3.6(2.9)	2.5(3.1)

In this experiment, we developed an efficient method for sample delivery and data collection for multicrystals. A Kapton membrane was utilized for crystal growth and sealing up the microplate. A complete dataset can be obtained after merging multiple datasets and the structure can be solved. Comparing the in situ microplates with the single nylon loop, the data collected using the in situ microplates at cryogenic temperature have the same good quality as the data collected using a single nylon loop, but the data collected using the in situ microplates at room temperature demonstrate worse resolution and signal-to-noise ratio, which is because the quality of crystals is affected at room temperature (Table 1). Moreover, electronic density comparation shows that there is no significant difference between results from nylon loop at 100 K and those from microplate B at 100 K. However, we do observe that there is a slight disappearance of electronic density obtained from microplate A at room temperature (F38 blue), compared with those from the nylon loop at 100 K (F38 green) or microplate B at 100 K (F38 red) (Figure 6). Our signal-to-noise analysis and electronic density analysis show that the quality of crystals is affected by radiation damage at room temperature. However, the discrepancy does not significantly affect the results.

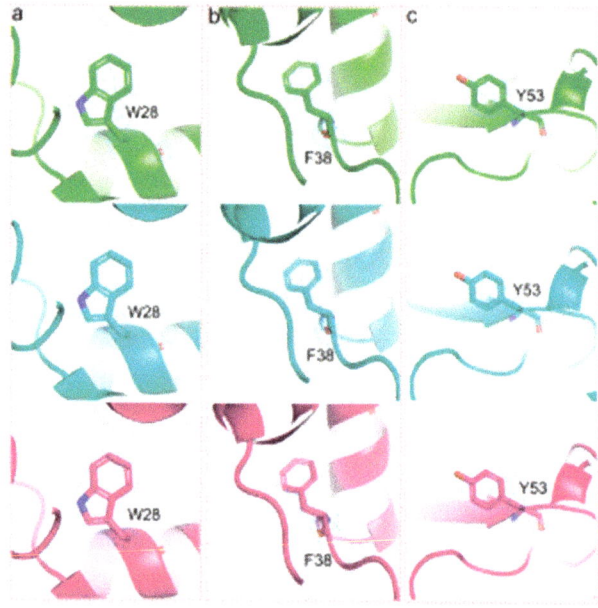

Figure 6. Enlargement of the three typical lysozyme residues and view of the extra electron density observed after partial refinement using the data set collected from crystal mounted by the Nylon loop at 100 K (green) and microplate A at room temperature (blue) and microplate B at 100 K (red). (**a**) View of Lysozyme W28 residue. (**b**) View of Lysozyme F38 residue. (**c**) View of Lysozyme Y57 residue.

In summary, this new method based on microplates realizes the in situ growth and simultaneous sample loading of multiple crystals, and it also realizes the rapid localization of crystals and the efficient data collection. On the premise that the in situ device has little effect on the data quality and low background scattering, we obtained comparable data quality to that of the traditional method with the nylon loop.

The in situ microplate including plate A and plate B has three advantages compared to the common commercially available in situ plate. First of all, except for the side parts of the microplate, the in situ microplate can almost perform a 360-degree data collection on the crystal without deviating from the optical path, and it can also be rotated at a full angle for rapid centering. The second is that the microplates are easy to manufacture and operate. The manual loading operation of the in situ microplate is the same as that of the ordinary nylon loop, which means that extra motors are not required, and it will not be limited by extra motors in most cases. The high-resolution structure of the protein can be obtained, and the manufacture of the in situ plate is fully customizable, which is suitable for most beamline stations. Third, the in situ microplates can support the cultivation of one protein under multiple crystallization conditions and the in situ cultivation of multiple proteins. This is not only suitable for the integration of multiple data sets of microcrystals, but also for multiple sets of data collection for multiple large crystals. Our microplate can save time for changing samples, and has the function of screening crystallization conditions.

However, the in situ microplates also have the following problems. When the in situ microplates are used to screen protein crystallization conditions, few crystallization conditions can be screened at a time, and it is only suitable for the fine screening of protein crystallization conditions. The size of these in situ microplates is affected by the limitation of the motor of the SSRF beamline BL18U1. In order to reduce costs, we use low-precision 3D printing technology and choose white resin materials, so far, we manually add protein samples to each well, therefore, the size of the protein wells are larger than chip device, the distance between the two protein wells is also longer than chip device. If the researcher needs to screen more protein crystallization conditions, the in situ microplates can also be designed to have more protein wells and crystallization chambers by adopting mechanical spotting or using higher-precision processing technology, but the design of the in situ microplates needs to meet requirements mentioned in the previous section.

Furthermore, the automation level of the fixed target serial crystallography method based on the thin film in this research is far from enough, and the automatic sample delivery and data collection of crystals is still not fully automatic. In the future, the automatic data collection of crystals should be further improved. At the same time, these methods have not yet resolved protein crystals of unknown structure. These methods should be used to further test protein crystals of unknown structure to improve these methods.

4. Conclusions

In recent years, there has been a growing interest in collecting X-ray diffraction data directly from the crystallization plates to eliminate the crystal harvesting step. The use of the microplates and the grid scanning at SSRF enables efficient data collection and crystallization condition screening. Hundreds of crystals can be mounted on the goniometer head simultaneously by microplates, which bypasses the tedious step of mounting single crystals by a user or robot, and microcrystals can be efficiently aligned to the beam position for data collection. Additionally, the experiments can be conducted at room temperature and cryogenic temperature. Furthermore, this method is particularly attractive for fragile or optically invisible microcrystals. Our microplates are suitable for data collection for solving structures of multiple crystals at room temperature or cryogenic temperature.

Author Contributions: Conceptualization, Z.W.; Data curation, M.L. and H.W.; Funding acquisition, J.H.; Methodology, M.L. and L.Y.; Project administration, Q.W. and J.H.; Resources, B.S., H.Z., F.Y., Q.W. and J.H.; Software, Z.W.; Supervision, L.Y., B.S., H.Z., F.Y. and Q.W.; Validation, M.L.; Visualization, H.W.; Writing—original draft, M.L.; Writing—review and editing, Z.W. All authors have read and agreed to the published version of the manuscript.

Funding: This research was funded by The National Key Research and Development Program of China (Grant No. 2017YFA0504901).

Acknowledgments: The authors thank the teams at the BL18U1 beamline of the Shanghai Synchrotron Radiation Facility for the beamtime allocated to this project and for their help in experiments.

Conflicts of Interest: The authors declare no conflict of interest.

References

1. Mcpherson, A. In situ X-ray crystallography. *J. Appl. Crystallogr.* **2010**, *33*, 397–400. [CrossRef]
2. Heymann, M.; Opthalage, A.; Wierman, J.L.; Akella, S.; Szebenyi, D.M.E.; Gruner, S.M.; Fraden, S. Room-temperature serial crystallography using a kinetically optimized microfluidic device for protein crystallization and on-chip X-ray diffraction. *IUCrJ* **2014**, *1*, 349–360. [CrossRef] [PubMed]
3. Li, L.; Mustafi, D.; Fu, Q.; Tereshko, V.; Chen, D.L.L.; Tice, J.D.; Ismagilov, R.F. Nanoliter microfluidic hybrid method for simultaneous screening and optimization validated with crystallization of membrane proteins. *Proc. Nat. Acad. Sci. USA* **2006**, *103*, 19243–19248. [CrossRef] [PubMed]
4. Perry, S.L.; Guha, S.; Pawate, A.S.; Henning, R.; Kosheleva, I.; Šrajer, V.; Kenis, P.J.A.; Ren, Z. In situserial Laue diffraction on a microfluidic crystallization device. *J. Appl. Crystallogr.* **2014**, *47*, 1975–1982. [CrossRef]
5. Kisselman, G.; Qiu, W.; Romanov, V.; Thompson, C.M.; Lam, R.; Battaile, K.P.; Pai, E.F.; Chirgadze, N.Y. X-CHIP: An integrated platform for high-throughput protein crystallization and on-the-chip X-ray diffraction data collection. *Acta Crystallogr. Sect. D Boil. Crystallogr.* **2011**, *67*, 533–539. [CrossRef]
6. Bingel-Erlenmeyer, R.; Olieric, V.; Grimshaw, J.P.A.; Gabadinho, J.; Wang, X.; Ebner, S.G.; Isenegger, A.; Schneider, R.; Schneider, J.; Glettig, W.; et al. SLS Crystallization Platform at Beamline X06DA—A Fully Automated Pipeline Enablingin SituX-ray Diffraction Screening. *Cryst. Growth Des.* **2011**, *11*, 916–923. [CrossRef]

7. Pinker, F.; Brun, M.; Morin, P.; Deman, A.-L.; Chateaux, J.-F.; Oliéric, V.; Stirnimann, C.; Lorber, B.; Terrier, N.; Ferrigno, R.; et al. ChipX: A Novel Microfluidic Chip for Counter-Diffusion Crystallization of Biomolecules and in Situ Crystal Analysis at Room Temperature. *Cryst. Growth Des.* **2013**, *13*, 3333–3340. [CrossRef]
8. Pineda-Molina, E.; Daddaoua, A.; Krell, T.; Ramos, J.L.; Delgado-López, J.M.; Gavira, J.A. In situ X-ray data collection from highly sensitive crystals of Pseudomonas putida PtxS in complex with DNA. *Acta Crystallogr. Sect. F Struct. Boil. Cryst. Commun.* **2012**, *68*, 1307–1310. [CrossRef]
9. Yadav, M.K.; Gerdts, C.J.; Sanishvili, R.; Smith, W.W.; Roach, L.S.; Ismagilov, R.F.; Kuhn, P.; Stevens, R.C. In situ data collection and structure refinement from microcapillary protein crystallization. *J. Appl. Crystallogr.* **2005**, *38*, 900–905. [CrossRef]
10. Solvas, X.C.I.; Demello, A. Droplet microfluidics: Recent developments and future applications. *Chem. Commun.* **2011**, *47*, 1936–1942. [CrossRef]
11. Maeki, M.; Yoshizuka, S.; Yamaguchi, H.; Kawamoto, M.; Yamashita, K.; Nakamura, H.; Miyazaki, M.; Maeda, H. X-ray diffraction of protein crystal grown in a nano-liter scale droplet in a microchannel and evaluation of its applicability. *Anal. Sci.* **2012**, *28*, 65. [CrossRef] [PubMed]
12. Roedig, P.; Duman, R.; Sanchez-Weatherby, J.; Vartiainen, I.; Burkhardt, A.; Warmer, M.; David, C.; Wagner, A.; Meents, A. Room-temperature macromolecular crystallography using a micro-patterned silicon chip with minimal background scattering. *J. Appl. Crystallogr.* **2016**, *49*, 968–975. [CrossRef] [PubMed]
13. Mueller, C.; Marx, A.; Epp, S.W.; Zhong, Y.; Kuo, A.; Balo, A.R.; Soman, J.; Schotte, F.; Lemke, H.T.; Owen, R.L.; et al. Fixed target matrix for femtosecond time-resolved and in situ serial micro-crystallography. *Struct. Dyn.* **2015**, *2*, 054302. [CrossRef]
14. Murray, T.D.; Lyubimov, A.Y.; Ogata, C.M.; Vo, H.; Uervirojnangkoorn, M.; Brünger, A.T.; Berger, J.M. A high-transparency, micro-patternable chip for X-ray diffraction analysis of microcrystals under native growth conditions. *Acta Crystallogr. Sect. D Boil. Crystallogr.* **2015**, *71*, 1987–1997. [CrossRef] [PubMed]
15. Huang, C.Y.; Olieric, V.; Ma, P.K.; Panepucci, E.; Diederichs, K.; Wang, M.; Caffrey, M. In meso in situ serial X-ray crystallography of soluble and membrane proteins. *Acta Crystallogr. Sect. D Boil. Crystallogr.* **2015**, *71*, 1238–1256. [CrossRef]
16. Roedig, P.; Vartiainen, I.; Duman, R.; Panneerselvam, S.; Stübe, N.; Lorbeer, O.; Warmer, M.; Sutton, G.; Stuart, D.I.; Weckert, E.; et al. A micro-patterned silicon chip as sample holder for macromolecular crystallography experiments with minimal background scattering. *Sci. Rep.* **2015**, *5*, 10451. [CrossRef]
17. Zarrine-Afsar, A.; Barends, T.R.M.; Müller, C.; Fuchs, M.R.; Lomb, L.; Schlichting, I.; Miller, R.J.D. Crystallography on a chip. *Acta Crystallogr. Sect. D Boil. Crystallogr.* **2012**, *68*, 321–323. [CrossRef]
18. Li, B.; Huang, S.; Pan, Q.Y.; Li, M.-J.; Zhou, H.; Wang, Q.-S.; Yu, F.; Sun, B.; Chen, J.-Q.; He, J.-H. New design for multi-crystal data collection at SSRF. *Nucl. Sci. Tech.* **2018**, *29*, 21. [CrossRef]
19. Baxter, E.L.; Aguila, L.; Alonso-Mori, R.; Barnes, C.O.; Bonagura, C.A.; Brehmer, W.; Brünger, A.T.; Calero, G.; Caradoc-Davies, T.T.; Chatterjee, R.; et al. High-density grids for efficient data collection from multiple crystals. *Acta Crystallogr. Sect. D Struct. Boil.* **2016**, *72*, 2–11. [CrossRef]
20. Michalska, K.; Tan, K.; Chang, C.; Li, H.; Hatzos-Skintges, C.; Molitsky, M.; Alkire, R.; Joachimiak, A. In situ X-ray data collection and structure phasing of protein crystals at Structural Biology Center 19-ID. *J. Synchrotron Radiat.* **2015**, *22*, 1386–1395. [CrossRef]
21. McPhillips, T.M.; McPhillips, S.E.; Chiu, H.J.; Cohen, A.E.; Deacon, A.M.; Ellis, P.J.; Garman, E.; Gonzalez, A.; Sauter, N.K.; Phizackerley, R.P.; et al. Blu-Ice and the Distributed Control System: Software for data acquisition and instrument control at macromolecular crystallography beamlines. *J. Synchrotron Radiat.* **2002**, *9*, 401–406. [CrossRef] [PubMed]
22. Owen, R.L.; Rudino-Pinera, E.; Garman, E.F. Experimental determination of the radiation dose limit for cryocooled protein crystals. *Proc. Natl. Acad. Sci. USA* **2006**, *103*, 4912–4917. [CrossRef]
23. Krug, M.; Weiss, M.S.; Heinemann, U.; Mueller, U. XDSAPP: A graphical user interface for the convenient processing of diffraction data using XDS. *J. Appl. Crystallogr.* **2012**, *45*, 568–572. [CrossRef]
24. Foadi, J.; Aller, P.; Alguel, Y.; Cameron, A.D.; Axford, D.; Owen, R.L.; Armour, W.; Waterman, D.G.; Iwata, S.; Evans, G. Clustering procedures for the optimal selection of data sets from multiple crystals in macromolecular crystallography. *Acta Crystallogr. Sect. D Boil. Crystallogr.* **2013**, *69*, 1617–1632. [CrossRef]
25. Collaborative, C.P. The CCP4 suite: Programs for protein crystallography. *Acta Crystallogr. Sect. D Biol. Crystallogr.* **1994**, *50*, 760. [CrossRef]

26. Adams, P.D.; Grosse-Kunstleve, R.W.; Hung, L.W.; Ioerger, T.R.; McCoy, A.J.; Moriarty, N.W.; Read, R.J.; Sacchettini, J.C.; Sauter, N.K.; Terwilliger, T.C. PHENIX: Building new software for automated crystallographic structure determination. *Acta Crystallogr. Sect. D Boil. Crystallogr.* **2002**, *58*, 1948–1954. [CrossRef] [PubMed]
27. Emsley, P.; Cowtan, K. Coot: Model-building tools for molecular graphics. *Acta Crystallogr. Sect. D Biol. Crystallogr.* **2004**, *60*, 2126–2132. [CrossRef] [PubMed]

© 2020 by the authors. Licensee MDPI, Basel, Switzerland. This article is an open access article distributed under the terms and conditions of the Creative Commons Attribution (CC BY) license (http://creativecommons.org/licenses/by/4.0/).

Article

The Steps from Batchwise to Continuous Crystallization for a Fine Chemical: A Case Study

Christian Melches, Hermann Plate, Jürgen Schürhoff and Robert Buchfink *

GEA Messo GmbH, 47229 Duisburg, Germany; Christian.Melches@gea.com (C.M.); Hermann.Plate@gea.com (H.P.); Juergen.Schuerhoff@gea.com (J.S.)
* Correspondence: Robert.Buchfink@gea.com; Tel.: +49-2065-903-327

Received: 15 May 2020; Accepted: 22 June 2020; Published: 24 June 2020

Abstract: Many processes to produce fine chemicals and precursors of pharmaceuticals are still operated in batchwise mode. However, recently, more producers have taken a change to continuous operation mode into consideration, performing studies and trials on such a change, while some have even already exchanged their production mode from batchwise to continuous operation. In this paper, the stepwise development from an initial idea to industrial implementation via laboratory testing and confirmation is revealed through the example of an organic fine chemical from the perspective of a crystallization plant manufacturer. We begin with the definition of the objectives of the project and a brief explanation of the advantages of continuous operation and the associated product properties. The results of the laboratory tests, confirming the assumptions made upfront, are reported and discussed. Finally, the implementation of an industrial plant using a draft tube baffled (DTB) crystallizer and the final product properties are shown. Product properties such as crystal size distribution, crystal shape, related storage stability and flowability have successfully been improved.

Keywords: fine chemicals; continuous crystallization; crystal shape; process design; DTB crystallizer; scale up

1. Introduction

The worldwide demand for pharmaceuticals, food and feed additives and their precursors is growing due to a growing global population and demographic changes. Crystallization is a major unit operation with regards to the separation and especially the purification of products of the pharmaceuticals, food and fine chemicals industry. Nowadays, many of these chemicals are produced in a batchwise operation mode [1]. The main disadvantages of batchwise operations are the innate system batch-to-batch variability and a lower process efficiency compared to continuous crystallization processes [2,3]. Schaber et al. [4] found savings of 9–40% of the production costs using continuous crystallization processes. In relation to expiring patents, competitiveness requires optimized process design with regards to operational costs and investments and/or beneficial product properties like crystal size, crystal size distribution, crystal shape and therefore product storage stability and free-flowing ability.

In general, several main requirements exist with regards to crystallization processes, which partly influence one another (see Figure 1). For the crystallization of APIs, (Active Pharmaceutical Ingredients) additional requirements like polymorphism and chirality may exist, which are not applicable for the fine chemical examined within this case study.

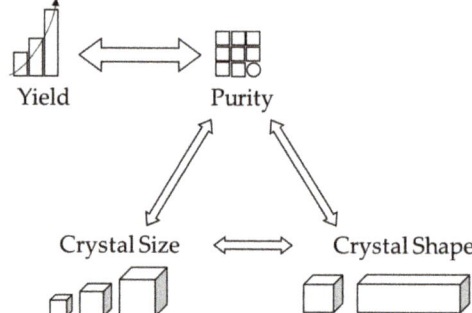

Figure 1. Main requirements for industrial crystallization processes.

From plant manufacturers or engineering companies' perspectives, all these requirements are defined by the customer or by the market in which the product is used and the crystallization process as well as the chosen equipment is to be designed to meet those requirements.

The phase diagram is thermo-dynamically fixed; however, side compounds or impurities are well known to have an influence on the solubility of organic products [5]. The solubility of the organic substance in this case study was suppressed by increasing the impurity concentrations expressed by the concentration factor shown and discussed further in Section 3.1.1.

Continuous evaporative crystallization processes with recycled mother liquor are mainly characterized by the concentration factor α, which is, hereafter, defined as the ratio of the final impurity concentration in the purge c^i_{Purge} and initial impurity concentration in the feed c^i_{Feed}.

$$\alpha = \frac{c^i_{Purge}}{c^i_{Feed}} \tag{1}$$

The yield of a process and the purity of the product show opposing trends and, unless one applies an additional process step or technology, increasing both at the same time is not possible [6]. Therefore, it is crucial to understand whether both requirements—yield and purity—can be fulfilled at the same time for a single-stage process or if an additional step needs to be added. Different process options like first crop–second crop or re-crystallization will be discussed in Section 4 of this paper.

The crystal size and crystal size distribution are other important properties for crystalline products that could be affected by retention time, temperature and impurity concentrations [6,7]. Most of the studies for the crystal sizes derived from continuous crystallization processes are performed using Mixed Suspension and Mixed Product Removal (MSMPR) crystallizers [8–10]; however, for the crystallization of inorganic substances, different crystallizer types have been developed in order to increase the crystal size.

The draft tube baffled (DTB) crystallizer, as a Cleared Suspension Mixed Product Removal (CSMPR) type was designed to increase the crystal size by limiting the mechanical energy input to the suspension (secondary nucleation limitation) and by an efficient crystal fines destruction in the outer heating circuit [11].

Another important requirement is the shape of the product crystals. The crystal shape determines the major properties like bulk density, dust formation, storage ability and free-flowing ability on the one hand, and directly influences the purity of a crystalline product by changing the final moistures of continuous separation, e.g., by centrifugation, on the other [12]. Differences in crystal shape result from the different growth rates of the specific faces of a crystal [13]. It is well known from the literature that even traces of impurities could change the crystal shape by adsorption to specific faces of a crystal for inorganic [14] as well as organic [15] products. The theoretical basis of this will not be discussed further due to our focus on the industrial implementation of the results; however, reference is made to the literature discussing the main theories of impurity-induced change in crystal shape [13,16].

A further major advantage of using a DTB crystallizer for a continuous crystallization process is the possibility of influencing crystal shape next to crystal size and crystal size distribution. By applying an adequate retention time, it is possible to mechanically shape the crystals by abrading their edges. The resulting fines of such desired secondary nucleation are redissolved in the outer heating circuit. Comparable considerations are taken into account by Kwon et al. using a fines trap, which is comparable to the clarification zone of the DTB crystallizer [17].

Comparable objectives were defined for the case study presented here, which was elaborated in cooperation with a well-known international chemical producer. The main objective was a change from existing batchwise crystallization to a continuous crystallization process.

Furthermore, the possibility of improving the above listed product properties was part of this study. In particular, the former market product showed a strong tendency to build agglomerates during storage and transport, which was related to the broad crystal size distribution and the elongated crystal shape of the product (see Figure 2). While changing the process from batchwise to continuous operation, an improvement in product properties was a major reason why we chose a proper crystallizer type and specific, well-defined process parameters.

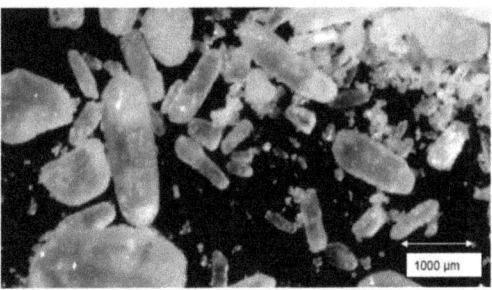

Figure 2. Optical characterization of commercial market product (broad crystal size distribution and elongated crystal shape).

Intense laboratory trials were performed in the GEA Messo GmbH (Duisburg, Germany) in-house research and development center using original feed samples supplied by the production facilities. A two-step approach for laboratory development was applied.

The initial step involves multi-stage batchwise evaporation/crystallization trials to observe the effect of increasing concentration factors α on important physical and chemical parameters like densities and boiling point elevations, on the solubility of the product substance and on the crystal size/crystal size distribution. The second step comprises continuous crystallization tests in a bench-scale DTB crystallizer, applying process parameters defined based on the results of step 1 before fixing all relevant process parameters and scaling up to the industrial plant.

In particular, for continuously operated crystallization plants aiming at a high yield, the effect of the accumulating impurities present in the feed solution are crucial for both process and product design and, as such, were tested. It was observed that the accumulating impurities have an effect on the solubility of the product's substance, while also changing the shape of the crystals from cubic shapes to increasingly needle-like shapes. In particular, for continuous crystallization processes, a balance must be found between the yield of the process, defined by the final purge, on the one hand, and the required product purity, which also includes properties like crystal size or crystal shape, on the other.

2. Materials and Methods

2.1. Discontinuous Multi-Stage Crystallization

The experimental setup for the discontinuous multi-stage crystallization (see Figure 3a) consists of a double-jacketed round-bottom durane glass beaker with a volume of two liters. The energy for the evaporation is transferred by using hot water, which is provided by an external thermostat circuit (Julabo MA12, Seelbach, Germany). A top-mounted motor-driven stirrer (IKA Eurostar 100 control, Stauffen, Germany) is used for the adequate mixing of the process liquor and the crystals. The condensate section consists of a glass-made surface condenser with an intermediate receiver, "Anschütz-Thiele", and a graduated receiver. A membrane vacuum pump (Gardner Denver, Ilmenau, Germany) provides the necessary pressure underneath and regulates the operating pressure and therefore the temperature in the evaporation chamber.

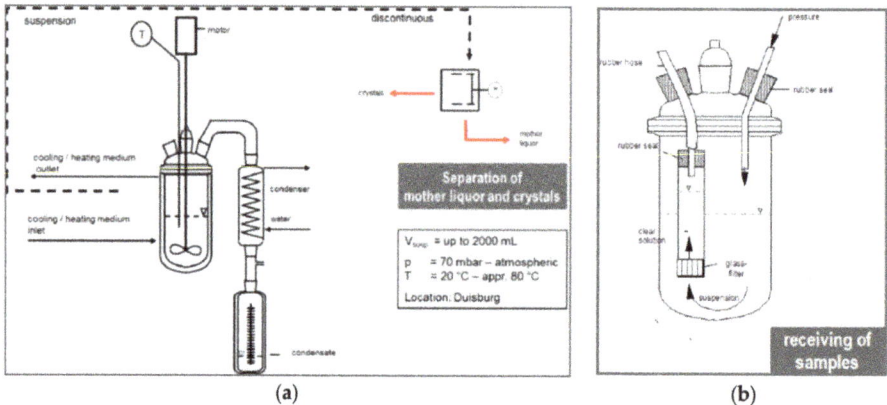

Figure 3. (a) Experimental setup for discontinuous multi-stage crystallization tests; (b) sampling procedure.

The process temperature of 60 °C was monitored by an integrated temperature probe (PT-100) and was adjusted by the corresponding pressure defined by the vacuum pump. After reaching saturation, a defined amount of seed crystals was added to the solution and evaporative crystallization continued at a constant temperature until the desired concentration factor α was reached. For de-supersaturation purposes, the suspension was further agitated for 1h at the selected process temperature.

Subsequently, a mother liquor sample was taken through a glass filter by pressure filtration (see Figure 3b) and, finally, the suspension was separated into wet crystals and mother liquor by centrifugation, applying the conditions summarized in Table 1.

Table 1. Parameters of solid–liquid separation by centrifugation used for the laboratory test (lab sieve drum centrifuge SIEVA2, Hermle Labortechnik GmbH, Wehingen, Germany.

Parameter	Value	Unit
Drum size diameter	140	mm
Paper filter (inlet)	PP/Heidland	-
Pore size	7	µm
Separation time	120	s
Rotation speed	5500	rpm
G-force	2200	g

2.2. Continuous Crystallization in Bench-Scale Draft Tube Baffled (DTB) Crystallizer

In order to ensure the proper transfer of the findings of our discontinuous tests into the design of the industrial plant, an intermediate confirmation step was executed, using continuous crystallization in a lab-scale DTB crystallizer. The general setup is shown in Figure 4.

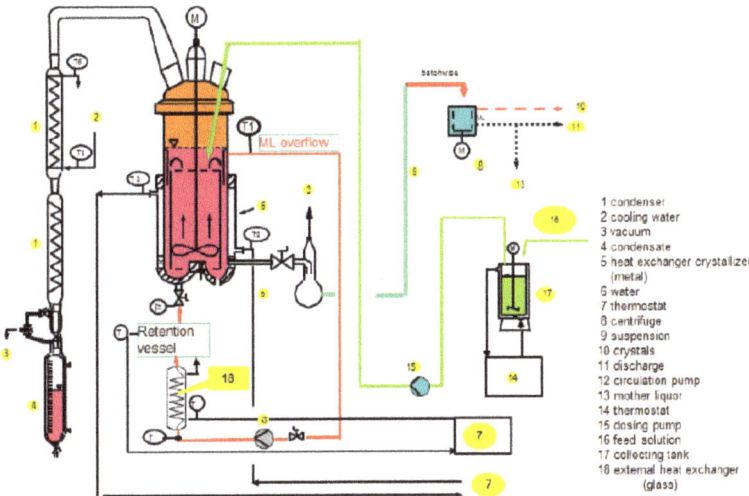

Figure 4. Experimental setup for continuous crystallization tests (laboratory scale draft tube baffled (DTB) crystallizer).

The DTB crystallizer is a Cleared Suspension Mixed Product Removal (CSMPR) type, which is a specific crystallizer type developed to produce coarser crystals compared to Mixed Suspension and Mixed Product Removal (MSMPR) type crystallizers such as, e.g., the Forced Circulation (FC) crystallizer.

The bench-scale DTB crystallizer used for the continuous test work has essentially the same setup as the industrial unit that will be applied; however, it is built of glass. The main circulation is realized by an agitator (top-mounted) within the central pipe. The mother liquor overflow is taken from an internal clarification zone and contains mainly fine crystals generated by secondary nucleation. The mother liquor is transported by an external circulation pump through the heat exchanger, which introduces the required heat for evaporation provided by a thermostat. Due to the temperature and therefore the solubility increase, the fines are dissolved within the mother liquor before re-entering the crystallizer. The setup of the condensation system is equal to that used for the discontinuous tests described in Section 2.1.

The feed is tempered by a thermostat and is continuously added to the top of the crystallizer, while the suspension is removed discontinuously from the bottom of the crystallizer by inducing a proper vacuum. The solid–liquid separation by centrifugation is performed with equal parameters, as described in Section 2.1.

Optical characterization was done using microscope Zeiss Photomikroskop II (Oberkochen, Germany).

The process solution was prepared according to heat and mass balance, which were based on the results of the discontinuous multi-stage crystallization trials. After setting the temperature and pressure, seed crystals were introduced and the continuous operation was maintained for at least 10 hours.

3. Results

3.1. Discontinuous Multi-Stage Crystallization

The main purpose of the discontinuous multi-stage crystallization trials was to observe the effect of increasing the concentration factor, α, on the most important parameters listed below, which were used to design the industrial process and equipment:

- The solubility of the product (relevant for process yield);
- The purity of the product (> requested product purity);
- The crystal shape of the product;
- The boiling point elevations, densities and viscosities (not shown here).

3.1.1. Effect of Concentration Factor α on Solubility of Product Substance

With increasing concentration factors, the concentration of impurities or side compounds in the mother liquor increase accordingly (if not co-crystallizing or volatile). This normally has an impact on the solubility of the product. Figure 5 shows the observed depressive effect on product solubility of an increasing concentration factor.

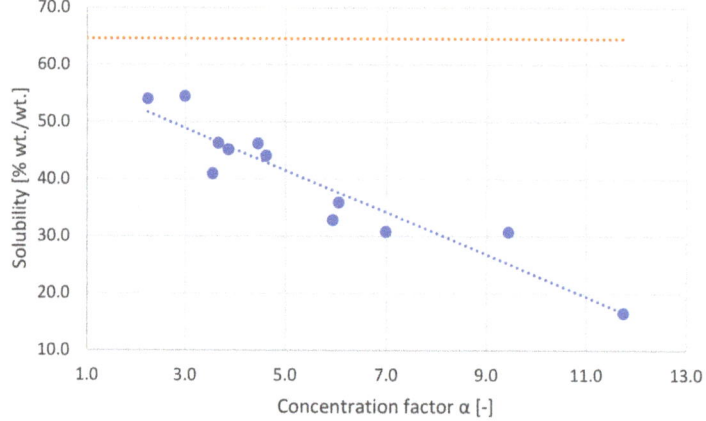

Figure 5. Solubility of the product as function of the concentration factor of soluble impurities/byproducts during multi-stage crystallization at 60 °C in aqueous solution (blue) and the solubility of the product in pure water at 60 °C (orange).

The solubility of the product is decreasing linearly in this matrix, with an increasing concentration factor corresponding to the concentration of impurities/byproducts. This information is crucial to close the overall heat and mass balance, as it affects the potential process yield. The extrapolated curve is not going to affect the solubility of the pure product substance in water, as the feed solution used for the tests contained impurities (feed solution is determined by $\alpha = 1$). No data between $1 < \alpha < 2$ are available, as the feed solution was undersaturated and needed to be pre-concentrated before any crystallization take place.

3.1.2. Effect of Concentration Factor α on the Purity of the Product

To maximize the yield of a continuous crystallization process, the concentration factor should be set as high as possible. However, with an increasing level of impurities, the achievable purity decreases. The requested purity on a dry basis was defined as >99.3%.

A critical concentration factor could be observed, above which the requested purity could no longer be achieved in a single-stage process. Under these apparent conditions, and considering the washing of the crystal cake with pure water in a centrifuge (15% compared to the solids), the critical concentration factor was identified to be ~3.8 (Figure 6).

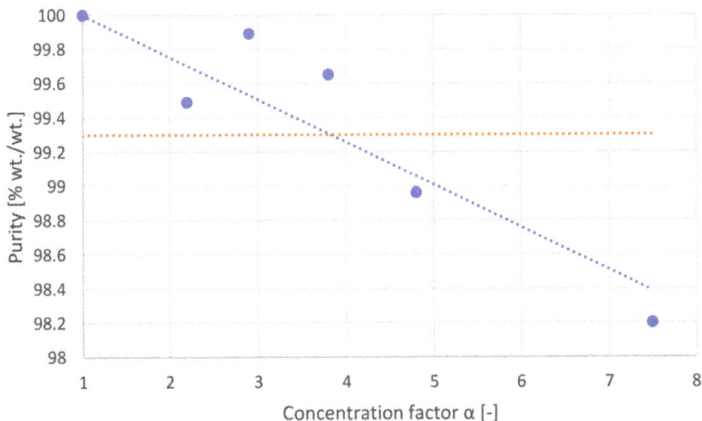

Figure 6. Purity of the product as function of the concentration factor of soluble impurities/byproducts after multi-stage crystallization with crystal washing (15% compared to solids) (blue) and the requested purity (orange).

3.1.3. Effect of Concentration Factor α on Crystal Shape

The main source of impurities within the product are derived from the adherent mother liquor, which could be reduced to a certain extent by the washing of the crystal cake in a centrifuge. The amount of adherent mother liquor is defined by the physical parameters of the solution, like density and (especially) viscosity on the one hand, and on the crystal size, shape and surface conditions of the solid crystals on the other.

Further to its effect on the final moisture of the cake, the shape of the crystalline product is very important for parameters such as the tendency to build up dust, free-flowing ability and storage stability.

It was observed that an increasing impurity level led to a change in the crystal shape from a compact shape to an increasingly needle-like shape (Figure 7a–d). As the tests were executed with the original feed solution, containing a defined matrix of impurities, no comparison to the product crystallized from a pure solution is available. Identifying the responsible impurity from the matrix of different impurities was not within the scope of the test, as this is fixed by upstream processes and cannot easily be adjusted.

The needle-like shape of the product crystals creates various difficulties, as follows:

- Undesired crystal shape, storage ability and angle of repose;
- Increased crystal breakage during solid–liquid separation, drying and bagging (dust formation);
- Increased moisture content after solid–liquid separation (negative impact on crystal purity) [18].

It was observed that a concentration factor of $\alpha = 3$ already led to an elongated crystal shape at a still sufficient purity. The elongated shape is the defining parameter, with regards to limiting the concentration factor of the industrial unit.

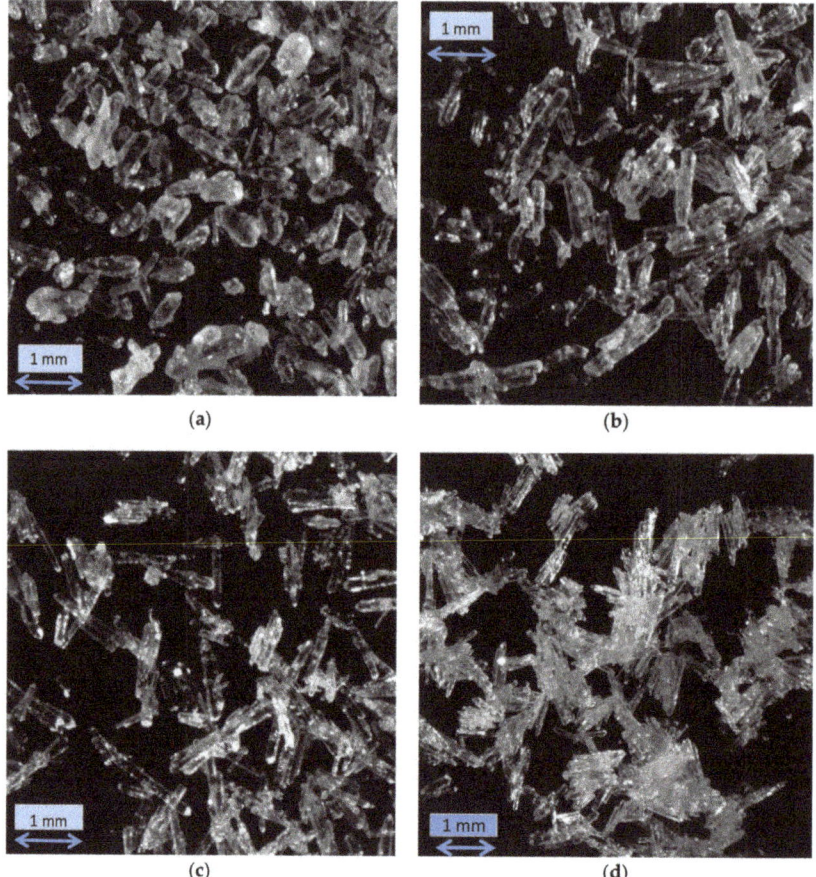

Figure 7. (**a**) Crystals after washing for a concentration factor of two. (**b**) Crystals after washing for a concentration factor of three. (**c**) Crystals after washing for a concentration factor of four. (**d**) Crystals after washing for a concentration factor of five.

3.2. Continuous Crystallization in DTB Crystallizer

Based on the results of the multi-stage discontinuous crystallization trials, the final heat and mass balance, as well as the crystallizer design, were defined, to serve as a basis for the continuous crystallization tests. Due to the negative impact of the present impurities on the crystal shape, and due to the required product purity, the concentration factor within the continuous crystallization was limited to 2.5 for the continuous tests.

The major parameters to be confirmed are the achievement of the crystal size, crystal size distribution and the purity of the final product.

The purity of the crystalline product (after washing with 15% water based on the solid amount) on a dry basis was analyzed to be above 99.7%. This is in accordance with the limit of 99.3% purity on a dry basis, which was defined by the customer.

Three major product aspects were observed for the product from the continuous laboratory-scale DTB crystallizer, which are as follows:

- The size of the product crystals increased to d´ > 1400μm (see Figure 8) without nucleation (no cycling behavior expected for the industrial DTB at this crystal size [19]), while d´ is defined as $d_{36.8}$;
- The increased retention time resulted in a rounded crystal shape (the rounding effect was induced mainly by the attrition caused by the internal circulation pump);
- The dissolving of fine crystals (produced by attrition) in the outer heating circuit resulted in a narrower crystal size distribution.

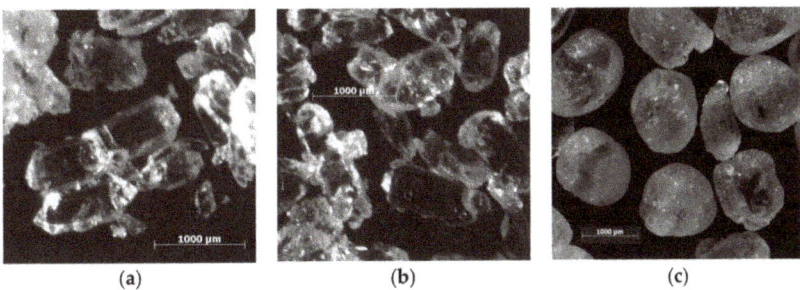

Figure 8. (a) Crystals after 1.5h of continuous crystallization in bench scale DTB (d´ = 956μm); (b) crystals after 6h (d´ = 1149 μm); (c) crystals after 10h (d´ = 1435 μm).

3.3. Implementation of Industrial DTB Crystallizer Unit

After the confirmation of the initial findings by continuous crystallization trials, the process, including the heat and mass balance, the design of equipment and further relevant basic engineering deliverables, was elaborated by GEA Messo GmbH. The industrial-scale continuous crystallization plant was then successfully installed at the customer's site (Figure 9).

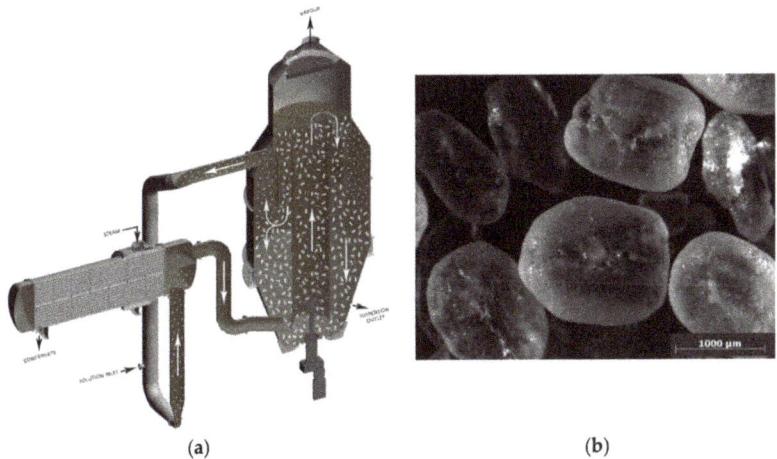

Figure 9. (a) Scheme of an industrial DTB crystallizer, by GEA Messo GmbH. (b) Crystals from the industrial DTB crystallizer (crystal size d´ = 1980 μm).

The performance of the crystallization plant was fully achieved with regards to product purity as well as product yield on the one hand, and product parameters like crystal size and crystal size distribution on the other.

4. Discussion

A variety of the literature deal with the heat and mass balance of industrial crystallization processes [13,16,20,21].

Figure 10 shows a schematic and simplified mass balance for an evaporative continuous crystallization process with an exemplary concentration factor of 10. Both the yield of the process as well as the purity of the product are a function of the concentration factor α. For the chosen example, the dependencies of the process yield and product purity are shown in Figure 11. Depending on the requirements defined by a producer or the market, only a certain working range for α is acceptable because, otherwise, the yield would be too low ($\alpha < \alpha_{critical}$) on the one hand, or the purity would be to low ($\alpha > \alpha_{critical}$) on the other.

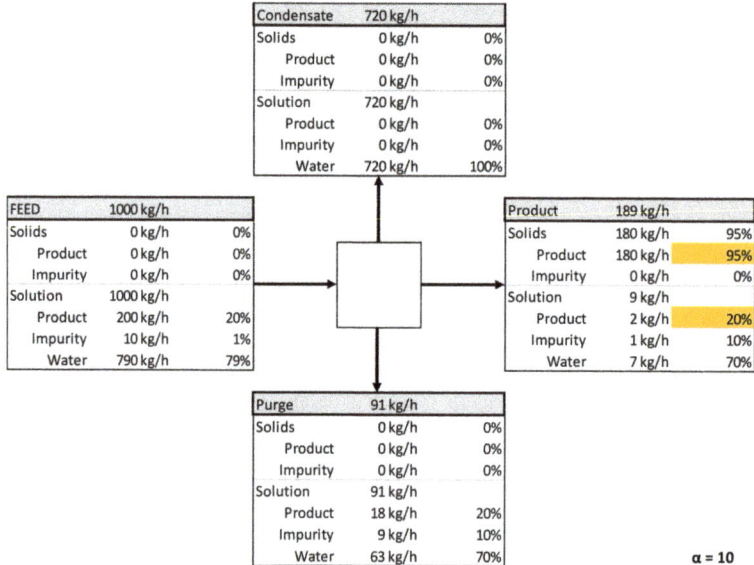

Figure 10. Simplified block balance of an evaporative crystallization process (concentration factor of 10).

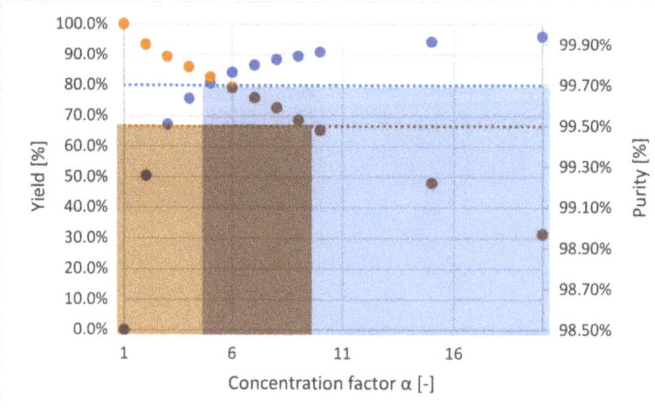

Figure 11. Theoretical definition of a concentration factor range to cover both a requested purity of the final product (orange curve) and a requested yield of the process (blue curve) (from $\alpha = 5 \ldots 10$).

From the results of the multi-stage discontinuous tests, it can be clearly seen that the concentration factor would need to be limited to maximum $\alpha = 4$ to avoid endangering the requested purity of 99.3% (on a dry basis). However, due to the strong effect on crystal shape (needle-like shape), it was decided that we would keep the concentration factor α even lower, below 2.5.

This limits the process yield for single-stage processes, which was not crucial for the case study presented here, as the client could recycle the purge from crystallization into an upstream process outside of the battery limit of the crystallization.

When there is no overlap for the concentration factor with regards to yield and purity, and intensive cake washing within solid–liquid separation is not enough to achieve a critical yield, there are, nevertheless, various concepts to achieve such a critical yield.

Two major concepts should be introduced here: the so-called first crop–second crop concept and the re-crystallization concept [21].

For the first crop–second crop concept, the product is crystallized to a certain concentration factor in the first crop crystallization, respecting the critical value of α with regards to purity. In order to increase the yield of the overall process, the mother liquor of the first crop crystallization is further concentrated, and an out-of-specification product is generated in the second crop crystallization step. This impure product is dissolved in the feed solution (if undersaturated) or by the addition of a solvent, and is recycled for the first crop crystallization. The final overall process purge is taken from the second crop crystallization.

For the re-crystallization concept, an out-of-specification product (raw product) is produced by an initial crystallization step, applying high concentration factors. The product is subsequently totally dissolved in the solvent and completely re-crystallized in a second crystallization step, producing a product with the requested purity (pure product). The purge is taken from the first raw crystallization step.

From plant manufacturing or engineering companies' perspectives, the concentration factor α is the most crucial factor for continuous evaporative crystallization processes, as it enables the balance of the outer process.

$$\dot{m}_{Feed} = \dot{m}_{Product} + \dot{m}_{Condensate} + \dot{m}_{Purge} \qquad (2)$$

The concentration factor, as one possible dimensionless characterization of the grade of concentration, is defined between one, standing for no evaporation and therefore no crystallization occurrence, and infinity, standing for a complete evaporation of all solvents, leaving a dry and solid product next to the solvent removed by evaporation.

By increasing the concentration factor, the concentration of impurities increases in the mother liquor of the crystallization, which causes several effects in relation to our case study:

1) The solubility of the product linearly decreases with the increasing concentration factor (the concentration of impurities in the mother liquor). This is a well-known thermodynamic effect, which is documented for many inorganic and organic substances. No further parameters known to influence the solubility, like temperature or pH-value, were investigated during this study, while the specific species in the impurity spectrum, which are responsible for the observed effect, were not identified.

2) The purity of the product linearly decreases with the increasing concentration factor, which was expected according to the theoretical approach discussed above. Similar results were presented by Alvarez et al., introducing a distribution coefficient (DC), defining the ratio of impurities in the product and the impurities in the mother liquor, showing a linear dependency [10].

3) With an increasing concentration factor and therefore an increasing concentration of impurities in the mother liquor, the shape of the product crystals changed to an increasingly needle-like shape. This is related to the interaction of the impurities with the specific surface of the crystal, changing its growth rate. Theoretically, those slow-growing faces determine the shape of the product crystals, as the fastest growing surfaces will disappear over time [21].

It must be highlighted here that all parts of the study were carried out to gain specific required knowledge on the relevant process and product parameters that will enable an engineering company to scale up this process to an industrial plant. Neither a complete parameter study, nor repetitions satisfying statistical approaches, could be performed.

5. Conclusions

The pathway from an initial idea to industrial implementation was successfully shown for an organic model substance, containing the following main steps:

- The initial idea to change from batchwise-operated to continuously operated crystallization, assessing the feasibility of producing a desired product with regards to crystal size, crystal size distribution and crystal shape;
- Discontinuous multi-stage crystallization trials to identify the major physical and chemical parameters, checking the effect of the increasing concentration factor on product purity, crystal shape and physical and chemical parameters (e.g. solubility), alongside the definition of the critical concentration factor;
- Continuous crystallization trials using a laboratory DTB crystallizer to confirm the findings of the discontinuous tests and, finally, to generate the mass and heat balance, as well as the design of the industrial crystallizer unit;
- The implementation of the industrial unit at a customer production site, based on the results of the above described laboratory tests.

This stepwise approach was successfully performed for the given model project, and for several other organic products, as well as inorganic products. Furthermore, the demonstrated stepwise approach could be used as a blueprint for any other crystallization project involving a change from batchwise to continuous operation mode, by considering specific product properties like crystal shape or crystal size distribution, or to increase the yield of an existing crystallization process.

Author Contributions: Conceptualization, C.M., H.P. and R.B.; methodology, H.P.; validation, C.M., H.P. and J.S.; formal analysis, C.M., H.P.; investigation, H.P., J.S.; writing—original draft preparation, C.M., H.P. and R.B.; writing—review and editing, C.M., H.P. All authors have read and agreed to the published version of the manuscript.

Funding: This research received no external funding.

Conflicts of Interest: The authors declare no conflict of interest.

References

1. Sen, M.; Rogers, A.; Singh, R.; Chaudhury, A.; John, J.; Ierapetritou, M.G.; Ramachandran, R. Flowsheet optimization of an integrated continuous purification-processing pharmaceutical manufacturing operation. *Chem. Eng. Sci.* **2013**, *102*, 56–66. [CrossRef]
2. Chen, J.; Sarma, B.; Evans, J.M.B.; Myerson, A.S. Pharmaceutical Crystallization. *Cryst. Growth Des.* **2011**, *11*, 887–895. [CrossRef]
3. Plumb, K. Continuous processing in the pharmaceutical industry changing the mindset. *Chem. Eng. Res. Des.* **2006**, *83*, 730–738. [CrossRef]
4. Schaber, S.; Gerogiorgis, D.I.; Ramachandran, R.; Evans, J.M.B.; Barton, P.I.; Trout, B.L. Economic Analysis of Integrated Continuous and Batch Pharmaceutical Manufacturing: A Case Study. *Ind. Eng. Chem. Res.* **2011**, *50*, 10083–10092. [CrossRef]
5. Capellades, G.; Wiemeyer, H.; Myerson, A.S. Mixed-Suspension, Mixed-Product Removal Studies of Ciprofloxacin from Pure and Crude Active Pharmaceutical Ingredients: The Role of Impurities on Solubility and Kinetics. *Cryst. Growth Des.* **2019**, *19*, 4008–4018. [CrossRef]
6. Variankaval, N.; Cote, A.S.; Doherty, M.F. From form to function: Crystallization of active pharmaceutical ingredients. *AIChE J.* **2008**, *54*, 1682–1688. [CrossRef]

7. Vetter, T.; Burcham, C.L.; Doherty, M.F. Regions of attainable particle sizes in continuous and batch crystallization processes. *Chem. Eng. Sci.* **2014**, *106*, 167–180. [CrossRef]
8. Zhang, D.; Xu, S.; Du, S.; Wang, J.; Gong, J. Progress of Pharmaceutical Continuous Crystallization. *Engineering* **2017**, *3*, 354–364. [CrossRef]
9. Su, Q.; Nagy, Z.; Rielly, C. Pharmaceutical crystallization processes from batch to continuous operation using MSMPR stages: Modelling, design, and control. *Chem. Eng. Proc.* **2015**, *89*, 41–53. [CrossRef]
10. Alvarez, A.J.; Singh, A.; Myerson, A.S. Crystallization of Cyclosporine in a Multistage Continuous MSMPR Crystallizer. *Cryst. Growth Des.* **2011**, *11*, 4392–4400. [CrossRef]
11. Wöhlk, W.; Hofmann, G. Types of crystallizers. *Int. Chem. Eng.* **1984**, *24*, 419–431.
12. Garg, J.; Arora, S.; Garg, J. Spherical crystallization: An overview. *Int. J. Pharm. Technol.* **2014**, *4*, 1909–1928.
13. Myerson, A.S. *Handbook of Industrial Crystallization*, 2nd ed.; Butterworth-Heinemann: Woburn, MA, USA, 2002.
14. Buchfink, R. Effects of impurities on an industrial crystallization process of ammonium sulfate. Ph.D. Thesis, Martin-Luther-University, Halle (Saale), Germany, 2 May 2011.
15. Winn, D.; Doherty, M.F. Modeling crystal shapes of organic materials grown from solution. *AIChE J.* **2000**, *46*, 1348–1367. [CrossRef]
16. Mullin, J.W. *Crystallization*, 3rd ed.; Butterworth-Heinemann: Oxford, UK, 1993.
17. Kwon, J.S.-I.; Nayhouse, M.; Christofides, P.D.; Orkoulas, G. Modeling and control of crystal shape in continuous protein crystallization. *Chem. Eng. Sci.* **2014**, *107*, 47–57. [CrossRef]
18. Wakeman, R. The influence of particle properties on filtration. *Sep. Purif. Rev.* **2007**, *58*, 234–241. [CrossRef]
19. Hofmann, G.; Wang, S.; Widua, J.; Wöhlk, W. Zyklische Korngrößenschwankungen in Massenkristallisatoren. In Proceedings of the Fachausschuss Kristallisation, Strasbourg, France, 27–28 March 2000. (In German).
20. Hofmann, G. *Kristallisation in der Industriellen Praxis*; Wiley-VCH: Weinheim, Germany, 2004. (In German)
21. Beckmann, W. *Crystallization Basic Concepts and Industrial Applications*; Wiley-VCH: Weinheim, Germany, 2013.

© 2020 by the authors. Licensee MDPI, Basel, Switzerland. This article is an open access article distributed under the terms and conditions of the Creative Commons Attribution (CC BY) license (http://creativecommons.org/licenses/by/4.0/).

Article

Systematic Investigations on Continuous Fluidized Bed Crystallization for Chiral Separation

Erik Temmel [1,2], Jonathan Gänsch [1,*], Andreas Seidel-Morgenstern [1,3] and Heike Lorenz [1]

1. Max Planck Institute for Dynamics of Complex Technical Systems, Sandtorstraße 1, 39106 Magdeburg, Germany; seidel@mpi-magdeburg.mpg.de (A.S.-M.); lorenz@mpi-magdeburg.mpg.de (H.L.)
2. Sulzer Chemtech Ltd., Gewerbestraße 28, 4123 Allschwil, Switzerland; erik.temmel@sulzer.com
3. Otto von Guericke University Magdeburg, Institute of Process Engineering, 39106 Magdeburg, Germany
* Correspondence: gaensch@mpi-magdeburg.mpg.de

Received: 27 April 2020; Accepted: 11 May 2020; Published: 14 May 2020

Abstract: A recently developed continuous enantioseparation process utilizing two coupled fluidized bed crystallizers is systematically investigated to identify essential correlations between different operation parameters and the corresponding process performance on the example of asparagine monohydrate. Based on liquid phase composition and product crystal size distribution data, it is proven that steady state operation is achieved reproducibly in a relatively short time. The process outputs at steady state are compared for different feed flow rates, supersaturations, and crystallization temperatures. It is shown that purities >97% are achieved with productivities up to 40 g/L/h. The size distribution, which depends almost exclusively on the liquid flow rate, can be easily adjusted between 260 and 330 µm (mean size) with an almost constant standard deviation of ±55 µm.

Keywords: fluidized bed; continuous; preferential crystallization; chiral separation; racemate resolution; enantiomer; asparagine monohydrate

1. Introduction

Recently, continuous crystallization has again become the focus of many research activities. Efforts have been made to develop design rules based on comparing it with the corresponding batch process [1,2] or to elucidate basic mechanisms, for example, the impurity incorporation and carryover [3]. Additionally, novel concepts, applying to example slug-flow [4], oscillatory baffled [5,6], Couette-Taylor [7,8] or fluidized bed crystallizers [9–12], as well as columns with static mixers [1] or coiled flow inverters [13], have been developed lately. These concepts exploiting tubular crystallizers in general are not superior nor applicable for every substance system or separation problem. They have specific benefits and drawbacks [1] compared to the well-known mixed-suspension, mixed-product removal (MSMPR) concept but have given new input to the topic of continuous crystallization and have widened the field of application for the food, agriculture, and pharmaceutical industries.

Fluidized bed crystallizers (FBCs) have recently been successfully exploited in wastewater treatment to recover phosphate [9,10], sulfates and magnesia [11], or boron [12] from diluted aqueous solutions. All of these studies aimed at a full conversion of the pollutant ions in the liquid phase to form a salt and utilized afterward the surface of fluidized, inert particles to collect the fine precipitate. FBCs can be also applied, however, for separations where the substance of interest has a similar or the same concentration as the impurity, as is the case for enantioseparations [14].

This process is then based on Preferential Crystallization, which is often applied as an efficient and inexpensive option for the production of pure enantiomers from the racemic, i.e., 50:50, mixture [2,15]. The classical variant is, nevertheless, an unstable process carried out in the metastable zone of the respective ternary substance system, consisting in this case of two enantiomers, D and L, dissolved in a solvent. Seeds of the desired enantiomer are utilized to selectively remove this species from the liquid phase. Hence, the dissolved mass of the seeded component decreases while the mass fraction of the counter-enantiomer increases. Consequently, the supersaturation of the counter-enantiomer also increases in most cases, which forces this species to nucleate after a certain induction time. A complete separation with the maximum yield is, hence, only achievable if the process is interrupted at the right time shortly before the nucleation.

To increase robustness, productivity, and yield of the classical variant, two opposite Preferential Crystallizations can be carried out in two vessels only connected via a continuous liquid phase exchange [16]. In this way, the selective removal of the enantiomers in the respective crystallizer is counterbalanced and a racemic liquid phase composition maintained, which reduces the risk of nucleation.

The application of fluidized bed crystallizers together with the coupled Preferential Crystallization principle, as shown in Figure 1, provides two additional benefits. Firstly, the flow rate of the feed entering the columns from the bottom drags small crystals out of the process at the top. Hence, contamination of the product can be avoided to a certain extend even if the counter enantiomer nucleates since the nuclei follow the liquid flow. Secondly, a product classification can be achieved by a conical shape of the columns. Then the liquid velocity profile varies over the height of the crystallizers and only a specific crystal size is present at the product outlet. This process allows, hence, for a continuous production of pure enantiomers with a high productivity and a specific, adjustable product crystal size. In previous studies, the feasibility of steady-state operation of this process was shown without considering the process dynamics [17] and only based on the liquid phase composition for a specific operation point [14,17]. However, no systematic investigation also based on the periodically harvested product crystals has been conducted.

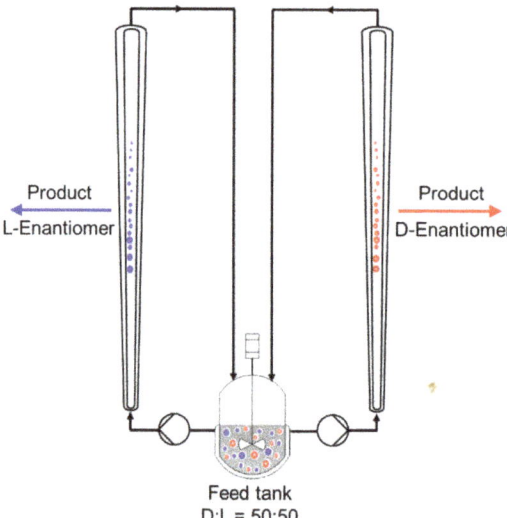

Figure 1. Principle scheme of a coupled Preferential Crystallization carried out in two fluidized bed crystallizers for the separation of the D- and L-enantiomer from the racemic (50:50) mixture provided in the feed tank (adapted from [2]).

In this work we, first, prove the achievability of steady-state operation on the example of D-/L-asparagine monohydrate based on the liquid phase composition as well as the periodically harvested product crystal size distribution. Furthermore, the required startup time to reach the steady-state will be estimated for different operation conditions and the high reproducibility of the results will be demonstrated. On this basis, consecutively the dependence of productivity, purity, and yield on the process conditions, feed flow rate, supersaturation, and crystallization temperature, will be systematically studied.

2. Materials and Methods

2.1. Substances

For the fluidized bed crystallization experiments, racemic asparagine monohydrate (rac Asn·H$_2$O) and the respective enantiopure D- and L-Asn·H$_2$O were purchased from Sigma Aldrich (Purity >99%, Steinheim, (Baden-Württemberg), Germany). Both were used without further treatment. Deionized water (Millipore, Milli-Q Advantage A10) was used as solvent and for washing of the crystalline product together with ethanol, which was purchased from VWR Chemicals (Purity >99.7%, Fontenay-sous-Bois, France).

Solubility data of the D-/L-Asn·H$_2$O/ water system was mainly determined in [18]. The accuracy was confirmed by additional measurements in [19] where also the parameters of polynomial (Equation (1)) were estimated to describe the saturation concentration, x_{sat}, as a function of temperature and composition. As shown in Figure 2, asparagine monohydrate forms a conglomerate, as a requirement for the application of Preferential Crystallization directly from a supersaturated racemic liquid phase.

$$x_{sat,i}\left(T, \frac{x_j}{x_{Solvent}}\right) = 0.0104 + 1.0584 \times 10^{-4} \cdot T + 2.4432 \times 10^{-5} \cdot T^2 + 0.0312 \cdot \frac{x_j}{x_{Solvent}}. \quad (1)$$

where

$$T = \text{temperature } [°C]$$

$$x = \text{mass fraction } [-]$$

for

$$i = \text{D-Asn·H}_2\text{O, L-Asn·H}_2\text{O and } j = \text{D-Asn·H}_2\text{O, L-Asn·H}_2\text{O} \neq i$$

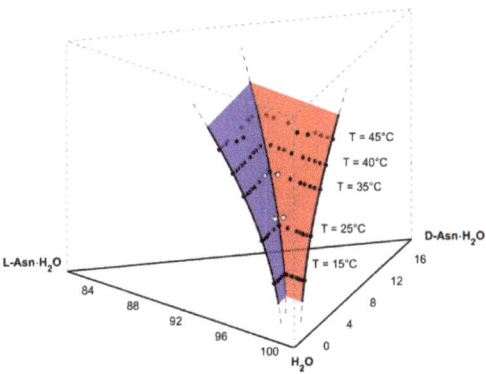

Figure 2. Upper 20% section of the ternary phase prism of the asparagine monohydrate enantiomers in water at different temperatures. All axes are given in mass fractions × 100 (wt%). Black and white dots—solubility data from [18] and [19], respectively. Red and blue surface—fitted solubility surface of D-Asn·H$_2$O and L-Asn·H$_2$O, respectively. Reprinted with permission from [19]. Copyright 2020 American Chemical Society.

The selective removal of the seeded enantiomer during Preferential Crystallization leads to an altering of the liquid phase composition during the course of the separation process. Hence, the driving forces of both enantiomers change and a different supersaturation calculation also need to be applied.

Thus, Equation (2) is utilized in the present study to describe the supersaturation, S, as a function of temperature and composition.

$$S_i(T, x_i) = \frac{x_i}{x_{sat,i}\left(T, \frac{x_j}{x_{Solvent}}\right)} \quad (2)$$

for

$$i = \text{D-Asn·H}_2\text{O, L-Asn·H}_2\text{O and } j = \text{D-Asn·H}_2\text{O, L-Asn·H}_2\text{O} \neq i$$

2.2. Experimental Setup

All separation experiments were performed in the setup shown in Figure 3. The plant consisted of two tubular crystallizers (C1 and C2, Figure 3) with an individual volume of approximately 0.5 L and a total height of 1130 mm. Each double-jacketed tubular crystallizer was composed of a conical section in the lower part and a cylindrical section in the upper part of the columns. Thus, the fluid velocity of a liquid phase passing through the crystallizers changes with the height of the columns due to the increasing diameter in the conical section. At the transition between the upper and lower section, where the diameter is the largest, the fluid velocity becomes constant again over the height of the cylindrical part.

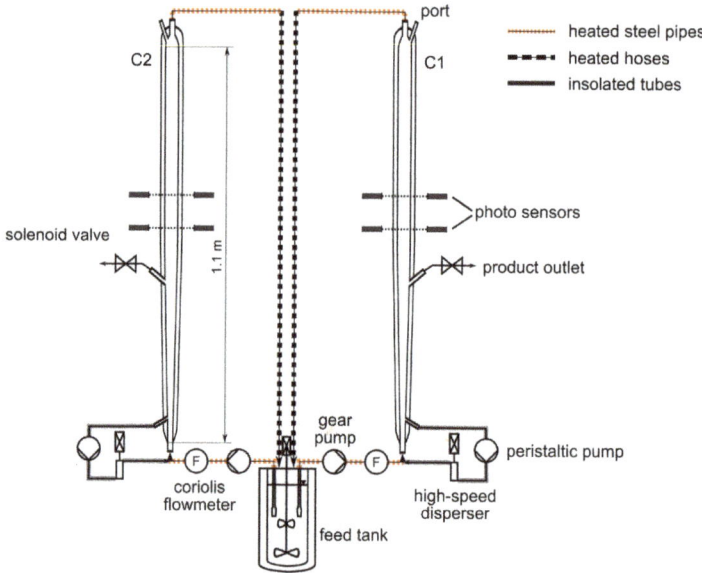

Figure 3. Process scheme with the main components of the utilized continuous fluidized bed plant.

The temperature inside the columns was measured at the inlet, in the middle, and at the outlet via resistance thermometers (Pt100s). Temperature control was ensured by two thermostats connected to the respective middle Pt100s. In this study, the middle temperature is referred to as the crystallization temperature for the sake of simplicity, even though a temperature gradient will be present over the height of the columns due to the fluid flow.

Both crystallizers were coupled via heated steel pipes and flexible heated hoses with a double-jacketed feed tank on the scale of 7.5 L. A propeller-type stirrer was utilized in this tank during the separation experiments to suspend a certain excess mass of racemic solid phase in a solution saturated at a defined initial temperature. This excess of solid phase served to re-saturate the depleted mother liquors from the crystallizers during the process to ensure constant initial conditions at the inlet of the columns. The temperature within the feed tank was again measured by a Pt100 and controlled by a thermostat.

Clear saturated racemic solution was withdrawn during the process from the feed tank via glass filters and was pumped continuously from the bottom through the tubular crystallizers using gear pumps. The rotation speed of these gear pumps was controlled by Coriolis mass flowmeters to ensure a constant volumetric flowrate.

To harvest the product, each column had a product outlet at a height of 365 mm, close to the end of the conical section. The product removal was realized periodically due to the scale of the plant, using solenoid valves, which were controlled by photoelectric barriers as described in the next section.

Due to the mild mixing conditions of the particulate phase, secondary nucleation will not be sufficient to counterbalance the loss of crystals of the periodically continuous product removal. Hence, a continuous seeding strategy was necessary, which also supports Preferential Crystallization by providing enantiopure crystals. Thus, the largest particles present at the bottom of the crystallizer columns were withdrawn via peristaltic pumps into a bypass. They were pumped, subsequently, through high-speed dispersers, which were utilized as mills working on the rotor-stator principle. Afterward, the ground crystals were fed back as seed material to the process.

2.3. Experimental Procedures and Operation Parameters

In the following the experiments performed are described with their operation parameters and objectives, and performance parameters for process evaluation are introduced (Section 2.3.1). Since starting and operation procedure is crucial for achieving a successful continuous racemate resolution, the procedure along each experiment, especially the crystallization progress and the periodical product removal, are explained and the used analytics are depicted (Section 2.3.2)

2.3.1. Study of Operation Parameters and Process Evaluation

Altogether, seven experiments were planned and carried out for 8 h in this first attempt to systematically investigate the influence of the operation conditions on the process performance. In continuation of the previous study [14], a saturation temperature of 35 °C was chosen for most of the experiments (T_{sat}, Table 1). The first three processes (Exps. 1–3, Table 1) were used to determine the operation window of the process with respect to the crystallization temperature (T_{crys}, Table 1).

Table 1. Process conditions and objectives of all experiments.

Exp.	T_{sat} [°C]	T_{crys} [°C]	F [L/h]	Initial Seed Crystals Sieve Fraction [µm]	m_{Seed} [g]	Objective of Investigation
1	35	27	10	90–125	6.28	steady-state, operation window
2	35	30	10	212–300	15	steady-state, operation window
3	35	31, 32	10	90–125	6.28	operation window
4	35	30	10, 12	212–300	15	volumetric flowrate, residence time
5	35	30	12, 14	212–250	15	volumetric flowrate, residence time
6	24.6	20	12	250–355	20	crystallization temperature
7	35	30	12	250–355	20	steady-state

The data of Exp. 2 were additionally utilized to determine the required time to reach steady-state operation. It was found that after two product withdrawals (process time approximately 2 h) all process characteristics became constant. Hence, some of the later experiments were split into two periods, where two sets of process parameters (compared, for example, to the volume flow of Exp. 4) were tested for 4 h each, to enhance the time efficiency of the investigation. Exps. 4 and 5 were carried out to evaluate stepwise the influence of the volume flow rate (F, Table 1) between 10 and 14 L/h. In

Exp. 6, the supersaturation was kept constant but the saturation and crystallization temperature were reduced by approximately 10 K to investigate the influence of reduced crystallization kinetics. In Exp. 7 the central test point was run again, T_{sat} = 35 °C, T_{crys} = 30 °C, and F = 12 L/h, to evaluate the standard deviation of all process characteristics for a longer time period.

Every reached operation point was evaluated based on the normalized volume related product crystal size distributions, q_3 (Equation (3)), their mean sizes, $\overline{L_3}$ (Equation (4)), and their respective standard deviation, s_{L3} (Equation (5)), as well as the achieved yields, Y, and productivities, Pr (Equations (6) and (7), respectively) calculated from the product masses and the time window between two withdrawals.

$$q_3(z_k) = \frac{\mu_k}{\Delta z_k} \tag{3}$$

$$\overline{L_3} = \sum_{k=1}^{N} z_k \cdot \mu_k \tag{4}$$

$$s_{L3} = \sqrt{\sum_{k=1}^{N} \left(z_k - \overline{L_3}\right)^2 \cdot \mu_k} \tag{5}$$

where

μ_k = mass fraction of sieve class k [-]
z_k = characteristic length of sieve class k [μm]
Δz_k = width of sieve class k [μm]

$$Y = \frac{m_{prod}}{m_{theo}} \tag{6}$$

$$Pr = \frac{m_{prod}}{\Delta t \cdot V_{tot}} \tag{7}$$

where

m_{prod} = mass of withdrawn product crystals [g]
m_{theo} = theoretical maximum of product mass [g]
Δt = time window between two withdrawals [h]
V_{tot} = total volume of tubular crystallizer [L]
(C1 = 0.478 L, C2 = 0.511 L)

2.3.2. Start and Operation Procedure, Crystallization Progress, and Analytics Used

The starting point of all experiments was the preparation of the feed suspension. Therefore, the solvent was filled into the feed tank and circulated via the gear pumps through the columns and via the peristaltic pumps through the comminution bypasses. The required racemic solute mass for saturating the aqueous solution at the desired temperature of each experiment (Table 1, T_{sat}) plus 200 g racemic excess solid was afterward filled in the feed tank, and the suspension heated to T_{sat}, which was kept constant throughout the whole experiments. To avoid any risk of nucleation within the columns during the preparation step, the coolant temperature of the double jackets was set to one Kelvin above T_{sat}. After the mother liquor within the feed tank reached T_{sat}, the coolant temperature of the double jackets was set to one Kelvin below T_{sat}. In parallel, the heating hoses were set to 40 °C to prevent any nucleation or clocking within the tubes: all other steel pipes (Figure 3) were appropriately heat traced to prevent clocking. The liquid flow was stopped for seeding each column from the top (Figure 3, port) with a certain mass (Table 1, m_{Seed}) of one of the pure, crystalline enantiomer once the respective temperatures were reached. In all experiments, crystallizer C1 was seeded with D-Asn·H_2O and crystallizer C2 with L-Asn·H_2O. Immediately after seed addition, the pumps were started again to fluidize the present crystals and to prevent nucleation in the stagnant fluid phase. After this initial seeding, both crystallizers were subsequently cooled to the planned operation temperature (Table 1, T_{crys}). The fluidized seed crystals in both columns started to grow according to the present supersaturation. Hence, they settled constantly due to their increasing mass towards the crystallizers' bottoms. All crystals, which reach the fluid phase inlet, were withdrawn into the inline seeding bypass, where they were ground. The generated crystal fragments were fed back to the process merged with the liquid inlet stream.

If the process conditions are chosen appropriately, the crystal fragments will grow (Figure 4a) large enough to settle within the cylindrical part and thus stay in the process. Then the overall crystal mass increases and the process progresses. The growing seeds start to settle, subsequently, on the top of the initial solid mass and the crystal bed increases in height.

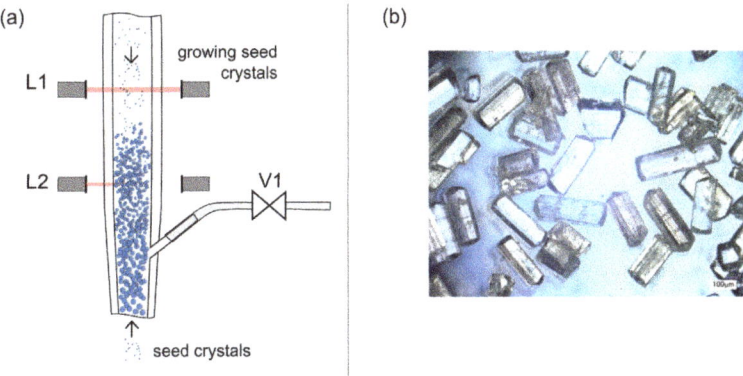

Figure 4. (**a**) Schematic detail drawing of the progressing process at the height of the product outlet. The seed crystals move through the fluidized crystal bed until they are grown to a size at which they settle at the top of the bed. (**b**) Microscope photograph of washed product crystals.

The product withdrawal was initiated when the crystal bed first blocks the upper photoelectric barrier (L1, Figure 4a). The solenoid valve (V1, Figure 4a) was opened at this point and the crystal product is harvested until the lower photoelectric barrier (L2, Figure 4a) was not blocked by the fluidized crystal phase anymore. Afterward, the valves were closed and the crystallization progressed until the crystal bed height was large enough for the next product withdrawal. This periodic procedure continued and the process progressed towards its dynamic steady-state. Immediately after the product withdrawal, the product crystals were filtered and washed with an ethanol-water mixture (40 wt% EtOH, 60 wt% H_2O). The dried product crystals (Figure 4b) were afterwards analyzed via high performance liquid chromatography (HPLC), X-ray powder diffraction (XRPD), and sieve analysis to evaluate each operation point with respect to process performance parameters (purity, yield, productivity, and size distribution, Equations (3)–(7)). The specifications of the applied analytics are given in Table 2.

Table 2. Specifications of the applied analytics.

Analysis	Specification
HPLC	Crownpak CR(+) (4 × 150 mm, particle size 5 µm) column with Dionex Ultimate 3000 system (Thermo Scientific). Perchloric acid/ water (pH = 1) as the eluent. The flow rate, UV wave length, column temperature and injection volume were 0.4 ml/min, 200 nm, 5 °C and 1 µL, respectively.
XRPD	X'Pert Pro diffractometer (PANalytical GmbH, Germany), 2-theta range of 5–40°, step size 0.017°, step time 50 s
Sieve analysis	Retsch, AS 200 digit, amplitude 1.8 mm, sieving time 20 min, sieve mesh widths: 0, 90, 125, 180, 212, 250, 300, 355, 400, 500, 630, 710, 800 µm.

The recycled, concentration-depleted mother liquor causes partial dissolution of the solid racemic excess within the feed tank. To minimize additional distortions of the feed concentration, and thus of the process, a piecewise addition of new solid excess material is required. Since the average product crystal mass of one withdrawal is approximately 20 g per column, subsequently, 50 g of racemic solid excess material was added to the feed tank after every product withdrawal.

3. Results and Discussion

3.1. Continuous Operation and Steady-State Analysis

Along the continuous operation, quantities of the liquid phase, such as volumetric flowrate and local temperatures, were measured and, as seen in Figure 5a,b, controlled appropriately. Figure 5 also depicts the temporal evolution of the fluid density (Figure 5c), measured via the Coriolis mass flowmeter, and the D-Asn enantiomeric proportion of the liquid phase within the feed tank and at the crystallizer outlet (Figure 5d), determined via HPLC offline samples. As seen in Figure 5c,d, these measured quantities also show no discernible trends during the whole experiment. Thus, the applied operation procedure successfully prevents fluctuations or depletions of the feed concentration. Furthermore, the depleted D-Asn enantiomeric proportion at the crystallizer outlet indicates the selective removal of the seeded enantiomer due to the Preferential Crystallization.

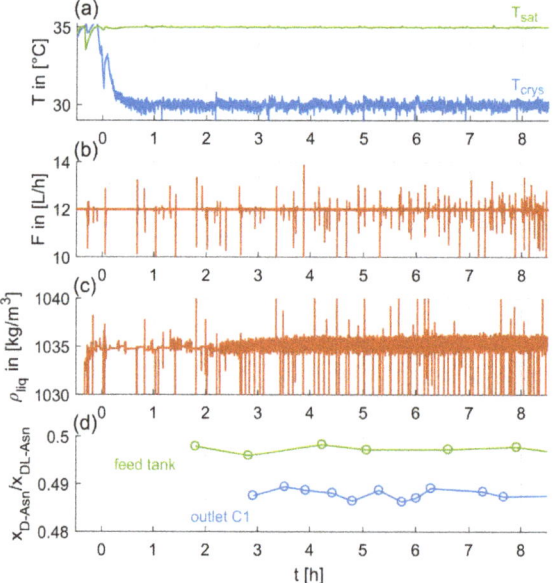

Figure 5. Evolution of the saturation temperature, T_{sat}, crystallization temperature, T_{crys} (**a**), volumetric flowrate (**b**), liquid phase density of the feed solution, ρ_{liq} (**c**), and enantiomeric proportion of the D-Asn·H_2O in the liquid phase at the crystallizer outlet and within the feed tank (**d**), of Exp. 7 (C1) over time, with t = 0 indicating the initial seeding point. Spikes occurring in the flowrate and solution density courses can be attributed to air bubbles passing the Coriolis mass flowmeter (data shown are raw data without any curve treatment).

The temporal evolution of the above introduced process performance parameters, product crystal mean size, $\overline{L_3}$, standard deviation of the product crystal size distribution, s_{L3}, and productivity, Pr, are illustrated in Figure 6 for Exp. 7. As shown in Figure 6, all process performance characteristics reached almost constant values after approximately 2 h process time. During the following 6 h operation time the mean product crystal size varies between 315 to 321 µm with a respective standard deviation between 57 and 60 µm. The calculated productivities range in the same time between 30 and 37 g/L/h, while the product purity exceeds 98% for the whole experiment.

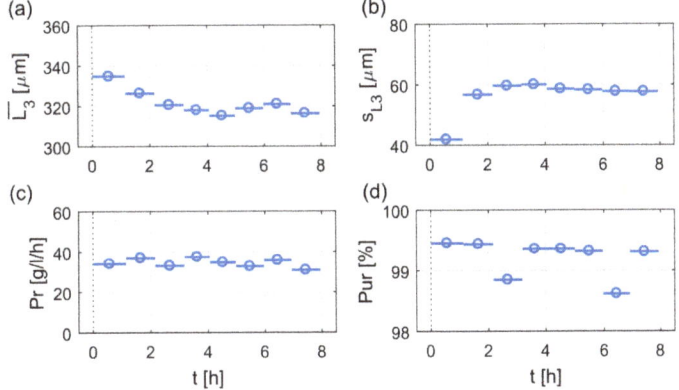

Figure 6. Evolution of mean product crystal size, $\bar{L}_3$ (**a**), its respective standard deviation, s_{L3} (**b**), productivity, Pr (**c**), and product purity, Pur (**d**), of Exp. 7 (C1) over time, with t = 0 indicating the initial seeding point. Illustrated are the values of the process characteristics for every product withdrawal as points with their corresponding time window as line.

3.2. Reproducibility of Steady-State Results

The central point (DoE, design of experiments) of the investigated operation window (operation parameters, T_{sat} = 35 °C, T_{crys} = 30 °C and F = 12 L/h) was evaluated during Exps. 4, 5, and 7. The corresponding operation parameters were used for the first period of experiments 4 and 7 (4 h), though different seed crystal sizes (Exp. 4: 212–300 µm, Exp. 7: 250–355 µm) were added for initialization. In Exp. 5, another operation point was tested before going to the central point. Figure 7 depicts the steady-state product crystal size distribution, q3 (Equation (3)), and the steady-state productivity, Pr, for both crystallizers, C2 and C1 (Figure 7a,b) for the three experiments. The average mean values, $\bar{L}_3$, and standard deviations, s_{L3}, of the steady-state product crystal size distribution are also given. The comparison of Exps. 4 and 5 shows that differences between the product crystal sizes are below 4% and thus negligible. In contrast, the scattering range of the productivity values for each evaluated operation point is approximately 8–10 g/L/h (≈ 30%). Responsible for this broad scattering are the dynamics of the periodic product withdrawal, since during each withdrawal an unknown and varying crystal mass remains within the outlet ports of the crystallizer. This crystal mass is removed from the tube after each withdrawal via a cleaning step with pure solvent and is, thus, not quantifiable. Several experimentations to capture and quantify this crystal mass led to the assumption that it varies between 1 and 3 g, which consequently reduces the evaluated productivity by approximately 2–10 g/L/h. However, the productivity values of Exps. 4 and 5 scatter within a common range, considering the previously mentioned deviation.

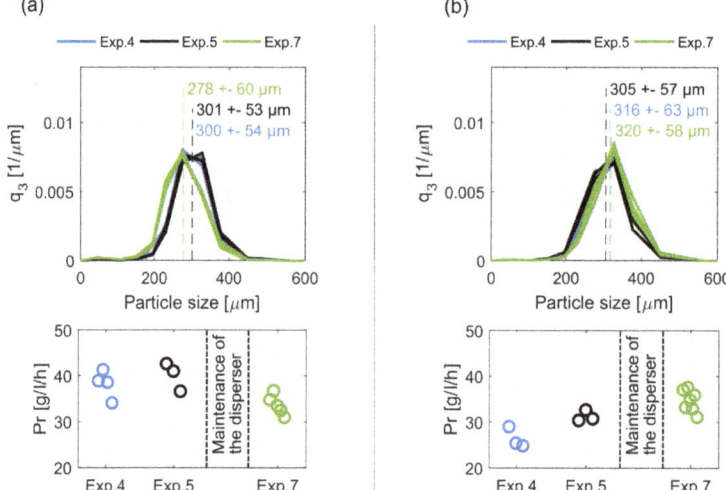

Figure 7. Steady-state crystal size distributions, q_3, with their respective mean values and standard deviations, as well as the corresponding productivities, Pr, of crystallizer C2 (**a**) and C1 (**b**) for Exps. 4, 5, 7.

As indicated in Figure 7, the high-speed dispersers were maintained before Exp. 7. During the process, the maintained disperser tools were placed in cylindrical steel vessels, through which the suspended crystals of the bypass flow. Since the disperser tools do not cover the whole cross section area of these vessels, crystals carried by the surrounded flow might not be comminuted. Thus, the maintenance could cause changes in the amount and size distribution of the generated seed crystals and affect the cyclic steady-state and its reproducibility. As seen in Figure 7, the product crystal size distributions and the productivities of Exp. 7 slightly differ from the respective results of Exps. 4 and 5, which shows exceptionally good reproducibility.

Comparing the results of crystallizer C2 with the corresponding data of crystallizer C1 (Figure 7a,b) shows that C2 produces somewhat smaller product crystals with higher productivities. This trend was persistent throughout all experiments. It indicates that both crystallizers differ in their hydrodynamics due to small differences resulting from their hand manufacturing. Furthermore, the total volume of C2 is 511 ml and, thus, significantly larger than C1 with a total volume of 478 ml. Due to the classifying effect of the conically shaped crystallizers, a larger diameter of C2 would lead to smaller product crystals assuming similar flow rates. These smaller crystals have a higher specific crystal surface, which enhance the total crystallization rate and, thus, the productivity.

The comparison of Exps. 4, 5, and 7 prove that the utilized pilot plant generates reproducible results. Furthermore, the three experiments clearly prove that, when starting from different initial conditions, the same operation points can be independently reproduced. The reproducibility of a further operation point (operation parameters, T_{sat} = 35 °C, T_{crys} = 30 °C, and F = 10 L/h) was also verified (see Appendix A, Figure A1).

3.3. Influence of Volumetric Flowrate

To study the influence of the volumetric flowrate and the resulting residence time of the liquid phase, three operation points with different volumetric flowrates, F = 10, 12, and 14 L/h, were investigated during Exps. 2, 4, 5, and 7. A constant saturation temperature (35 °C) and a constant crystallization temperature (30 °C), were ensured during all experiments. It was observed that at the lowest flowrate, F = 10 L/h, incrustations at the top of both crystallizers occurred and spread significantly faster than at the other operation points. The incrustation layer detached after a certain time, and settled towards the crystallizer bottom and the milling bypasses, where they were comminuted. The subsequently detected decrease of the product purity indicates that the incrustations are induced by nucleation of the counter-enantiomer. To achieve a successful racemate resolution at lower flowrates and, thus, higher residence time of the liquid phase, the risk of nucleation could be reduced by setting lower supersaturation.

Figure 8a depicts the mean product crystal size distributions with their respective mean values and standard deviations for the three volumetric flowrates, from Exps. 4 and 5. As seen, the product crystal size distributions differ regarding the obtained product crystal size; with increasing flowrate the mean crystal size increases significantly from 267 to 300 and 330 µm. As expected, higher volumetric flowrates and, thus, higher fluid velocities, cause larger product crystals. Thus, our results verify the results of former studies [17].

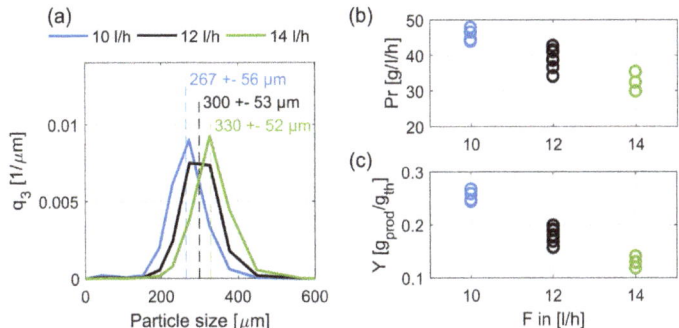

Figure 8. Mean steady-state crystal size distributions, q_3, with their respective mean values and standard deviations (**a**), productivities, Pr (**b**), and yields, Y (**c**), for three different volumetric flowrates, F = 10, 12, and 14 L/h. Given results are from crystallizer C2 for Exps. 4 and 5.

Figure 7b,c shows the influence of the volumetric flowrate on productivity and yield. As shown, a higher volumetric flowrate leads to lower productivity and yield. This can be correlated to the larger crystals and the resulting lower total crystal surface of the fluidized bed. A second reason is that the higher volumetric flowrate reduces the residence time of the seed crystals and, thus, reduces their growing period. Consequently, a higher amount of seed crystals is discharged at the top of the crystallizers. Comparing productivity and yield with respect to the total decrease shows that productivity decreases by approximately 33% whereby yield decreases by approximately 50%. This disproportion is attributed to the increasing throughput rate of the mother liquor, and consequently its decreased depletion.

3.4. Influence of Supersaturation

The first three experiments (Exps. 1–3, Table 1) were utilized to identify a suitable operation window and, hence, four different supersaturation values were tested. A constant volumetric flowrate (10 L/h) and saturation temperature (35 °C) were ensured during all three experiments. The crystallization temperature was set to T_{crys} = 27, 30, 31, and 32 °C, which corresponds to supersaturations of S = 1.40, 1.23, 1.18, and 1.13, respectively. Significant gain of the fluidized crystal bed height was not observed during 90 min operation time using the lowest driving force (1.13), and, thus, no product withdrawal could be realized. Previous studies [19] proved that crystal growth of Asn·H_2O still takes place at this supersaturation. The absent gain of the crystal bed height could be explained by the reduced growth rate of the seed crystals, which are mainly discharged at the top of the crystallizer as a consequence. Nucleation within the tubular crystallizers and a significant decrease of the product purity were observed at the highest supersaturation (1.40). Thus, the operation window is limited by a maximum supersaturation, where nucleation of the counter-enantiomer prevents continuous racemate resolution, as well as a minimal supersaturation, where seed crystals are mainly discharged due to insufficient growth.

The steady-state results of the applied supersaturations are presented in Figure 9 regarding the product crystal size distributions and their characteristics (Figure 9a), productivity (Figure 9b), and yield (Figure 9c). As depicted in Figure 9a, the influence of the supersaturation on the product crystal size is negligible, which again verifies the classifying effect of the fluidized bed. The standard deviation of the product crystal size distribution increases slightly at higher supersaturations. Figure 9b,c shows that the supersaturation clearly enhances both productivity and yield, which correlates with the growth kinetics (dotted lines) of the given substance system [19].

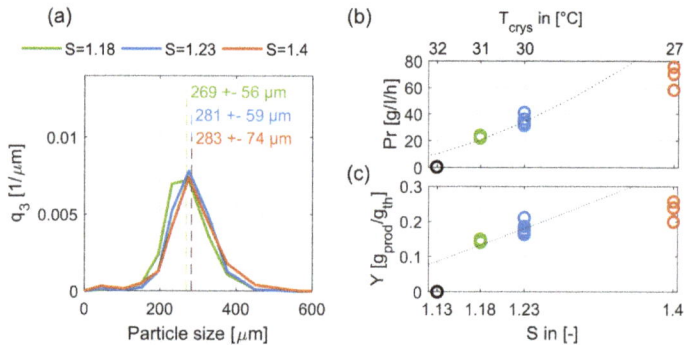

Figure 9. Mean steady-state crystal size distributions, q_3, with their respective mean values and standard deviations (**a**), productivities, Pr (**b**), and yields, Y (**c**), for four different supersaturations, S = 1.40, 1.23, 1.18, and 1.13, and the same saturation temperature and volumetric flowrate. The respective crystallization temperatures, T_{crys} = 27, 30, 31, and 32 °C, are indicated. The dotted lines (b and c) represent the change with respect to the operation point at T_{crys} = 30 °C, assuming the withdrawn product mass is proportional to the growth rate [19]. Given results are from crystallizer C1 for Exps. 1–3.

3.5. Influence of Crystallization Temperature

In Exp. 6 the crystallization temperature was reduced to T_{crys} = 20 °C. To guarantee an approximately constant supersaturation of 1.23, the saturation temperature was adjusted to 24.6 °C. Figure 10 depicts the steady-state results of Exp. 6 together with the respective results of Exp. 4 (T_{crys} = 30 °C, T_{sat} = 35 °C) for the same volumetric flowrate (12 L/h).

As seen in Figure 10a, an influence of the crystallization temperature on the product crystal size was not observed. At the lower crystallization temperature, the standard deviation decreases by almost 15%, hence, the selectivity of the size classifying effect is improved. Since the crystal growth rate decreases with the crystallization temperature, the crystal growth and the standard deviation show the same correlation as in the previous section. This correlation leads to the assumption that the crystal growth counteracts the size-classifying effect and thus its selectivity. The influences of the crystallization temperature on productivity and yield are depicted in Figure 10b,c. As expected, a lower crystallization temperature leads to both lower productivity and yield. The observed decreases of productivity and yield are in good agreement with the studied correlation between crystal growth kinetics (dotted lines) and crystallization temperature [19].

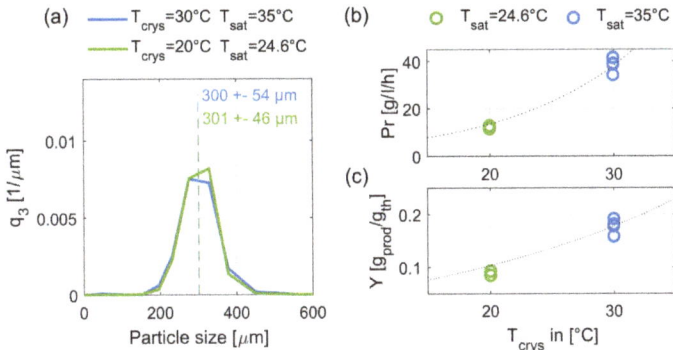

Figure 10. Mean steady-state crystal size distributions, q_3, with their respective mean values and standard deviations (**a**), productivities, Pr (**b**), and yields, Y (**c**), for two different crystallization temperatures, T_{crys} = 20 and 30 °C, at the same supersaturation and volumetric flowrate. The dotted lines (**b**) indicate the change with respect to the central point at T_{crys} = 30 °C, assuming the withdrawn product mass is proportional to the growth rate [19]. Given results are from crystallizer C2 for Exps. 4, 6.

4. Conclusions

In the present experimental parameter study, seven experiments were performed to investigate the continuous fluidized bed crystallization at twice 0.5 L scale for racemate resolution at its steady-state. Each experiment was conducted for 8 hours to ensure constant conditions. As verified, after a relatively short operation time (approximately 2 hours), the utilized pilot plant reaches a cyclic steady-state, where product crystals with constant crystal size distribution and productivity can be periodically withdrawn. The reproducibility of the steady-state results was proven and sensitivities of the utilized pilot plant on the steady-state results were identified. In particular, it was observed that changes of the high-speed disperser slightly effect the steady-state results in terms of productivity and product crystal size distribution. Nevertheless, the steady-state results were shown in the present study to have exceptionally good reproducibility.

It was proven that the steady-state product crystal size mainly depends on the volumetric flowrate, and thus can be easily adjusted (in this work between 260 and 330 µm). All products, withdrawn at steady-state, have a narrow crystal size distribution (standard deviation <60 µm) and a low fines content. Thus, the size classifying effect of the conically shaped tubular crystallizers and its selectivity is verified. Productivity and yield increase with supersaturation, whereby it was shown that the continuous racemate resolution is limited by a certain maximal and minimal supersaturation. At too low supersaturation, the seed crystals grow insufficiently and are mainly discharged at the top of the crystallizer. At too high supersaturation, the respective counter-enantiomer nucleates and, thus, contaminates the resolution product. Furthermore, the limitations regarding the supersaturation are not fixed values and also depend on the volumetric flowrate. In particular, the nucleation probability increases with higher supersaturation and lower volumetric flowrates. A decrease of productivity was observed at lower supersaturation and higher volumetric flowrates, and thus, higher fluid velocities. These observed correlations enhance the expectation that the location and width of the operation window and, thus, the process performance, are tunable via geometrical aspects of the conically shaped tubular crystallizer [20].

The utilized pilot plant enables continuous racemate resolution with enantiomer purities above 97% and productivities up to 40 g/L/h for each enantiomer, which is far above productivities documented by other studies [14,21]. Since the process was not optimized at all, productivities higher than 40 g/L/h are to be expected. Thus, the coupled fluidized bed crystallization was proven to be an excellent technology for continuous enantioseparation, which facilitates high purities and the robust production of both enantiomers simultaneously.

Author Contributions: Conceptualization, H.L. and A.S.-M.; methodology, E.T. and J.G.; investigation, formal analysis, validation, data curation and visualization, J.G.; writing—original draft preparation, E.T. and J.G.; writing—review and editing, E.T., H.L. and A.S.-M.; supervision, H.L. and A.S.-M.; project administration and funding acquisition, A.S.-M. All authors have read and agreed to the published version of the manuscript.

Funding: This research was funded by Deutsche Forschungsgemeinschaft (DFG) within the Research Program SPP 1679 "Dynamische Simulation vernetzter Feststoffprozesse".

Acknowledgments: The authors thank Jacqueline Kaufmann and Stefanie Leuchtenberg for their analytical support as well as Detlef Franz, Klaus-Dieter Stoll, Stefan Hildebrandt and Steve Haltenhof for their technical support.

Conflicts of Interest: The authors declare no conflict of interest. The funders had no role in the design of the study; in the collection, analyses, or interpretation of data; in the writing of the manuscript, or in the decision to publish the results.

Appendix A

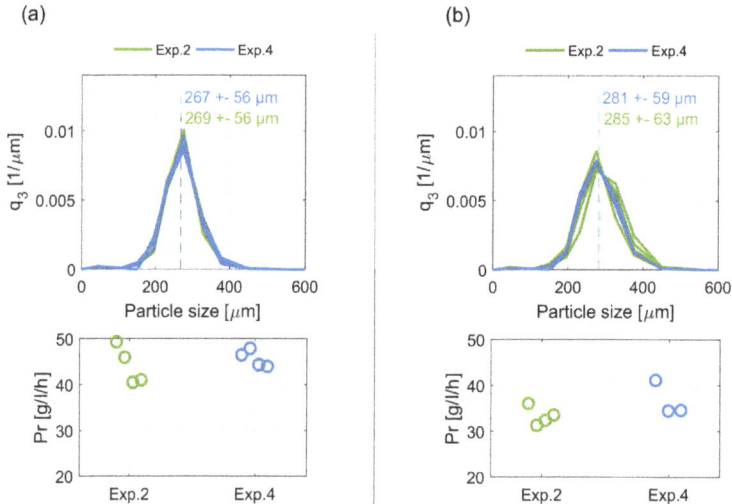

Figure A1. Steady-state crystal size distributions, q_3, with their respective mean values and standard deviations as well as the corresponding productivities, Pr, of crystallizer C2 (**a**) and C1 (**b**) for Exps. 2, 4.

References

1. Jiang, M.; Braatz, R.D. Designs of continuous-flow pharmaceutical crystallizers: Developments and practice. *Cryst. Eng. Comm.* **2019**, *21*, 3534–3551. [CrossRef]
2. Lorenz, H.; Temmel, E.; Seidel-Morgenstern, A. Continuous Enantioselective Crystallization of Chiral Compounds. In *The Handbook of Continuous Crystallization*, 1st ed.; Yazdanpanah, N., Nagy, Z.K., Eds.; Royal Society of Chemistry: Cambridge, UK, 2020; Volume 1, pp. 422–468. [CrossRef]
3. Darmali, C.; Mansouri, S.; Yazdanpanah, N.; Woo, M.W. Mechanisms and Control of Impurities in Continuous Crystallization: A Review. *Ind. Eng. Chem. Res.* **2018**, *58*, 1463–1479. [CrossRef]
4. Su, M.; Gao, Y. Air–Liquid Segmented Continuous Crystallization Process Optimization of the Flow Field, Growth Rate, and Size Distribution of Crystals. *Ind. Eng. Chem. Res.* **2018**, *57*, 3781–3791. [CrossRef]
5. Kacker, R.; Maaß, S.; Emmerich, J.; Kramer, H.J.M. Application of inline imaging for monitoring crystallization process in a continuous oscillatory baffled crystallizer. *AIChE J.* **2018**, *64*, 2450–2461. [CrossRef]

6. Onyemelukwe, I.; Benyahia, B.; Reis, N.M.; Nagy, Z.K.; Rielly, C. The heat transfer characteristics of a mesoscale continuous oscillatory flow crystalliser with smooth periodic constrictions. *Int. J. Heat Mass Transf.* **2018**, *123*, 1109–1119. [CrossRef]
7. Nguyen, A.-T.; Kim, W.-S. Influence of feeding mode on cooling crystallization of L-lysine in Couette-Taylor crystallizer. *Korean J. Chem. Eng.* **2017**, *34*, 2002–2010. [CrossRef]
8. Kim, J.-E.; Kim, W.-S. Synthesis of Core–Shell Particles of Nickel–Manganese–Cobalt Hydroxides in a Continuous Couette-Taylor Crystallizer. *Cryst. Growth Des.* **2017**, *17*, 3677–3686. [CrossRef]
9. Shih, Y.-J.; Abarca, R.R.; De Luna, M.D.G.; Huang, Y.-H.; Lu, M.-C. Recovery of phosphorus from synthetic wastewaters by struvite crystallization in a fluidized-bed reactor: Effects of pH, phosphate concentration and coexisting ions. *Chemosphere* **2017**, *173*, 466–473. [CrossRef]
10. Priambodo, R.; Shih, Y.-J.; Huang, Y.-H. Phosphorus recovery as ferrous phosphate (vivianite) from wastewater produced in manufacture of thin film transistor-liquid crystal displays (TFT-LCD) by a fluidized bed crystallizer (FBC). *RSC Adv.* **2017**, *7*, 40819–40828. [CrossRef]
11. Maharaj, C.; Chivavava, J.; Lewis, A. Treatment of a highly-concentrated sulphate-rich synthetic wastewater using calcium hydroxide in a fluidised bed crystallizer. *J. Environ. Manag.* **2018**, *207*, 378–386. [CrossRef] [PubMed]
12. Vu, X.; Lin, J.-Y.; Shih, Y.-J.; Huang, Y.-H. Reclaiming Boron as Calcium Perborate Pellets from Synthetic Wastewater by Integrating Chemical Oxo-Precipitation within a Fluidized-Bed Crystallizer. *ACS Sustain. Chem. Eng.* **2018**, *6*, 4784–4792. [CrossRef]
13. Hohmann, L.; Greinert, T.; Mierka, O.; Turek, S.; Schembecker, G.; Bayraktar, E.; Wohlgemuth, K.; Kockmann, N. Analysis of Crystal Size Dispersion Effects in a Continuous Coiled Tubular Crystallizer: Experiments and Modeling. *Cryst. Growth Des.* **2018**, *18*, 1459–1473. [CrossRef]
14. Binev, D.; Seidel-Morgenstern, A.; Lorenz, H. Continuous Separation of Isomers in Fluidized Bed Crystallizers. *Cryst. Growth Des.* **2016**, *16*, 1409–1419. [CrossRef]
15. Lorenz, H.; Seidel-Morgenstern, A. Processes to Separate Enantiomers. *Angew. Chem. Int. Ed.* **2014**, *53*, 1218–1250. [CrossRef] [PubMed]
16. Elsner, M.P.; Ziomek, G.; Seidel-Morgenstern, A. Simultaneous preferential crystallization in a coupled batch operation mode. Part II: Experimental study and model refinement. *Chem. Eng. Sci.* **2011**, *66*, 1269–1284. [CrossRef]
17. Binev, D.; Seidel-Morgenstern, A.; Lorenz, H. Study of crystal size distributions in a fluidized bed crystallizer. *Chem. Eng. Sci.* **2015**, *133*, 116–124. [CrossRef]
18. Petrusevska-Seebach, K. Overcoming Yield Limitations when Resolving Racemates by Combination of Crystallization and/or Chromatography with Racemization. Ph.D. Thesis, Otto-von-Guericke-Universität, Magdeburg, Germany, 2012.
19. Temmel, E.; Gänsch, J.; Lorenz, H.; Seidel-Morgenstern, A. Measurement and Evaluation of the Crystallization Kinetics of l-Asparagine Monohydrate in the Ternary l-/d-Asparagine/Water System. *Cryst. Growth Des.* **2018**, *18*, 7504–7517. [CrossRef]
20. Mangold, M.; Khlopov, D.; Temmel, E.; Lorenz, H.; Seidel-Morgenstern, A. Modelling geometrical and fluid-dynamic aspects of a continuous fluidized bed crystallizer for separation of enantiomers. *Chem. Eng. Sci.* **2017**, *160*, 281–290. [CrossRef]
21. Köllges, T.; Vetter, T. Design and Performance Assessment of Continuous Crystallization Processes Resolving Racemic Conglomerates. *Cryst. Growth Des.* **2018**, *18*, 1686–1696. [CrossRef]

© 2020 by the authors. Licensee MDPI, Basel, Switzerland. This article is an open access article distributed under the terms and conditions of the Creative Commons Attribution (CC BY) license (http://creativecommons.org/licenses/by/4.0/).

Article

Concept Study for an Integrated Reactor-Crystallizer Process for the Continuous Biocatalytic Synthesis of (S)-1-(3-Methoxyphenyl)ethylamine

Dennis Hülsewede [1], Erik Temmel [2], Peter Kumm [1] and Jan von Langermann [1,*]

[1] Biocatalytic Synthesis Group, Institute of Chemistry, University of Rostock, Albert-Einstein-Str. 3A, 18059 Rostock, Germany; dennis.huelsewede@uni-rostock.de (D.H.); peter.kumm@uni-rostock.de (P.K.)
[2] Sulzer Chemtech Ltd., Gewerbestraße 28, 4123 Allschwil, Switzerland; temmel@mpi-magdeburg.mpg.de
* Correspondence: jan.langermann@uni-rostock.de; Tel.: +49-381-4986456

Received: 27 February 2020; Accepted: 23 April 2020; Published: 27 April 2020

Abstract: An integrated biocatalysis-crystallization concept was developed for the continuous amine transaminase-catalyzed synthesis of (S)-1-(3-methoxyphenyl)ethylamine, which is a valuable intermediate for the synthesis of rivastigmine, a highly potent drug for the treatment of early stage Alzheimer's disease. The three-part vessel system developed for this purpose consists of a membrane reactor for the continuous synthesis of the product amine, a saturator vessel for the continuous supply of the amine donor isopropylammonium and the precipitating reagent 3,3-diphenylpropionate and a crystallizer in which the product amine can continuously precipitate as (S)-1-(3-methoxyphenyl)ethylammonium-3,3-diphenylpropionate.

Keywords: amine; biocatalysis; enzyme; process intensification; crystallization; enantioselective

1. Introduction

In recent years, biocatalytic synthesis reactions have made a significant impact in the scientific community and have even replaced existing chemical pathways in industrial processes. The main advantages are often higher stereo-, regio- and chemoselectivities, while mild reaction conditions and environmentally friendly solvents such as water can be applied. The high selectivity of biocatalysts results also frequently in less or even no side reactions, which itself yields higher process and atom efficiencies. In addition, recent scientific and technological advances in enzyme engineering allow the relatively fast design and production of tailor-made biocatalysts for a specific process [1–7]. Furthermore, downstream-processing from biocatalyst-based reaction systems remains an issue and is a major economic factor in the overall process. This problem originates mostly from the presence of water-soluble proteins, buffer salts, biocatalyst-based cofactors, remaining unreacted substrates and co-substrates, which have to be removed efficiently to ensure high product purities. The purification from such complex mixtures is typically achieved by multiple extractions and further purification steps [8].

In contrast, selective crystallization techniques provide a more selective product isolation approach from complex mixtures, especially aqueous solutions and was integrated in this study directly into the biocatalytic synthesis process [9]. In the presented study this is utilized at the synthesis of (S)-1-(3-methoxyphenyl)ethylamine, which is a valuable intermediate for the synthesis of rivastigmine, a highly potent drug for the treatment of early stage Alzheimer's disease (Scheme 1). Studies have shown that the (S)-enantiomer is more potent as the (R)-enantiomer and preferably the enantiomerically pure (S)-form should be administered to avoid complications [10–13].

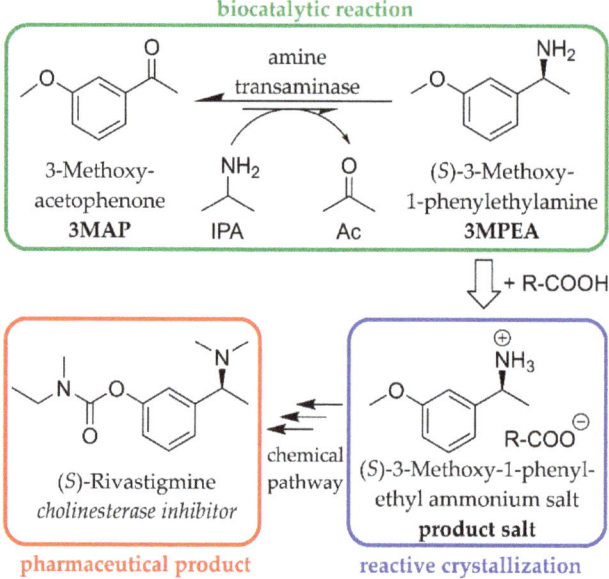

Scheme 1. Investigated biocatalytic transamination reaction and the inclusion of a reactive crystallization step for the synthesis of rivastigmine; R = –CH$_2$–CHPh$_2$.

The applied amine transaminase from *Ruegeria pomeroyi* catalyzes the transfer of the amine group from the donor amine isopropylamine (IPA) to the carbonyl compound 3-methoxyacetophenone (3MAP), forming (S)-1-(3-methoxyphenyl)ethylamine (3MPEA) and acetone (Ac) as a co-product [14,15]. The biocatalytic transformation forming 3MPEA itself is very enantioselective but suffers from an unfavorable reaction equilibrium [16]. These limitations in amine transaminase-catalyzed reactions are often overcome by classical (bio)chemical solutions to remove the co-product from the equilibrium or using specifically tailor-made donor amines to shift the reaction to the product side [17–28]. These chemically-driven options offer higher yields but require often a complex process control, additional (bio)catalytic reaction systems, additional chemicals and eventually lead to lower atom efficiencies [29]. As a crystallization-based alternative we apply our recently developed in situ-product crystallization approach in the amine-transaminase-catalyzed reaction [16,30]. This reactive crystallization approach removes the product amine 3MPEA from solution by forming an ammonium salt with a suitable carboxylate anion that exhibits a very low solubility. Additional chemical reactants or (bio)catalysts are not required and a simplified downstream-processing approach via filtration is possible.

2. Materials and Methods

2.1. Chemicals

All chemicals were obtained from Acros (Fair Lawn, NJ, USA), TCI Chemicals (Tokyo, Tokyo Prefecture, Japan), Aldrich (St. Louis, MO, USA), Alfa Aesar (Haverhill, MA, USA) and ABCR (Karlsruhe, Baden-Württemberg, Germany) and were used as received. Deionized water was used throughout this study.

2.2. Biocatalyst

Amine transaminase from *Ruegeria pomeroyi* (as ECS-ATA08) was obtained as whole cell lyophilizate (Enzymicals AG (Greifswald, Mecklenburg-Vorpommern, Germany)) and was used as received during this study. The activity of 364 U/g was determined using the conversion of 2.5 mmol/L 1-phenylethylamine and 2.5 mmol/L pyruvate to acetophenone and alanine in 50 mM phosphate buffer pH 8.0 with 0.025 mmol PLP and 0.25 %(v/v) DMSO at 25 °C. One unit of enzyme activity is defined as the formation of 1 mmol acetophenone per minute, which was observed spectrophotometrically at 245 nm with ε(acetophenone) = 11.852 L/(mol·cm).

2.3. Donor Amine Salt Synthesis

39.6 g 3,3-diphenylpropionic acid (3DPPA) were dissolved in 500 mL methyl tert-butyl ether (MTBE) in a 1 L round flask. Then 16.0 mL isopropylamine (IPA) were added to the stirred solution with a syringe and the resulting suspension was stirred overnight. MTBE and excess IPA were evaporated afterwards with a rotary evaporator to obtain the donor salt isopropylammonium 3,3-diphenylpropionate (IPA-3DPPA).

2.4. Solubility Measurements

Solubilities were measured in 10 mL 50 mM phosphate buffer with an excess of the respective solid salt to obtain a saturated solution. The resulting mixture was adjusted to the desired pH value and shaken for 7 days at 160 rpm and 30 °C. The pH was re-adjusted daily with NaOH and HCl, if required. The resulting mixture was then filtered to obtain a clear, saturated solution. For pH-dependent measurements a sample was diluted (typically 20×) for the absorption measurement of 3DPPA at 249 nm and compared with a calibration curve of 3DPPA in 50 mM phosphate buffer at the same pH (3DPPA concentration range: 3–0.5 mM, 1 cm cuvette path length). For temperature dependent solubility measurements, 25 mM 2-[4-(2-hydroxyethyl)piperazin-1-yl]ethanesulfonic acid buffer (HEPES) at pH 7.5 was used. After 7 days a 1 mL sample was taken, evaporated and the amount of salt determined gravimetrically with the subtraction of the amount of HEPES.

2.5. Reactor Setup

Two Plane flange vessels with a tempering jacket and a bottom drain valve of 250 mL (Pfaudler GmbH, Germany) were used as crystallizer and saturator. The membrane reactor consists of two Teflon chambers separated by a polyvinylidene fluoride transfer membrane (PVDF) with a cut-off of 0.2 µm (Bio-Rad Laboratories GmbH, Germany) (Figure 1). The vessels were connected with PharMed - BPT and Tygon LMT - 55 tubes (Saint - Gobain, France), tempered with the cooling bath thermostat Huber CC-K6 attached to a Pilot ONE control panel (both by Peter Huber Kältemaschinenbau AG, Germany) and stirred with overhead stirrers MICROSTAR 7.5 control. The membrane reactor was stirred with magnetic stirrers (IKA Werke, Germany). Two peristaltic pumps—Hei-FLOW Precision 01 (Heidolph Instruments, Germany)—were used to transfer the mother liquors from the saturator to the membrane reactor and from the crystallizer to the saturator. To ensure a constant filling level in the membrane reactor, an overflow was integrated from which the reaction solution can flow into the lower positioned crystallizer. In order to keep the solid salts in the saturator and the crystallizer, 11 µm filters have been placed at the inlet of the tubes.

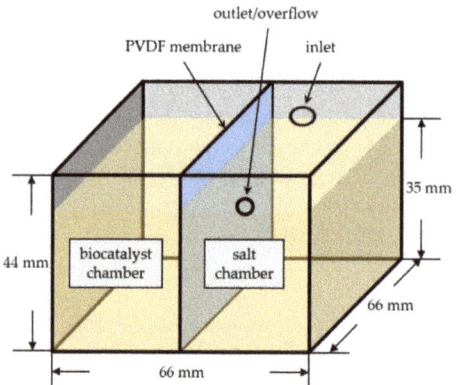

Figure 1. Design of the membrane reactor with polyvinylidene fluoride membrane (PVDF).

2.6. Membrane Reactor Procedure

In the membrane reactor, 857 µL IPA (100 mM) and 596 mg HEPES buffer (25 mM) were added to 100 mL water and the pH was adjusted to 7.5. This solution was divided equally into both chambers of the reactor. After the addition of 2.85 g donor salt (100 mM) and 566 mg product salt (26 mg for saturation and 540 mg as seed crystals) to the salt chamber, the reactor was stirred until the dissolved salts were homogeneously distributed in both chambers. To start the reaction, 124 mg pyridoxal 5′-phosphate (PLP) (5 mM) and 1.37 mL substrate 3′-methoxyacetophenone (3MAP) (100 mM) were added to the salt chamber and 1.56 g amine transaminase (ATA) (6 U/mL) to the biocatalyst chamber. During the reaction the formed product (S)-1-(3-methoxyphenyl)ethylamine (3MPEA) can diffuse through the membrane and result in crystal growth with 3DPPA on the product salt seed crystals, while the cells containing the ATA are retained by the membrane. Every 24 h samples were taken from both chambers of the reactor and measured by gas chromatography.

2.7. Triple Vessel Procedure

5.14 mL IPA (100 mM) and 3.57 g HEPES buffer (25 mM) were added to 600 mL water and the pH was adjusted to 7.5. From this solution, 80 mL were added to each of the two chambers of the membrane reactor and the remainder divided between the saturator and the crystallizer. 10.3 g donor salt ($\hat{=}$ 60 mM, if it would be fully dissolved) was then added to the saturator and 1 g product salt was added to the crystallizer (159 mg for saturation and 841 mg as seed crystals) (see Figure 2). By switching on the pumps (2.5 L/h) and stirrers (200 rpm), the partially dissolved salts can be distributed throughout the system, while crystalline salt is retained by the filters in the respective vessels. To start the reaction, 741 mg PLP (5 mM) and 8.24 mL 3MAP ($\hat{=}$ 100 mM) were added to the saturator and 1.56 g ATA (1 U/mL) to the biocatalyst chamber of the membrane reactor. Samples were taken regularly from the biocatalyst chamber of the membrane reactor and measured by gas chromatography, to re-adjust the 3MAP concentration to the initial concentration of 100 mM, based on the observed conversion. As long as solid donor salt is still present in the saturator, it is not necessary to add additional material.

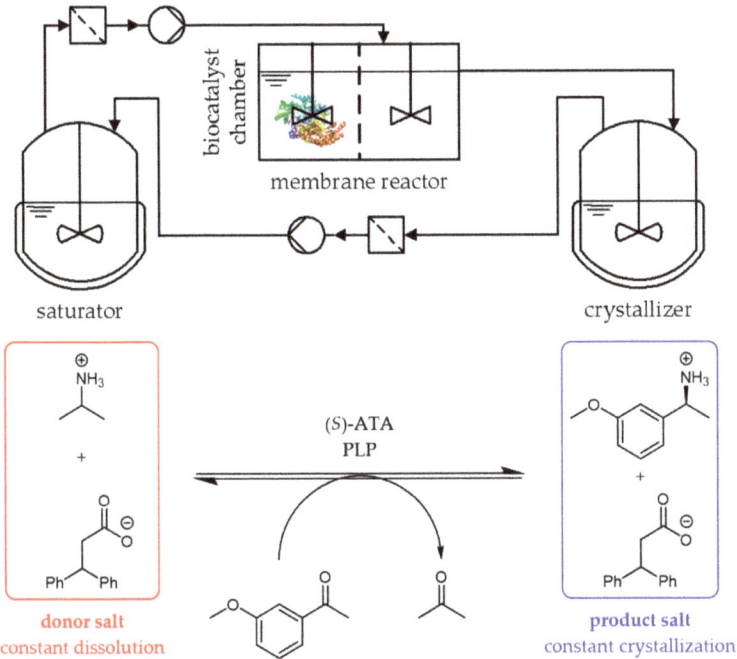

Figure 2. Reaction scheme of the continuous transaminases-catalyzed reaction with in situ donor salt dissolution (left) and product salt crystallization (right) in separated vessels.

2.8. Sampling

Samples of the suspension (500 µL) were taken periodically and 50 µL conc. NaOH was added to completely solve the undissolved salts and quench the reaction. Then 500 µL cyclopentylmethyl ether (CPME) was added to extract substrate and product, mixed by a vortex mixer and then centrifuged (2 min, 3000 rpm) to improve phase separation. From the organic layer, 200 µL was taken and was diluted with 800 µl CPME, combined with 200 µL of a 25 mM n-decane solution in CPME (internal standard) and subsequently analyzed by gas chromatography.

2.9. Chromatography

Concentration was measured with a Trace 1310 gas chromatograph by Thermo Scientific (Dreieich, Germany), equipped with a 1300 flame ionization detector and an Agilent Capillary HP-5 19091 J-433 (0.25 mm × 30 m × 0.25 µm). n-Decane was used as internal standard in all measurements. Temperatures of injector and detector were set to 250 °C. Temperature profile started at 90 °C, followed by a heating rate of 2 K/min to 100 °C, 20 K/min to 130 °C, 2 K/min to 138 °C and 20 K/min to 160 °C.

2.10. X-ray Powder Diffraction (XRPD)

Solid samples were measured via x-ray powder diffraction (XRPD) to discriminate the solid composition of donor salt and/or product salt. Powder x-ray diffraction data were collected on a Stoe Stadi-P with germanium-monochromatised Cu-Kα-radiation (λ = 1.5418 Å) in horizontal transmission/Debey-Scherrer geometry. The x-rays were detected with a position sensitive detector in the 2Theta range from 5 to 35°. The 40 kV high voltage and 40 mA current were generated by a Seifert high voltage generator (ID 3003). The equipment was controlled and the raw data were handled with the software STOE WinXPOW (version 2.25, 2009). The position of the 2Theta and ω-circle were adjusted with the (111)-reflex of crystalline silicon (2Theta = 28.44°). All samples were measured as flat preparation between two layers of poly acetate foil. The sample was spun around its center during the measurvent and the ω-circle was also spun with 1/2 2Theta.

2.11. Nuclear Magnetic Resonance (NMR)

^{1}H NMR and ^{13}C NMR spectra were recorded with a Bruker AVANCE 250 II, AVANCE 300 III and AVANCE 500. Chemical shifts are reported in parts per million relative to the solvent peak as an internal reference. Splitting patterns are indicated as follows: s, singlet; d, doublet; t, triplet; m, multiplet (see Appendix A).

3. Results and Discussion

Initial studies of the integrated use of crystallization in amine transaminase-catalyzed reactions included a direct application of a carboxylic acid, which yields the crystallization of the product amine salt but specifically avoids the crystallization of the donor amine isopropylamine as its salt (donor salt) [16]. This application of this concept results in a moderate apparent shift of the reaction equilibrium towards the product side but is unfortunately often limited to batch reactions and relatively low substrate concentrations due to a continuously increasing amount of solid product salt stopping the entire reaction.

In this study we present an alternative continuous approach towards this reactive crystallization, which intentionally includes the presence of the originally undesired donor amine salt isopropylammonium 3,3-diphenylpropionate. The donor amine salt dissolves continuously and thus release stoichiometric amounts of isopropylammonium and 3,3-diphenylpropionate into solution. Any excess beyond the solubility limit remains as a dispersed solid phase in the reaction mixture. This basically limits the amount of amine in solution to an absolute minimum, in contrast to conventional approaches using high excesses of isopropylamine in solution, which may cause limited enzyme stability [30,31]. Similarly, the substrate 3MAP is continuously dissolved in the solution up to its solubility limit, ensuring a constant 3MAP concentration in the aqueous solution throughout the process. Consequently, the reaction equilibrium in solution is based on the aqueous phase concentration since it is only accessible by the biocatalyst. The conversion towards the products leads to a continuous removal of the educts from the aqueous solution, which is adjusted to the original concentration due to the above mentioned solubility equilibrium. The reaction cycle is closed by the final continuous crystallization of the product amine salt, which removes in stoichiometric amounts the dissolved 3,3-diphenylpropionate anion. The only byproduct is acetone, which evaporates quite easily from solution due to its high vapor pressure at 30 °C [30]. Applications at large scale will require additional solutions to remove acetone effectively from solution to avoid a full stop of the in situ-product crystallization and the inhibition of the biocatalyst, for example, via stripping with an inert gas.

The presence of two solid salt phases requires a separation into two vessels to avoid an undesired mixing. In this work we present a triple vessel system, which separates both solid phases and the catalyst from each other, which enables the above mentioned continuous reaction mode (Figure 2). A membrane reactor is applied to retain the biocatalyst (amine transaminase from *Ruegeria pomeroyi*) behind a polyvinylidene fluoride membrane (PVDF), while the filtered mother liquor is pumped via peristaltic pumps through a crystallizer for product salt crystallization, a saturator for donor salt saturation and eventually back into the membrane reactor to close the loop. An exception is the connection between the membrane reactor and the crystallizer, which is directly fed by an overflow from the higher positioned membrane reactor.

3.1. Salt Solubilities

The solubility difference between both salts, donor salt and product salt, is the main parameter within the shown reaction mode. The donor salt must have a significantly higher solubility than the product salt, which will only then crystallize selectively from solution. Using the 3,3-diphenylpropionate (3DPPA) as the anion results in a solubility difference of approx. 50 mM between the donor salt (IPA-3DPPA) and the product salt (3MPEA-3DPPA) (Figure 3). For the investigated biocatalytic reaction system the concentration of the donor salt remains for above pH 7 and 30 °C, at >50 mM, while the product salt is considerably less soluble at approx. 5 mM, depending on the chosen pH in solution (Figure 3A). These results are comparable with the model product amine salt 1-phenylethylammonium 3,3-diphenylpropionate in an earlier study [30]. Please note that the shown concentrations may be altered by the presence of other salts such as other buffer components, impurities and especially the additionally used isopropylamine. Changes in temperature will also affect the solubilities of these two main salts, however the observed effect is relatively small. As shown in Figure 3B at pH 7.5 no significant effect is visible and the donor salt remains strongly more soluble than the product amine salt. The choice of temperature is fortunately mostly controlled by the temperature optimum of the biocatalyst itself, which limits the choice of reaction temperature to a narrow range at 30 °C.

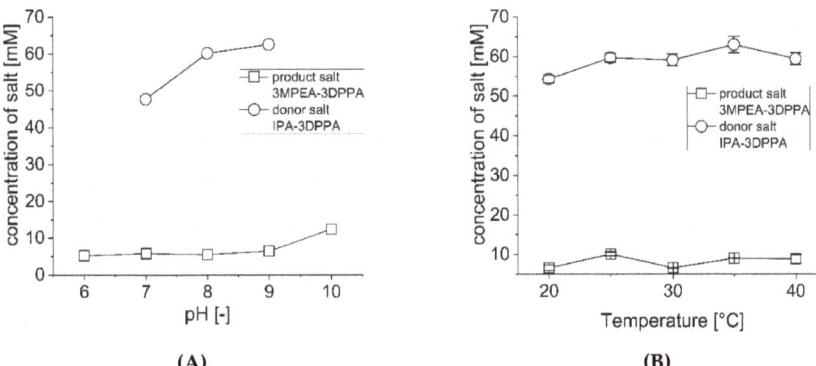

Figure 3. pH- and temperature-dependent solubility profile of the donor salt IPA-3DPPA and product salt 3MPEA-3DPPA; (**A**) pH dependency in 50 mM phosphate buffer at 30 °C; (**B**) temperature dependency in 25 mM HEPES buffer pH 7.5.

3.2. Single Membrane Reactor

The central membrane reactor is the key component within the triple vessel concept. The applied membrane primarily retains the biocatalyst (whole *E. coli* cells) in the biocatalyst chamber from the remaining solution and both solid salts, IPA-3DPPA and 3MPEA-3DPPA, in the salt chamber (Figure 4). During the reaction, IPA-3DPPA is continuously consumed as the amount of product salt increases and eventually accumulates as the only solid phase.

The dissolved reactants diffuse freely between both chambers and a relevant diffusion limitation was not observed (full equilibrium conditions can be achieved within ca. 10 min). This approach provides an alternative to classical encapsulation and immobilization approaches and thus prevents undesired deactivation or diffusion problems of the biocatalytic reaction system [32–36]. The applied PVDF transfer membrane is fully biocompatible and was described by Wachtmeister et al. in 2014 for a lyase-catalyzed reaction [37]. In addition, the use of a membrane reactor offers a simple adjustment of the biocatalytic synthesis system without interfering with the solid salt phases, including the addition or a full exchange of the biocatalyst during the reaction. In addition, the accumulation of inactivated biocatalyst in combination with an undesired mixing with the product salt is prevented, which simplifies downstream processing enormously to a simple filtration step.

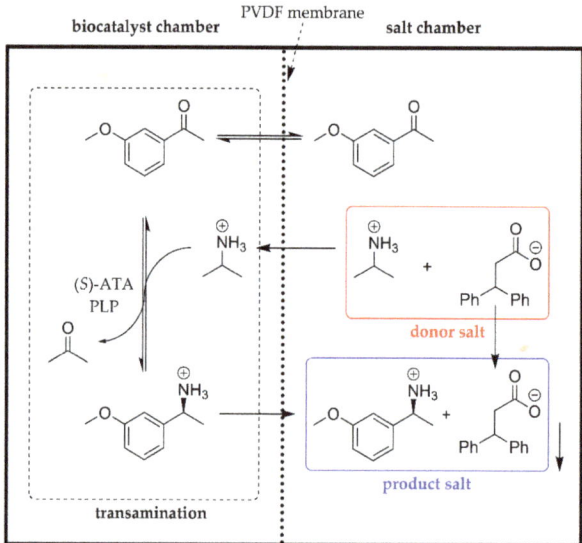

Figure 4. Reaction flow in the membrane reactor.

As shown in Figure 5, a batch experiment of the membrane reactor, without any connection to a saturator and crystallizer, allows the conversion of 55 mM 3MAP to the corresponding product amine salt 3MPEA-3DPPA. The product salt accumulates exclusively in the salt chamber, while the donor salt IPA-3DPPA is consumed in parallel in the salt chamber. A small amount of the product amine is always present in the biocatalyst chamber, which relates to the solubility limit of the product salt in solution. Eventually the reaction stops at approx. 55 mM product concentration due to an accumulation of acetone in the aqueous phase due to the absence of an active acetone removal step, which equals the equilibrium position of this biochemical reaction. Due to absence of an any observable substrate and product inhibition at the chosen reaction conditions, the only rate determining step is the available catalytic activity of the biocatalyst, which leaves a lot of room for optimization. Transport through the membrane and crystal growth are always significantly faster and did not limit the overall process.

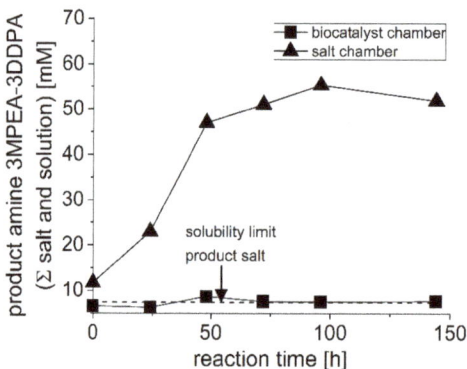

Figure 5. 3MPEA concentration curve in the chambers of the membrane reactor, 30 °C, 25 mM HEPES buffer pH 7.5, 100 mM 3MAP, 100 mM IPA-3DPPA, 100 mM additional isopropylamine and 5 mM PLP (within entire solution).

3.3. Combined Triple Vessel Concept

To permanently separate the salts from each other, the salt chamber of the membrane reactor is replaced by a flow-through chamber connected to two other vessels (see Figure 2). The donor salt is placed in the so-called saturator, which is placed directly before the membrane reactor. Here it can dissolve the donor salt continuously up to the solubility limit to keep the isopropylammonium and 3,3-diphenylpropionate concentrations in solution constant, which is later lowered by the biocatalytic reaction and the product crystallization step. The product salt is formed in the crystallizer, which is positioned after the membrane reactor and facilitates a constant crystal growth due to a slightly oversaturated product salt solution coming from the membrane reactor. Throughout the reaction donor salt is constantly consumed by dissolution in the saturator and equally product salt collected by crystal growth in the crystallizer. In the solution all concentrations are at an almost steady state except the above mentioned small oversaturation of the product salt and small undersaturation of the donor salt directly after the membrane reactor. In contrast to the single membrane reactor, an accumulation of acetone and thus limited equilibrium conversion seems not to be present here due the high surface area of the vessels and probably the applied tubing, which are permeable to acetone. In total, 1 g of biocatalytically produced product amine salt was obtained throughout the reaction. By dissolving the product salt in a basic solution, the product amine can be extracted into cyclopentylmethyl ether (CPME). By adding HCl, the hydrochloride of the product amine can be precipitated and filtered off, which was described in an earlier study [30]. The system remained fully stable over 33 h with a constant process productivity of 1.2 g/(L·d) (Figure 6). After 33 h an undesired product salt crystallization occurred in the biocatalyst chamber, which decreased overall productivity significantly. The crystal morphology between donor salt (platelets) and product salt (needles) is clearly different (Figure 7).

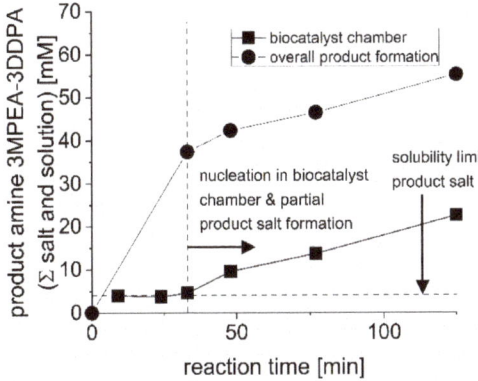

Figure 6. Product formation in the combined triple vessel concept; 30 °C, 25 mM HEPES buffer pH 7.5, 100 mM 3MAP, 60 mM IPA-3DPPA, 100 mM additional isopropylamine and 5 mM PLP (within entire solution).

The product salt was easily obtained from the reaction solution by a simple filtration and a single rinse with distilled water. The solid phase does not contain any cell or protein residue. The purity was determined by NMR with >99.5% and the enantiomeric excess was determined by high-performance liquid chromatography with >99.5%. XRPD analysis also showed that no donor salt formation occurred in the crystallizer and similarly no product salt was found in the saturator (Figure 8).

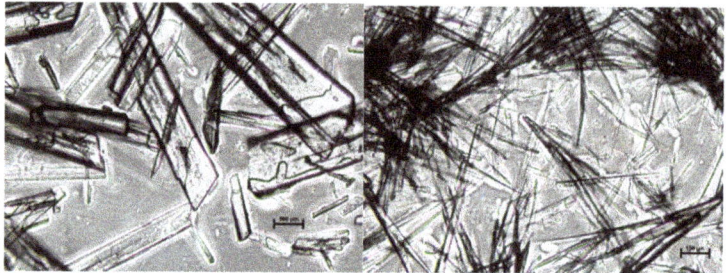

Figure 7. Crystal morphology of donor salt (left) and product salt (right).

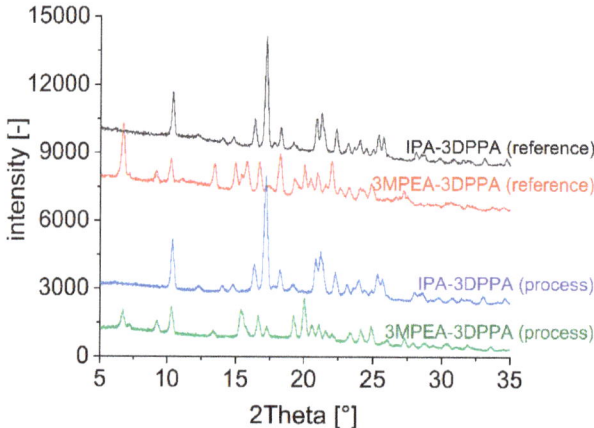

Figure 8. X-ray powder diffraction (XRPD) -analysis of donor and product salt; reference substances and samples from combined triple vessel concept.

4. Summary and Conclusions

In this study we reported the development of a continuously operated amine transaminase-catalyzed reaction, which is based on the integration of a reactive crystallization step for the in situ-removal of the product amine as a product salt. The presented concept involves the use of a membrane reactor, which retains the whole cell biocatalyst and two separate vessels for the application and collection of the donor salt (saturator) and product salt (crystallizer). The saturator provides a constant concentration of the required donor amine salt isopropylammonium 3,3-diphenylpropionate simultaneous to the crystallizer that collects the product amine salt (S)-1-(3-methoxyphenyl)ethylammonium 3,3-diphenylpropionate.

In conclusion, the shown triple vessel concept with its central membrane reactor and final crystallizer allows to overcome the very unfavorable chemical reaction equilibrium of the amine transaminase-catalyzed reaction in a continuously operated vessel concept. A fully stoichiometric reaction was achieved, which is not obtained in classical reaction concepts using isopropylamine as donor amine. The spatial separation of biocatalyst, saturator and crystallizer allows a full control of these components, including its separate removal and recycling after usage. The shown concept achieves very high product purity by the integrated crystallization step with only very few downstream-processing steps. The application of the membrane reactor provides a localization of the biocatalyst and prevents the use of potentially harmful immobilization techniques. The herein achieved space-time-yield of 1.2 g/(L·d) is directly correlated with the applied biocatalyst activity, which will increase in parallel with higher biocatalyst loadings. Future studies will also target the optimization of the shown concept in order to improve process productivity in such a continuous reaction mode. This primarily includes techniques to prevent the undesired nucleation by an optimized reactor design and the use of purified enzyme within the biocatalyst chamber.

Author Contributions: Conceptualization, J.v.L. and D.H.; methodology, D.H., P.K. and E.T.; formal analysis, D.H. and J.v.L.; investigation, D.H. and J.v.L.; writing—original draft preparation, D.H.; writing—review and editing, J.v.L. and E.T.; visualization, D.H. and J.v.L.; supervision, J.v.L.; project administration, J.v.L.; funding acquisition, J.v.L. All authors have read and agreed to the published version of the manuscript.

Funding: Funding by German Research Foundation (DFG, grant LA 4183/1-1), the Central Innovation Program SME of the Federal Ministry for Economic Affairs and Energy (ZIM, grant number 16KN073233) and the Leibniz ScienceCampus Phosphorus Research Rostock (grant CryPhos), is gratefully acknowledged.

Acknowledgments: The authors thank Hubert Bahl and Ralf-Jörg Fischer for their continuous support in microbiology, Dirk Michalik and Heike Borgwaldt for their assistance with NMR measurements, Martin Köckerling and Florian Schröder for their assistance with XRPD-measurements and Sandra Diederich and Rike Thomsen for technical and experimental support. We acknowledge financial support by Deutsche Forschungsgemeinschaft and Universität Rostock within the funding programme Open Access Publishing.

Conflicts of Interest: The authors declare no conflict of interest.

Abbreviation

3DPPA	3,3-Diphenylpropionic acid
3MAP	3′-Methoxyacetophenone
3MPEA	(S)-1-(3-Methoxyphenyl)ethylamine
3MPEA-3DPPA	(S)-1-(3-Methoxyphenyl)ethylammonium 3,3-diphenylpropionate (product salt)
ATA	Amine transaminase
CPME	Cyclopentylmethyl ether
HEPES	2-[4-(2-Hydroxyethyl)piperazin-1-yl]ethanesulfonic acid
IPA	Isopropylamine
IPA-3DPPA	Isopropylammonium 3,3-diphenylpropionate (donor salt)
NMR	Nuclear magnetic resonance
PLP	Pyridoxal 5′-phosphate
PVDF	Polyvinylidene fluoride transfer membrane
XRPD	X-ray powder diffraction

Appendix A

Appendix A.1. NMR-Data

Appendix A.1.1. Donor Salt IPA-3DPPA

^{1}H-NMR (373.2 K, DMSO-d6, 500.13 MHz, δ in ppm): 7.29–7.09 (m, 10H, Ar H), 4.47 (t, J = 7.7 Hz,1H, CH), 3.07 (m, J = 6.4 Hz, 1H, CH), 2.86 (d, J = 7.7 Hz, 2H, CH$_2$), 1.05 (d, J = 6.4 Hz, 6H, CH$_3$), NH$_3^+$ not given

^{13}C-NMR (373.1 K, DMSO-d6, 125.76 MHz, δ in ppm): 172.8 (CO$_2$), 144.7 (Ar), 127.7 (Ar), 127.2 (Ar), 125.4 (Ar), 46.9 (CH), 41.9 (CH$_2$), 41.6 (CH), 23.2 (CH$_3$)

Appendix A.1.2. Product Salt 3MPEA-3DPPA

^{1}H-NMR (373.1 K, DMSO-d6, 500.13 MHz, δ in ppm): 7.32–6.74 (m, 14H, Ar H), 5.16 (s, 3H, NH$_3^+$), 4.46 (t, J = 7.8 Hz, 1H, CH), 4.01 (q, J = 6.7 Hz, 1H, CH), 3.76 (s, 3H, O–CH$_3$), 2.97 (d, J = 7.8 Hz, 2H, CH$_2$), 1.29 (d, J = 6.7 Hz, 3H, CH$_3$)

^{13}C-NMR (373.2 K, DMSO-d6, 125.76 MHz, δ in ppm): 172.0 (CO$_2$), 159.0 (Ar), 149.2 (Ar), 143.9 (Ar), 128.5 (Ar), 127.8 (Ar), 127.1 (Ar), 125.5 (Ar), 117.7 (Ar), 111.5 (Ar), 111.3 (Ar), 54.6 (O–CH$_3$), 50.0 (CH), 46.5 (CH), 40.1 (CH2), 24.9 (CH$_3$)

References

1. Narancic, T.; Davis, R.; Nikodinovic-Runic, J.; O' Connor, K.E. Recent developments in biocatalysis beyond the laboratory. *Biotechnol. Lett.* **2015**, *37*, 943–954. [CrossRef] [PubMed]
2. Choi, J.-M.; Han, S.-S.; Kim, H.-S. Industrial applications of enzyme biocatalysis: Current status and future aspects. *Biotechnol. Adv.* **2015**, *33*, 1443–1454. [CrossRef] [PubMed]

3. Reetz, M.T. Biocatalysis in organic chemistry and biotechnology: Past, present, and future. *J. Am. Chem. Soc.* **2013**, *135*, 12480–12496. [CrossRef] [PubMed]
4. Munoz Solano, D.; Hoyos, P.; Hernaiz, M.J.; Alcantara, A.R.; Sanchez-Montero, J.M. Industrial biotransformations in the synthesis of building blocks leading to enantiopure drugs. *Bioresour. Technol.* **2012**, *115*, 196–207. [CrossRef]
5. Clouthier, C.M.; Pelletier, J.N. Expanding the organic toolbox: A guide to integrating biocatalysis in synthesis. *Chem. Soc. Rev.* **2012**, *41*, 1585–1605. [CrossRef]
6. Bornscheuer, U.T.; Huisman, G.W.; Kazlauskas, R.J.; Lutz, S.; Moore, J.C.; Robins, K. Engineering the third wave of biocatalysis. *Nature* **2012**, *485*, 185–194. [CrossRef]
7. Wohlgemuth, R. The locks and keys to industrial biotechnology. *New Biotechnol.* **2009**, *25*, 204–213. [CrossRef]
8. Ni, Y.; Holtmann, D.; Hollmann, F. How Green is Biocatalysis? To Calculate is To Know. *ChemCatChem* **2014**, *6*, 930–943. [CrossRef]
9. Hülsewede, D.; Meyer, L.-E.; von Langermann, J. Application of In Situ Product Crystallization and Related Techniques in Biocatalytic Processes. *Chem. Eur. J.* **2019**, *25*, 4871–4884. [CrossRef]
10. Enz, A. Phenylcarbamate for the Inhibition of Acetylcholinesterase. Patent DE3805744C2, 24 February 1988.
11. Farlow, M.R.; Cummings, J.L. Effective pharmacologic management of Alzheimer's disease. *Am. J. Med.* **2007**, *120*, 388–397. [CrossRef]
12. Emre, M. Rivastigmine in Parkinson's Disease Dementia. *CNS Drugs* **2006**, *20*, 748–750. [CrossRef]
13. Fuchs, M.; Koszelewski, D.; Tauber, K.; Kroutil, W.; Faber, K. Chemoenzymatic asymmetric total synthesis of (S)-Rivastigmine using omega-transaminases. *Chem. Commun.* **2010**, *46*, 5500–5502. [CrossRef] [PubMed]
14. Steffen-Munsberg, F.; Vickers, C.; Kohls, H.; Land, H.; Mallin, H.; Nobili, A.; Skalden, L.; van den Bergh, T.; Joosten, H.-J.; Berglund, P.; et al. Bioinformatic analysis of a PLP-dependent enzyme superfamily suitable for biocatalytic applications. *Biotechnol. Adv.* **2015**, *33*, 566–604. [CrossRef] [PubMed]
15. Mallin, H.; Höhne, M.; Bornscheuer, U.T. Immobilization of (R)- and (S)-amine transaminases on chitosan support and their application for amine synthesis using isopropylamine as donor. *J. Biotechnol.* **2014**, *191*, 32–37. [CrossRef]
16. Hülsewede, D.; Tänzler, M.; Süss, P.; Mildner, A.; Menyes, U.; Langermann, J. von. Development of an in situ-Product Crystallization (ISPC)-Concept to Shift the Reaction Equilibria of Selected Amine Transaminase-Catalyzed Reactions. *Eur. J. Org. Chem.* **2018**, *18*, 2130–2133. [CrossRef]
17. Payer, S.E.; Schrittwieser, J.H.; Kroutil, W. Vicinal Diamines as Smart Cosubstrates in the Transaminase-Catalyzed Asymmetric Amination of Ketones. *Eur. J. Org. Chem.* **2017**, *2017*, 2553–2559. [CrossRef]
18. Satyawali, Y.; Ehimen, E.; Cauwenberghs, L.; Maesen, M.; Vandezande, P.; Dejonghe, W. Asymmetric synthesis of chiral amine in organic solvent and in-situ product recovery for process intensification: A case study. *Biochem. Eng. J.* **2017**, *117*, 97–104. [CrossRef]
19. Heintz, S.; Borner, T.; Ringborg, R.H.; Rehn, G.; Grey, C.; Nordblad, M.; Kruhne, U.; Gernaey, K.V.; Adlercreutz, P.; Woodley, J.M. Development of in situ product removal strategies in biocatalysis applying scaled-down unit operations. *Biotechnol. Bioeng.* **2017**, *114*, 600–609. [CrossRef]
20. Rehn, G.; Ayres, B.; Adlercreutz, P.; Grey, C. An improved process for biocatalytic asymmetric amine synthesis by in situ product removal using a supported liquid membrane. *J. Mol. Catal. B Enzym.* **2016**, *123*, 1–7. [CrossRef]
21. Gomm, A.; Lewis, W.; Green, A.P.; O'Reilly, E. A New Generation of Smart Amine Donors for Transaminase-Mediated Biotransformations. *Chem. Eur. J.* **2016**, *22*, 12692–12695. [CrossRef]
22. Börner, T.; Rehn, G.; Grey, C.; Adlercreutz, P. A Process Concept for High-Purity Production of Amines by Transaminase-Catalyzed Asymmetric Synthesis: Combining Enzyme Cascade and Membrane-Assisted ISPR. *Org. Process Res. Dev.* **2015**, *19*, 793–799. [CrossRef]
23. Green, A.P.; Turner, N.J.; O'Reilly, E. Chiral amine synthesis using omega-transaminases: An amine donor that displaces equilibria and enables high-throughput screening. *Angew. Chem. Int. Ed.* **2014**, *53*, 10714–10717. [CrossRef] [PubMed]
24. Cassimjee, K.E.; Branneby, C.; Abedi, V.; Wells, A.; Berglund, P. Transaminations with isopropyl amine: Equilibrium displacement with yeast alcohol dehydrogenase coupled to in situ cofactor regeneration. *Chem. Commun.* **2010**, *46*, 5569–5571. [CrossRef] [PubMed]

25. Rehn, G.; Adlercreutz, P.; Grey, C. Supported liquid membrane as a novel tool for driving the equilibrium of omega-transaminase catalyzed asymmetric synthesis. *J. Biotechnol.* **2014**, *179*, 50–55. [CrossRef] [PubMed]
26. Höhne, M.; Bornscheuer, U.T. Biocatalytic Routes to Optically Active Amines. *ChemCatChem* **2009**, *1*, 42–51. [CrossRef]
27. Simon, R.C.; Richter, N.; Busto, E.; Kroutil, W. Recent Developments of Cascade Reactions Involving ω-Transaminases. *ACS Catal.* **2014**, *4*, 129–143. [CrossRef]
28. Koszelewski, D.; Lavandera, I.; Clay, D.; Rozzell, D.; Kroutil, W. Asymmetric Synthesis of Optically Pure Pharmacologically Relevant Amines Employing ω-Transaminases. *Adv. Synth. Catal.* **2008**, *350*, 2761–2766. [CrossRef]
29. Slabu, I.; Galman, J.L.; Lloyd, R.C.; Turner, N.J. Discovery, Engineering, and Synthetic Application of Transaminase Biocatalysts. *ACS Catal.* **2017**, *7*, 8263–8284. [CrossRef]
30. Hülsewede, D.; Dohm, J.-N.; von Langermann, J. Donor Amine Salt-Based Continuous in situ- Product Crystallization in Amine Transaminase-Catalyzed Reactions. *Adv. Synth. Catal.* **2019**, *361*, 2727–2733. [CrossRef]
31. Meng, Q.; Capra, N.; Palacio, C.M.; Lanfranchi, E.; Otzen, M.; van Schie, L.Z.; Rozeboom, H.J.; Thunnissen, A.-M.W.H.; Wijma, H.J.; Janssen, D.B. Robust ω-Transaminases by Computational Stabilization of the Subunit Interface. *ACS Catal.* **2020**, 2915–2928. [CrossRef]
32. Grabner, B.; Nazario, M.A.; Gundersen, M.T.; Loïs, S.; Fantini, S.; Bartsch, S.; Woodley, J.M.; Gruber-Woelfler, H. Room-temperature solid phase ionic liquid (RTSPIL) coated ω-transaminases: Development and application in organic solvents. *Mol. Catal.* **2018**, *452*, 11–19. [CrossRef]
33. Miložič, N.; Lubej, M.; Lakner, M.; Žnidaršič-Plazl, P.; Plazl, I. Theoretical and experimental study of enzyme kinetics in a microreactor system with surface-immobilized biocatalyst. *Chem. Eng. J.* **2017**, *313*, 374–381. [CrossRef]
34. Rehn, G.; Grey, C.; Branneby, C.; Lindberg, L.; Adlercreutz, P. Activity and stability of different immobilized preparations of recombinant E. coli cells containing ω-transaminase. *Process Biochem.* **2012**, *47*, 1129–1134. [CrossRef]
35. Sheldon, R.A. Enzyme Immobilization: The Quest for Optimum Performance. *Adv. Synth. Catal.* **2007**, *349*, 1289–1307. [CrossRef]
36. Uthoff, F.; Sato, H.; Gröger, H. Formal Enantioselective Hydroamination of Non-Activated Alkenes: Transformation of Styrenes into Enantiomerically Pure 1-Phenylethylamines in Chemoenzymatic One-Pot Synthesis. *ChemCatChem* **2017**, *9*, 555–558. [CrossRef]
37. Wachtmeister, J.; Jakoblinnert, A.; Kulig, J.; Offermann, H.; Rother, D. Whole-Cell Teabag Catalysis for the Modularisation of Synthetic Enzyme Cascades in Micro-Aqueous Systems. *ChemCatChem* **2014**, *6*, 1051–1058. [CrossRef]

© 2020 by the authors. Licensee MDPI, Basel, Switzerland. This article is an open access article distributed under the terms and conditions of the Creative Commons Attribution (CC BY) license (http://creativecommons.org/licenses/by/4.0/).

Article

TiO$_2$-Seeded Hydrothermal Growth of Spherical BaTiO$_3$ Nanocrystals for Capacitor Energy-Storage Application

Ming Li [1,2], Lulu Gu [1], Tao Li [2], Shiji Hao [2], Furui Tan [2], Deliang Chen [2,3,*], Deliang Zhu [1,*], Yongjun Xu [2], Chenghua Sun [2] and Zhenyu Yang [2,*]

1. College of Materials Science and Engineering, Shenzhen University, Shenzhen 518060, China; 2172341493@email.szu.edu.cn (M.L.); 21611200210@email.szu.edu.cn (L.G.)
2. School of Materials Science and Engineering, Institute of Science & Technology Innovation, Dongguan University of Technology, Dongguan 523808, China; 2016812@dgut.edu.cn (T.L.); haosj@dgut.edu.cn (S.H.); 2019126@dgut.edu.cn (F.T.); hnllxyj@dgut.edu.cn (Y.X.); chenghua.sun@monash.edu (C.S.)
3. School of Materials Science and Engineering, Zhengzhou University, Zhengzhou 450001, China
* Correspondence: dlchen@dgut.edu.cn (D.C.); dlzhu@szu.edu.cn (D.Z.); zyyang@dgut.edu.cn (Z.Y.)

Received: 17 February 2020; Accepted: 2 March 2020; Published: 14 March 2020

Abstract: Simple but robust growth of spherical BaTiO$_3$ nanoparticles with uniform nanoscale sizes is of great significance for the miniaturization of BaTiO$_3$-based electron devices. This paper reports a TiO$_2$-seeded hydrothermal process to synthesize spherical BaTiO$_3$ nanoparticles with a size range of 90–100 nm using TiO$_2$ (Degussa) and Ba(NO$_3$)$_2$ as the starting materials under an alkaline (NaOH) condition. Under the optimum conditions ([NaOH] = 2.0 mol L^{-1}, $R_{Ba/Ti}$ = 2.0, T = 210 °C and t = 8 h), the spherical BaTiO$_3$ nanoparticles obtained exhibit a narrow size range of 91 ± 14 nm, and the corresponding BaTiO$_3$/polymer/Al film is of a high dielectric constant of 59, a high break strength of 102 kV mm^{-1}, and a low dielectric loss of 0.008. The TiO$_2$-seeded hydrothermal growth has been proved to be an efficient process to synthesize spherical BaTiO$_3$ nanoparticles for potential capacitor energy-storage applications.

Keywords: spherical BaTiO$_3$ nanoparticle; hydrothermal synthesis; nanoscale TiO$_2$ seed; crystal growth; dielectric property

1. Introduction

Barium titanate (BaTiO$_3$) has been an important material in the manufacture of electronic components for many years due to its unique properties of high dielectric constant, high ferroelectricity, and piezoelectricity [1–5]. BaTiO$_3$-based ceramics are of a wide range of potential applications in ferroelectric random access memory (FRAM) [6], photoelectric humidity sensors [7], solid oxide fuel cells [8], superconductors [9], ferromagnets [10,11], high capacitance capacitors [12–14], pyroelectric detectors [15], and magneto resistors [16]. Especially, tetragonal BaTiO$_3$ ceramics are widely used in multi-layer ceramic capacitors (MLCC) [17], thermistors [18], and piezoelectric sensors [19]. Conventional methods used to prepare BaTiO$_3$ ceramics are solid-state reaction processes using TiO$_2$ and BaCO$_3$ as the raw materials at an elevated temperature of more than 1200 °C. The large size and low purity of the BaTiO$_3$ ceramics obtained by the solid-state reaction have limited their applications in nanotechnological fields.

The miniaturization of electronic components and nanotechnology makes it necessary to synthesize nanometer-scale BaTiO$_3$ materials, including nanowires [20] and nanoparticles [21], with scientific appeal and technical urgency. Device miniaturization and high dielectric constant can be achieved by controlling their microstructures and compositions, which are strongly dependent on the phase, uniformity, surface area, and size of the BaTiO$_3$ materials [22–24]. For the applications in MLCC, BaTiO$_3$ powders are usually used as dielectric fillers and blended with a polymer to a fabricate composite film with a compact and flexible surface. In order to manufacture a reliable BaTiO$_3$-based MLCC, high-quality BaTiO$_3$ powders with high purity, high crystallinity, high dispersibility, and uniform small size are the precondition. The BaTiO$_3$ fillers with a narrow particle-size distribution and suitable phases are in favor of obtaining a compact composite film with a lower content of pores, and the dense and homogeneous BaTiO$_3$ phase in polymer matrix can lead to higher dielectric properties of the composite films [25]. R.K.Goyal et al. found that the dielectric constants of the composite films filled with tetragonal BaTiO$_3$ powders are higher than those of the films with cubic BaTiO$_3$ fillers; whereas the effect of crystal phase on the dielectric losses presents an opposite trend that the composite filled with a cubic BaTiO$_3$ filler shows a lower dielectric loss than that of the tetragonal BaTiO$_3$ composite film [26]. Therefore, a high-quality BaTiO$_3$ filler is important for high performance composite dielectric films, and a recent investigation on the synthesis of BaTiO$_3$ nanocrystals via various processes has become one of the hot topics.

There have been a number of methods developed to prepare high-quality BaTiO$_3$ powders [27]. As mentioned above, the conventional route used to prepare BaTiO$_3$ powders is via a solid-state reaction between BaCO$_3$ and TiO$_2$ at a high temperature of 850–1400 °C [28]. This solid-state method is easy in operation and allows for mass production, but there are a number of serious drawbacks in the control of particle-size (morphology) and compositional purity. Ball-milling is usually used to mix BaCO$_3$ and TiO$_2$. It is not only time-consuming and labor-intensive but also easy to introduce impurities [29]. As an alternative to the solid-state process, various "wet chemical" methods, including sol-gel process [30,31], hydrothermal method [32], micro-emulsions [33], and oxalate process [34] have been developed to synthesize BaTiO$_3$ powders. These methods can produce high-purity, uniform, ultrafine BaTiO$_3$ powders. Because of the complexity of operation, multi-stage, and relatively high cost, most of these methods are mainly used at the laboratory level. It should be noted that the hydrothermal process is a promising method to synthesize BaTiO$_3$ powders with controllable morphology and chemical uniformity.

The hydrothermal method can use various processing conditions in the synthesis of BaTiO$_3$ powders including the sources of barium and titanium in an aqueous medium under crystallization or amorphous state, the hydrothermal temperature and time, and morphology-controlled agents. Because of the diversity of the factors that affect the synthesis of BaTiO$_3$ nanoparticles, hydrothermal methods are full of opportunities to improve their quality in phase composition, dimensions, and morphology. Li et al. [35] reported the synthesis of tetragonal BaTiO$_3$ nanocrystals using TiCl$_4$ (or TiO$_2$) as the source of titanium, BaCl$_2$ as the source of barium, and polymer(vinylpyrrolidone) (PVP) as the surfactant. Grendal et al. [36] used two titanium sources of amorphous titanium dioxide and a Ti-citrate complex solution to synthesize BaTiO$_3$ nanoparticles with a size range of 10–15 nm at different hydrothermal temperatures and times. Zhao et al. [37] used cetyltrimethylammonium bromide (CTAB), Ba(OH)$_2$·8H$_2$O, and tetrabutyl titanate as the precursors to synthesize BaTiO$_3$ nanocrystals via a self-assembly process. Ozen et al. [38] reported the hydrothermal synthesis of tetragonal BaTiO$_3$ nanocrystals from a single-source amorphous barium titanate precursor in a high concentration sodium hydroxide solution via a homogeneous dissolution-precipitation reaction. From the above cases, one can see that different hydrothermal parameters and growth mechanisms can effectively adjust the formation of BaTiO$_3$ nanocrystals. In addition, a single cubic phase of BaTiO$_3$ can be formed at a low alkalinity, and a tetragonal phase of BaTiO$_3$ is easily formed under a strong alkaline condition [39].

With the motivation of preparing cubic/tetragonal BaTiO$_3$ nanocrystals with a spherical morphology, this paper herein develops a TiO$_2$-seeded hydrothermal process to grow BaTiO$_3$ nanocrystals using Ba(NO$_3$)$_2$ and TiO$_2$(P25) as the barium and titanium sources, respectively. This synthesis is conducted under a strong alkaline NaOH aqueous solution (pH = 13.6), and the factors that affect the formation of BaTiO$_3$ nanocrystals are systematically investigated. The major influencing factors involve molar Ba/Ti ratios, hydrothermal temperature, and hydrothermal time, and their effects on the morphology, particle size, and phase composition of the BaTiO$_3$ nanoparticles are investigated. The possible growth mechanisms are discussed. The BaTiO$_3$/polymer/Al films containing the BaTiO$_3$ nanoparticles obtained under the optimum conditions are of a high dielectric constant of 59, a high break strength of 102 kV mm^{-1} and a low dielectric loss of 0.008. This work achieves this aim to seek optimum methods to synthesize spherical BaTiO$_3$ nanoparticles with potential applications in capacitor energy-storage and other electric devices.

2. Materials and Methods

2.1. Chemicals and Settings

Barium nitrate (Ba(NO$_3$)$_2$, analytical grade) was purchased from Tianjin Shengao Chemical Reagent Co., Ltd (Tianjin, China). Titanium dioxide (TiO$_2$, P25, chemically pure) was purchased from Degussa. Sodium hydroxide (NaOH, analytical grade) was purchased from Tianjin Komi Chemical Reagent Co., Ltd (Tianjin, China). Ethanol (analytical grade) was purchased from Tianjin Kaitong Chemical Reagent Co., Ltd (Tianjin, China). Distilled water was used in all the experiments. The drying oven (XMTD-8222) was purchased from Shanghai Jinghong Experimental Equipment Co., Ltd (Shanghai, China). The desktop high-speed centrifuge (H1850) was purchased from Hunan Xiangyi Centrifuge Co., Ltd (Xiangtan, China). The polymer (ceramic glue), a silicon-containing heat-resistant resin, was purchased from the IPINRU Chen Yu Technology Co., Ltd (Chengdu, China, Product No. CYN-01 with a curing temperature of ~220 °C). A silane coupling agent (KH550, NH$_2$CH$_2$CH$_2$CH$_2$Si(OC$_2$H$_5$)$_3$) was purchased from Guangzhou Yuantai Synthetic Material Co., Ltd (Guangzhou, China). Al foils (thickness = ~12 μm, tensile strength ≥ 180 MPa, ductility ≥ 15%) were purchased from Shenzhen Kejing Star Technology Co., Ltd (Shenzhen, China).

2.2. Growth of Spherical BaTiO$_3$ Nanoparticles

BaTiO$_3$ samples were synthesized via a hydrothermal process using TiO$_2$ (P25) nanoparticles as the Ti source and seeds. The synthetic process of the BaTiO$_3$ nanocrystals is shown in Figure 1. Teflon-lined autoclaves with a volume of 100 mL were used as the reaction vessel. Typically, 6.0 g of NaOH and 1.5 g of TiO$_2$ nanoparticles were first added into 75 mL of distilled water under magnet stirring; then a given amount of Ba(NO$_3$)$_2$ was added to the above suspension containing TiO$_2$ nanoparticles and NaOH under magnetic stirring. In the final suspensions, the molar ratios of Ba(NO$_3$)$_2$ to TiO$_2$ ($R_{Ba/Ti}$) were kept at 1.6–2.0, and the molar concentration of NaOH was about 2 mol L^{-1}. The pH values of the as-obtained suspensions before hydrothermal treatment were about 13.6. The prepared suspensions were then transferred into the Teflon-lined steel autoclaves. After carefully sealing, the autoclaves were heated in an oven at 150–210 °C for 2–16 h. After the hydrothermal reaction, the autoclaves were cooled naturally, and the solid samples were collected using a centrifugal machine (5000 rpm, 5 min), followed by washing with water for more than three times and drying at 120 °C for 24 h. The as-obtained BaTiO$_3$ solids were ground into powders using an agate mortar. These white powders, i.e., BaTiO$_3$ nanocrystals, were collected and used for characterization. The detailed processing parameters for the synthesis of BaTiO$_3$ nanocrystals are listed in Table 1. It was assumed that TiO$_2$ added was completely converted into BaTiO$_3$, and the theoretical mass could be calculated. The yield of BaTiO$_3$ was the ratio of the actual mass of the BaTiO$_3$ sample to their corresponding theoretical mass.

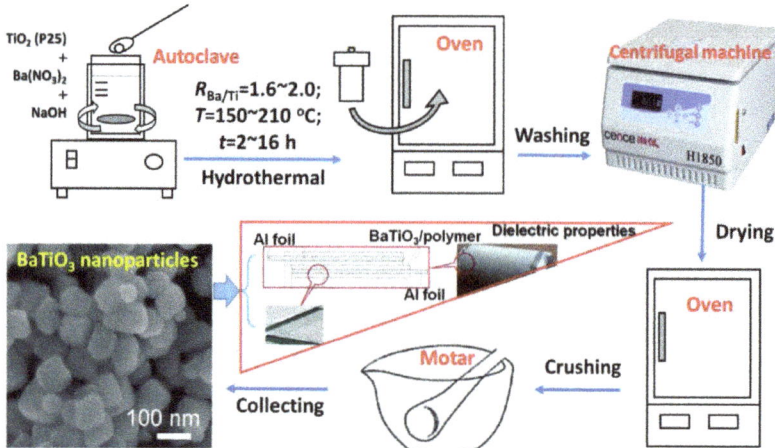

Figure 1. Schematic of the hydrothermal synthesis of BaTiO$_3$ nanocrystals using TiO$_2$ (P25) nanoparticles as the seeds and Ti source.

Table 1. A summary of experimental conditions for hydrothermal synthesis of BaTiO$_3$ nanoparticles.

Sample	[NaOH] (mol L^{-1})	$R_{Ba/Ti}$	Hydrothermal Temperature (°C)	Hydrothermal Duration (h)	Particle Size (nm)
S1	2	1.6	200	8	97 ± 15
S2	2	1.8	200	8	93 ± 24
S3	2	2.0	200	8	91 ± 22
S4	2	2.5	200	8	98 ± 26
S5	2	2.0	150	8	85 ± 15
S6	2	2.0	165	8	74 ± 13
S7	2	2.0	180	8	88 ± 10
S8	2	2.0	210	8	91 ± 14
S9	2	2.0	210	2	76 ± 17
S10	2	2.0	210	4	90 ± 15
S11	2	2.0	210	12	100 ± 20
S12	2	2.0	210	16	103 ± 20

2.3. Preparation of BaTiO$_3$/Polymer/Al (BPA) Films

To determine the possibility of the as-obtained BaTiO$_3$ nanocrystals to form a uniform film for capacitor energy-storage application, we chose sample S8 (in Table 1) as an example to prepare BaTiO$_3$/polymer/Al films (BPA films, Figure 2) using the similar method reported in our previous work [25]. Typically, the BaTiO$_3$ nanocrystals (S8) were mixed with a silicon-containing heat-resistant resin (CYN-01), and then some silane coupling agent (KH550) was added into the above mixture. Dimethylacetamide (DMAc, Guangzhou Jinhuada Chemical Reagent Co., Ltd., Guangzhou, China)) was used as the solvent. The mass ratio of M_{BaTiO3}:M_{DMAc}:$M_{Polymer}$:M_{KH550} was kept at 100:45:25:4. The as-prepared mixture was ultrasonically treated for 30 min for a uniform slurry. The above slurry was coated on an Al foil by a bar coater (T-300CA) and a coating rod (D10-OSP010-L0400) from Shijiazhuang Ospchina Machinery Technology Co., Ltd (Shijiazhuang, China)). The as-formed films were then dried in an oven at 220 °C for 10 min and finally used for the test of dielectric properties.

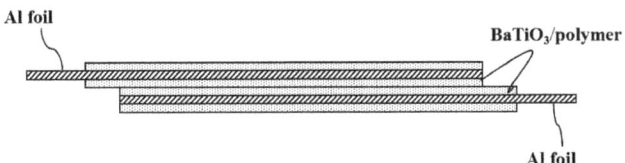

Figure 2. A schematic diagram of the BaTiO$_3$/polymer/Al film for capacitor cells.

2.4. Characterization of BaTiO$_3$ Nanocrystals and BPA Films

The X-ray diffraction (XRD) patterns of the BPA composite films and BaTiO$_3$ powders were recorded by a DX-2700BH X-ray diffractometer (Dandong, China) using Cu Kα irradiation. The morphologies and particle sizes of the BaTiO$_3$ samples were measured using a scanning electron microscope (SEM, Hitachi S-4800, Japan). The particle-size distribution was statistically analyzed according to the SEM images. The pH values of the suspensions were measured using a pH meter (PHS-2C). The yields of the BaTiO$_3$ samples were calculated according to the ratios of experimental BaTiO$_3$ mass to its theoretical mass on the basis of Ba conservation. Fourier-transform infrared (FT-IR) spectra were recorded on a Bruker–Equinox 55 spectrometer in a wavenumber range of 4000–400 cm^{-1} using the KBr technique. The dielectric constant (ε) and loss (tanδ) of the BPA films were measured using a high-precision high-voltage capacitor bridge (QS89, Shanghai Yanggao Capacitor Co., Ltd., Shanghai, China), and the frequency during dielectric performance test was kept at 10 Hz. The breakdown strengths of the BPA films were measured using a withstand voltage tester (GY2670A, Guangzhou Zhizhibao Electronic Instrument Co., Ltd., Guangzhou, China).

3. Results and Discussion

The TiO$_2$-seeded growth process of BaTiO$_3$ nanocrystals is shown in Figure 1. The commercially available TiO$_2$ (P25) nanoparticles, with a mixed phase of anatase and rutile and a size range of 20–25 nm, are used as the Ti source and seeds in the synthesis of BaTiO$_3$ nanocrystals via a conventional hydrothermal process in a strongly basic aqueous solution. In this synthesis, TiO$_2$ nanoparticles can first react with NaOH and form insoluble titanate species (e.g., Na$_2$TiO$_3$), which then act like the crystal nucleus to form BaTiO$_3$ nanocrystals by reacting with Ba^{2+} ions under the hydrothermal conditions. We systematically investigated the effects of molar ratios of Ba/Ti ($R_{Ba/Ti}$), hydrothermal temperature ($T/°C$) and time (t/h) on the phase, morphology and particle size of the BaTiO$_3$ nanocrystals.

3.1. Influence of Molar Ba/Ti Ratio

In order to verify the effect of the molar Ba/Ti ratio on the formation of BaTiO$_3$ nanoparticles, we synthesized a series of samples with various $R_{Ba/Ti}$ values from 1.6 to 2.5, and the other hydrothermal conditions were kept as the same: sodium hydroxide concentration [NaOH] = 2.0 mol L^{-1} (pH = 13.6), T = 200 °C, and t = 8 h. The typical results of these samples are shown in Figure 3.

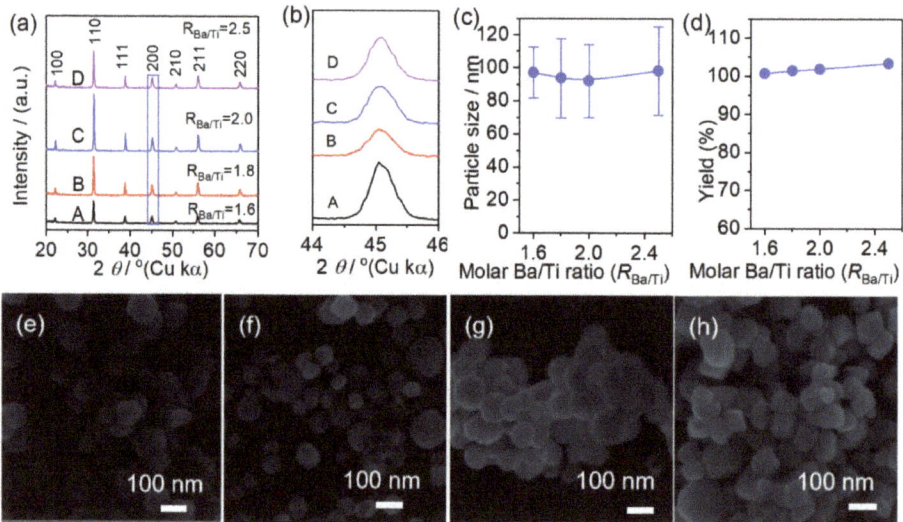

Figure 3. X-ray diffraction (XRD) patterns (**a**,**b**), particle sizes and yields (**c**,**d**), and scanning electron microscope (SEM) images (**e**–**h**) of the BaTiO$_3$ samples obtained with various molar Ba/Ti ratios ($R_{Ba/Ti}$ = 1.6–2.5) under hydrothermal conditions at 200 °C for 8 h ([NaOH] = 2.0 mol L^{-1}, pH ≈ 13.6).

Figure 3a,b shows the XRD patterns of the BaTiO$_3$ samples synthesized at different molar Ba/Ti ratios. From Figure 3a, one can find that all the samples show seven distinct peaks at around 21.98, 31.36, 38.64, 44.92, 50.58, 55.86 and 65.44°, corresponding to the (100), (110), (111), (200), (210), (211) and (220) reflections of the cubic BaTiO$_3$ phase, respectively, according to the JCPDS card no. 31-0174 [40]. No peaks belonging to other identifiable impurities can be found in all the samples obtained, indicating the as-obtained BaTiO$_3$ samples are pure. As Figure 3b shows, the peak at about 45° can be divided into two diffraction sub-peaks at 44.9 and 45.3°, attributable to the (200) and (002) reflections of the tetragonal BaTiO$_3$ species, respectively [41]. With the increase of the $R_{Ba/Ti}$ value from 1.6 to 2.5, the peaks near 45° become wider and wider, suggesting that a higher $R_{Ba/Ti}$ value is favorable in forming a tetragonal BaTiO$_3$ phase.

Figure 3c shows the plots of particle size dependent on the $R_{Ba/Ti}$ values. When $R_{Ba/Ti}$ = 1.6–1.8, the particle sizes are 90–100 nm (97 ± 15 nm for $R_{Ba/Ti}$ = 1.6 and 93 ± 24 nm for $R_{Ba/Ti}$ = 1.8), but the uniform degree is not high. Figure 3d shows the yields of BaTiO$_3$ samples synthesized with various $R_{Ba/Ti}$ values after hydrothermally treating at 200 °C for 8 h ([NaOH] = 2.0 mol L^{-1}). One can see that the yields of all the samples are close to 100%, indicating the complete conversion of TiO$_2$ to BaTiO$_3$ nanocrystals. The formation of a small amount of crystal water may make the BaTiO$_3$ yield a little larger than 100% according to the TiO$_2$ amount [42].

Figure 3e–h shows the typical SEM images of the BaTiO$_3$ samples obtained with various $R_{Ba/Ti}$ values ([NaOH] = 2.0 mol L^{-1}, T = 200 °C, t = 8 h). According to the SEM observations, when $R_{Ba/Ti}$ = 2.0 (Figure 3g), the particle size of the BaTiO$_3$ sample is 91 ± 22 nm, and it shows a more uniform solid spherical particle morphology. When $R_{Ba/Ti}$ = 2.5 (Figure 3h), the particle size of the BaTiO$_3$ sample is 98 ± 26 nm, and one can see that it shows obviously clean-cut crystal faces for the BaTiO$_3$ particles, suggesting a higher degree of crystallinity and favorable formation of the tetragonal BaTiO$_3$ phase.

Taking the results of XRD and particle-size distribution into account, we can tentatively conclude that a higher Ba/Ti ratio is more favorable in forming tetragonal BaTiO$_3$ nanocrystals with a more uniform size.

3.2. Influence of Hydrothermal Temperature

The effect of hydrothermal temperature on the synthesis of BaTiO$_3$ nanoparticles was investigated by changing the hydrothermal temperature from 150 to 210 °C under the conditions: $R_{Ba/Ti} = 2.0$, $t = 8$ h and [NaOH] = 2.0 mol L^{-1}, and Figure 4 shows their characterization results of XRD and SEM.

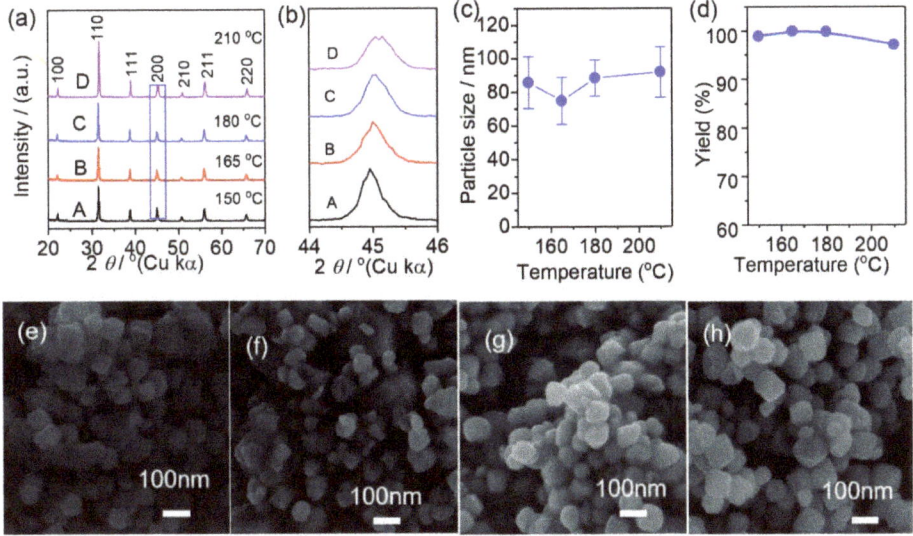

Figure 4. XRD patterns (**a**,**b**), particle sizes and yields (**c**,**d**), and SEM images (**e**–**h**) of the BaTiO$_3$ nanocrystals obtained with $R_{Ba/Ti} = 2.0$ under hydrothermal conditions at 150–210 °C for 8 h ([NaOH] = 2.0 mol L^{-1}, pH ≈ 13.6).

Figure 4a,b shows the typical XRD patterns of the BaTiO$_3$ samples obtained at different hydrothermal temperatures. From Figure 4a, one can see that all the BaTiO$_3$ samples show similar XRD patterns, all peaks of which can be attributed to the cubic BaTiO$_3$ phase (JCPDS card no. 31-0174), and no impure XRD peaks are found, indicating that the formation of pure BaTiO$_3$ crystals. The partially enlarged XRD patterns located in 2θ = 44–46° (Figure 4b) show the XRD peaks become wider and wider with the increase of hydrothermal temperature from 150 to 210 °C, and can be sub-divided to two peaks at 44.9° and 45.3°, attributable to the (200) and (002) reflections of the tetragonal BaTiO$_3$ phase.

Figure 4c shows the particle-size distribution plot versus hydrothermal temperature (T). When T = 150 °C, the particle sizes of the as-obtained BaTiO$_3$ nanocrystals are 85 ± 15 nm. When T = 165 °C, the particle size of the as-obtained BaTiO$_3$ is about 74 ± 13 nm, seeming to become smaller, but their uniformity is low. When the temperature increases to 180 °C, the particle size of the as-obtained BaTiO$_3$ is 88 ± 10 nm, and the morphology of the BaTiO$_3$ particles becomes relatively uniform. When T = 210 °C, the particle size of the as-obtained the BaTiO$_3$ sample is 91 ± 14 nm, just a slight increase. As Figure 4c shows, the particle sizes of the BaTiO$_3$ samples obtained at various hydrothermal temperatures are kept almost constant at about 80–90 nm.

Figure 4d shows the plot of the yield of the BaTiO$_3$ sample versus the hydrothermal temperature. One can see that during the hydrothermal temperature of 150–180 °C, the yield is close to 100%; when the hydrothermal temperature is 210 °C, the yield slightly decreases because of the complete dehydration reaction in the elevated temperature.

Figure 4e–h shows the typical SEM images of the BaTiO$_3$ samples synthesized under various hydrothermal temperatures for 8 h ([NaOH] = 2.0 mol L^{-1}, $R_{Ba/Ti} = 2.0$): (e) 150 °C, (f) 165 °C, (g) 180 °C, and (h) 210 °C. One can see that all the BaTiO$_3$ samples consist of spherical nanoparticles. With the increase of hydrothermal temperature, the as-obtained BaTiO$_3$ samples exhibit a higher degree of crystallinity indicated by the clean-cut crystal planes.

According to the XRD patterns (Figure 4a,b) and SEM images (Figure 4e–h), we find that a higher hydrothermal temperature is helpful to form tetragonal BaTiO$_3$ nanocrystals with more uniform spherical morphology. For safety's sake, the hydrothermal temperature is chosen as 210 °C for the synthesis of BaTiO$_3$ nanocrystals in the following investigation. Cautions: the working temperature limit of a PTFE hydrothermal reactor is usually about 220 °C, and a too high temperature will cause explosion.

3.3. Influence of Hydrothermal Time

The effect of hydrothermal time on the formation of BaTiO$_3$ nanocrystals (Figure 5) are investigated under the conditions: $R_{Ba/Ti} = 2.0$, $T = 210$ °C, [NaOH] = 2.0 mol L^{-1}, and $t = 2$–16 h. Figure 5a,b shows their XRD patterns. As Figure 5a shows, the XRD peaks of all the samples can be assignable to the cubic/tetragonal BaTiO$_3$ phase with no other identifiable impurity peaks. The partially enlarged XRD patterns in Figure 5b shows the details that the XRD peaks at around 45° become wider and wider as the hydrothermal time increases from 2 h to 16 h, indicating that the BaTiO$_3$ sample obtained with a longer hydrothermal time has more tetragonal BaTiO$_3$ species.

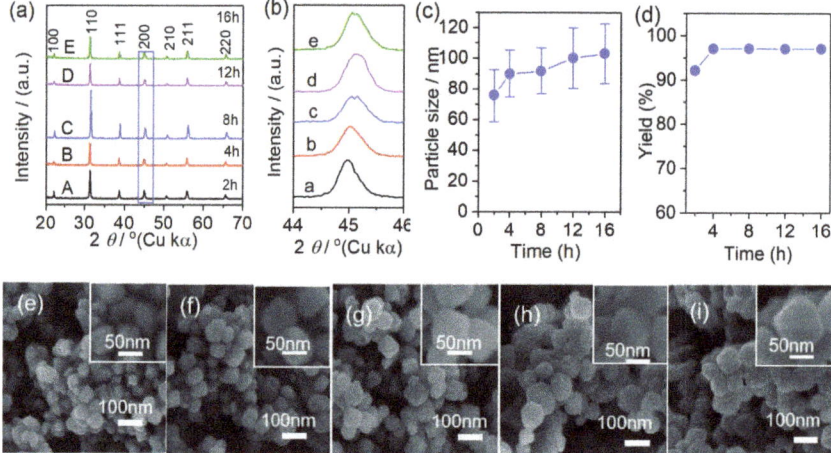

Figure 5. XRD patterns (**a,b**), particle sizes and yields (**c,d**), and SEM images (**e–i**) of the BaTiO$_3$ nanocrystals obtained with $R_{Ba/Ti} = 2.0$([NaOH]=2.0 mol L^{-1}, pH ≈ 13.6) by hydrothermally treating at 210 °C for various times ($t = 2$–16 h).

Figure 5c shows the BaTiO$_3$ sample gradually changes from small nanoparticles (~70 nm) to large ones (~100 nm) as the hydrothermal time is prolonged from 2 h to 16 h. Figure 5d shows the yield plot of the BaTiO$_3$ nanocrystals versus hydrothermal time. With a short hydrothermal time of 2 h, the BaTiO$_3$ yield is about 92% because of the incomplete reaction. When the hydrothermal time increases to 4–16 h, the yields of the BaTiO$_3$ samples is close to 98%.

Figure 5e–h shows the SEM images of the BaTiO$_3$ samples obtained with various hydrothermal times ($R_{Ba/Ti} = 2.0$, $T = 210$ °C, [NaOH] = 2.0 mol L^{-1}). The BaTiO$_3$ samples obtained with short hydrothermal times of 2–8 h, as shown in Figure 5e–g, exhibit a spherical shape; when the hydrothermal time increases to 12–16 h, as Figure 5h,i shows, the as-obtained BaTiO$_3$ samples take on a planar polyhedral morphology. It is interesting that the particle sizes of the BaTiO$_3$ samples are close to 100 nm and not changed obviously with the prolonging of hydrothermal time to 16 h. In addition, as

Figure 5i shows, the BaTiO$_3$ nanoparticles obtained by hydrothermal treating at 210 °C for 16 h are uniform in particle size and well dispersed.

Figure 6 shows the FT-IR spectra of the BaTiO$_3$ samples synthesized with different hydrothermal times ($R_{Ba/Ti}$ = 2.0, T = 210 °C, [NaOH] = 2.0 mol L^{-1}). The bands at 3431 and 1568 cm^{-1} can be attributed to the stretching mode of the adsorbed water molecules and O–H groups, indicating that the surfaces of the BaTiO$_3$ nanocrystals contain some adsorbed water and –OH groups. The weak band at 1400 cm^{-1} can be attributed to the stretching mode of the C–O groups because of the incorporation of CO$_2$ into the basic solution. The broad and strong absorption bands at 562 cm^{-1} is attributed to the normal vibration of Ti–O$_I$ stretching, and the weaker and sharper absorption bands near 438 cm^{-1} can be attributed to the normal vibration of Ti–O$_{II}$ bending. When the hydrothermal time is extended from 2 h to 16 h, the bands at 562 and 438 cm^{-1} become stronger and sharper, indicating that the BaTiO$_3$ nanocrystals with a high degree of crystallinity are formed. According to the XRD patterns (Figure 5a,b), SEM images (Figure 5e–i) and FT-IR spectra (Figure 6), the BaTiO$_3$ nanocrystals obtained by hydrothermal treating at 210 °C for more than 8 h are of uniform spherical morphologies with a size range of 95–100 nm and high degree of crystallinity. Therefore, the optimum hydrothermal parameters for the synthesis of BaTiO$_3$ nanocrystals can be $R_{Ba/Ti} \geq 2$, $T \geq 200$ °C, $t \geq 8$ h. The as-obtained BaTiO$_3$ nanocrystals are of a mixture of cubic and tetragonal phases and exhibit a uniform spherical particulate morphology with a size range of 90–100 nm. The as-obtained spherical BaTiO$_3$ nanocrystals show a high performance in ceramic capacitor for energy-storage applications.

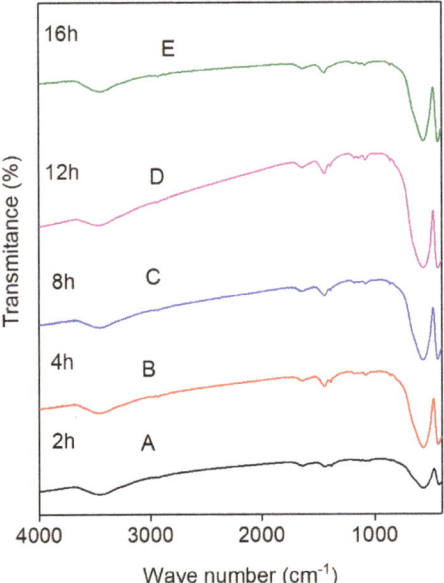

Figure 6. Typical FT-IR spectra of the BaTiO$_3$ nanocrystals obtained by hydrothermally treating at 210 °C for various times (2–16 h) with $R_{Ba/Ti}$ = 2.0 and [NaOH] = 2.0 mol L^{-1}.

3.4. Understanding of Growth Mechanism

In the hydrothermal synthesis of BaTiO$_3$ nanocrystals, TiO$_2$ (P25) nanoparticles are used as the solid-state Ti source and seeds for crystal growth. The possible growth mechanism of the BaTiO$_3$ nanocrystals by the hydrothermal process is shown in Figure 7. TiO$_2$ nanoparticles first react with OH$^-$ ions in a strong alkaline solution to form a soluble titanium hydroxide complex, which can form a negatively charged Ti–O chain. These negatively charged Ti–O chains attract positively charged Ba^{2+}

or BaOH$^+$ ions to form BaTiO$_3$ nuclei, on which the excess Ba^{2+} species continue to grow in the strong alkaline solution under the hydrothermal conditions for a long time. The possible reactions for the growth of BaTiO$_3$ nanocrystals can be described as follows:

$$TiO_2(P25) + OH^- \rightarrow TiO(OH)_2 \qquad (1)$$

$$TiO(OH)_2 + OH^- + H_2O \rightarrow Ti(OH)_6{}^{2-} \qquad (2)$$

$$Ti(OH)_6{}^{2-} + Ba^+ \rightarrow BaTiO_3 + H_2O \qquad (3)$$

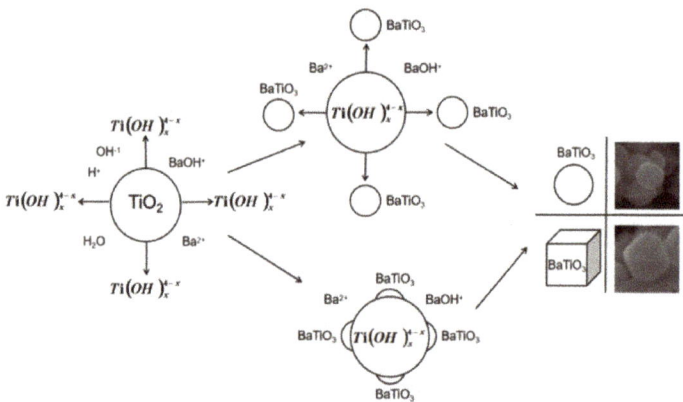

Figure 7. Possible growth mechanism for the synthesis of BaTiO$_3$ nanocrystals under the hydrothermal conditions using TiO$_2$ nanoparticles (P25) as the seeds and Ti source.

Using TiO$_2$ (P25) nanoparticles as the seeds and Ti source for the synthesis of BaTiO$_3$ nanocrystals, the negatively charged Ti–O chains are first formed on the surface of TiO$_2$ (P25) particles in the strong alkaline solution, and the whole TiO$_2$ (P25) nanoparticles are then gradually transformed to the [Ti(OH)$_x$]$^{4-x}$ species. The negatively charged Ti–O chains (i.e., [Ti(OH)$_6$]$^{2-}$) react with Ba^{2+} ions to form BaTiO$_3$ nanocrystals under hydrothermal conditions. The large spherical particles in situ formed on the TiO$_2$ (P25) nuclei may overcome the agglomeration because of their weak attraction to each other. The small particles can be self-regulated by the interaction of van der Waals torque (Casimir Torque) under high-temperature Brownian motion via the orientation attachment mechanism [43]. During the long hydrothermal reaction, smaller crystals dissolve and re-deposit on larger particles for orientation attachment and crystal extension via the Ostwald ripening process. Therefore, the growth mechanism for the formation of BaTiO$_3$ nanoparticles may involve the following steps: (1) TiO$_2$ (P25) nanoparticles are transformed to [Ti(OH)$_x$]$^{4-x}$ species in the strong alkaline solution; (2) Ba^{2+} ions reacts with [Ti(OH)$_x$]$^{4-x}$ species to form BaTiO$_3$ nanocrystals; (3) small BaTiO$_3$ nanocrystals grows to large ones via the Ostwald ripening process and the orientation attachment mechanism.

3.5. Dielectric Properties of the BPA Film with BaTiO$_3$ Nanoparticles

The spherical BaTiO$_3$ nanoparticles with a size range of 91 ± 14 nm (S8 in Table 1) obtained under the optimum conditions ([NaOH] = 2.0 mol L^{-1}, $R_{Ba/Ti}$ = 2.0, T = 210 °C and t = 8 h) were used to prepare BaTiO$_3$/polymer/Al (BPA) composite films to verify the feasibility of the BaTiO$_3$ sample in capacitor energy-storage applications.

The typical XRD patterns, SEM image and dielectric properties of the typical BPA films with the BaTiO$_3$ sample (S8) are shown in Figure 8. Figure 8a shows the XRD patterns of the BaTiO$_3$ sample, polymer/Al foil, and BPA film. According to the JCPDS card (No. 99-0005), the diffraction peaks at

2θ = 38.47°, 44.72°, and 65.09° correspond to the (111), (200), and (220) of the Al foil, respectively. The XRD pattern of the BPA film is a superposition of the BaTiO$_3$ sample and Al foil, and no other impurities are found in the BPA film. Figure 8b shows a typical SEM image of the BPA film. The film exhibits a uniform distribution of BaTiO$_3$ nanoparticles. Figure 8c gives the dielectric properties of the BPA films with spherical BaTiO$_3$ nanoparticles. As the statistical results show, the average dielectric constant of the BPA films reaches 59, the average dielectric loss reaches 0.008, and the average breakdown strength reaches 102 kV mm^{-1}. These electrical properties are much higher than those of the previous reports [44–49]. The TiO$_2$-seeded hydrothermal process is an efficient process to synthesize spherical BaTiO$_3$ nanoparticles for potential capacitor energy-storage applications.

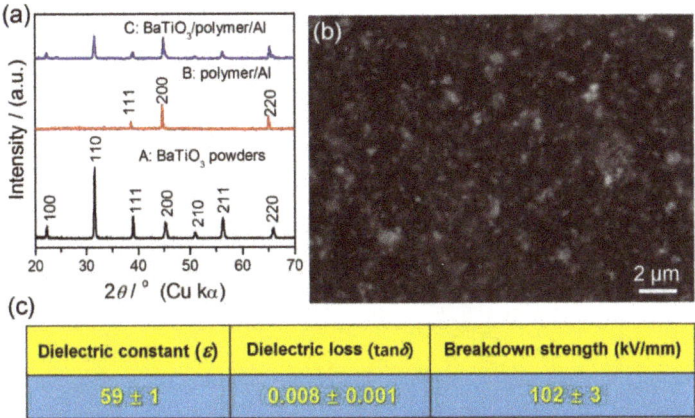

Figure 8. XRD, SEM, and dielectric properties of the typical BaTiO$_3$/polymer/Al films (BPA) films with BaTiO$_3$ nanocrystals (S8): (**a**) XRD patterns (A-BaTiO3 nanocrystals, B-Polymer/Al film and (C)-BaTiO$_3$/polymer/Al (BPA) film); (**b**) the typical SEM image of the BPA film; (**c**) typical electric properties of the BPA films.

We compared the dielectric constant, dielectric loss, and breakdown strength of the BPA films with those of the literature reports [25,31,44,46,49,50], and the results are shown in Table 2. One can find that the BPA films with the TiO$_2$-seeded BaTiO$_3$ nanocrystals exhibit an excellent balanced dielectric performance.

Table 2. Comparisons of dielectric constant, dielectric loss and breakdown strength of the composites containing BT particles.

Fillers	Polymer Matrix	Dielectric Constant	Break Strength	Dielectric Loss	Reference
BT microparticles	Resin	32	20.8 V/μm	0.014	[25]
BT microparticles	Resorcinol and formaldehyde	16.6	/	0.019	[31]
PDA coated BT nanoparticles (100 nm)/BN nanosheets	Poly(vinylidene fluoride -chlorotrifluoroethylene)	11.7	425 MV/m	0.10	[44]
Sphere-like TiO$_2$ nanowire clusters	Poly(vinylidene fluoride-co -hexafluoropylene)	11.9	160 kV/mm	0.048	[46]
CaCu$_3$Ti$_4$O$_{12}$@TiO$_2$ nanofibers	In suit prepared polyimide	5.85	236 kV/mm	0.025	[49]
PVP coated BT nanoparticles (100 nm)	Poly(vinylidene fluoride)	80.4	240 kV/mm	0.085	[50]
BT microparticles	Resin	59	102 kV/mm	0.008	This work

4. Conclusions

TiO$_2$ (P25) nanoparticle assisted hydrothermal process has been developed to synthesize BaTiO$_3$ nanocrystals in a strong alkaline solution (pH = 13.6) using TiO$_2$ (P25) and Ba(NO$_3$)$_2$ as the starting materials and NaOH as the mineralizer. The particle sizes, morphologies, and phases of the BaTiO$_3$ nanocrystals have been controlled by changing the molar Ba/Ti ratio, the hydrothermal temperature, and time. The XRD and SEM results indicate that a high Ba/Ti ratio (≥ 2.0), a high hydrothermal temperature (≥ 200 °C), and a long hydrothermal time (≥ 8 h) are favorable in forming a mixture of cubic/tetragonal BaTiO$_3$ nanocrystals with a uniform, well-dispersed spherical particulate morphology (90–100 nm). Under the optimum conditions ([NaOH] = 2.0 mol L^{-1}, $R_{Ba/Ti}$ = 2.0, T = 210 °C and t = 8 h), the as-obtained spherical BaTiO$_3$ nanoparticles have a narrow particle size range of 91 ± 14 nm. It should be emphasized that the particle size and morphology of the BaTiO$_3$ nanocrystals are kept relatively stable when the hydrothermal conditions change in a proper range, suggestive of a robust and efficient process toward spherical BaTiO$_3$ nanocrystals. The growth mechanism of the TiO$_2$-assisted hydrothermal process for the synthesis of BaTiO$_3$ nanocrystals has been attributed to the dissolution-crystallization, Oswald ripening, and oriented attachment process. The BaTiO$_3$/polymer/Al films containing the above BaTiO$_3$ nanoparticles are of a high dielectric constant of 59, a high break strength of 102 kV mm^{-1}, and a low dielectric loss of 0.008. The TiO$_2$-seeded hydrothermal process developed here is an efficient process to synthesize spherical BaTiO$_3$ nanoparticles for potential capacitor energy-storage applications.

Author Contributions: Conceptualization, D.C.; data curation, T.L. and F.T.; funding acquisition, D.C., C.S., and Z.Y.; investigation, D.Z. and Y.X.; methodology, F.T.; Software, S.H.; writing–original draft, M.L. and L.G.; writing–review and editing, M.L., L.G., T.L., and D.C. All authors have read and agreed to the published version of the manuscript.

Funding: This work was partly supported by the National Natural Science Foundation of China (Grant No. 51574205), the Natural Science Foundation of Guangdong Province (Grant No. 2018B030311022), Guangdong Innovation Research Team for Higher Education (Grant No. 2017KCXTD030), the Engineering Research Center of None-Food Biomass Efficient Pyrolysis & Utilization Technology of Guangdong Higher Education Institutes (Grant No.2016GCZX009), High-level Talents Project of Dongguan University of Technology (Grant No. KCYKYQD2017017), and Program from Dongguan University of Technology (Grant No. G200906-17).

Conflicts of Interest: The authors declare no conflict of interest.

References

1. Wang, C.; Yan, F.; Yang, H.; Yin, Y.; Wang, T. Dielectric and ferroelectric properties of SrTiO$_3$-Bi$_{0.54}$Na$_{0.46}$TiO$_3$-BaTiO$_3$ lead-free ceramics for high energy storage applications. *J. Alloys Compd.* **2018**, *749*, 605–611. [CrossRef]
2. Si, F.; Tang, B.; Fang, Z.X.; Li, H.; Zhang, S.R. A new type of BaTiO$_3$-based ceramics with Bi(Mg$_{1/2}$Sn$_{1/2}$)O$_3$ modification showing improved energy storage properties and pulsed discharging performances. *J. Alloys Compd.* **2020**, *819*, 153004. [CrossRef]
3. Liu, S.N.; Liu, C.C.; You, Y.; Wang, Y.J.; Wei, R.B. Fabrication of BaTiO$_3$-Loaded Graphene Nanosheets-Based Polyarylene Ether Nitrile Nanocomposites with Enhanced Dielectric and Crystallization Properties. *Nanomaterials* **2019**, *9*, 9121667. [CrossRef] [PubMed]
4. Li, Y.H.; Wu, J.; Wu, X.D.; Suo, H.; Shen, X.D.; Cui, S. Synthesis of bulk BaTiO$_3$ aerogel and characterization of photocatalytic properties. *J. Sol-Gel Sci. Technol.* **2019**, *90*, 313–322. [CrossRef]
5. Song, R.X.; Zhao, Y.; Li, W.L.; Yu, Y.; Sheng, J.; Li, Z.; Zhang, Y.L.; Xia, H.T.; Fei, W.D. High temperature stability and mechanical quality factor of donor-acceptor co-doped BaTiO$_3$ piezoelectrics. *Acta Mater.* **2019**, *181*, 200–206. [CrossRef]
6. Meeker, M.A.; Kundu, S.; Mauryn, D.; Kang, M.; Sosa, A.; Mudiyanselage, R.H.; Clavel, M.; Gollapudit, S.; Hudait, M.K.; Priya, S.; et al. The permittivity and refractive index measurements of doped barium titanate (BT-BCN). *Opt. Mater.* **2017**, *73*, 793–798. [CrossRef]

7. Sikarwar, S.; Sonker, R.K.; Shukla, A.; Yadav, B.C. Synthesis and investigation of cubical shaped barium titanate and its application as opto-electronic humidity sensor. *J. Mater. Sci.-Mater. Electron.* **2018**, *29*, 12951–12958. [CrossRef]
8. Hapani, D.S.; Singh, P.; Jha, P.K.; Singh, P. A comparative electrical conductivity behavior of $BaTiO_3$ and $CaTiO_3$ ceramics. *AIP Conf. Proc.* **2018**, *2009*, 020010.
9. Kabanov, V.V.; Piyanzina, I.I.; Tayurskii, D.A.; Mamin, R.F. Towards high-temperature quasi-two-dimensional superconductivity. *Phys. Rev. B* **2018**, *98*, 094522. [CrossRef]
10. Sasikumar, S.; Thirumalaisamy, T.K.; Saravanakumar, S.; Sivaganesh, D. Effect of neodymium doping in $BaTiO_3$ ceramics on structural and ferroelectric properties. *J. Mater. Sci.-Mater. Electron.* **2019**, *19*, 02670. [CrossRef]
11. Abdullah-Al, M.M.; Haque, A.; Pelton, A.; Paul, B.; Ghosh, K. Structural, electronic, and magnetic analysis and device characterization of ferroelectric-ferromagnetic heterostructure (BZT-BCT/LSMO/LAO) devices for multiferroic applications. *IEEE Trans. Magn.* **2018**, *54*, 2502908.
12. Akram, F.; Kim, J.C.; Khan, S.A.; Zeb, A.; Yeo, H.G.; Sung, Y.S.; Song, T.K.; Lee, S. Less temperature-dependent high dielectric and energy-storage properties of eco-friendly $BiFeO_3$-$BaTiO_3$-based ceramics. *J. Alloys Compd.* **2020**, *818*, 152878. [CrossRef]
13. Wang, H.X.; Liu, B.B.; Wang, X.H. Effects of dielectric thickness on energy storage properties of surface modified $BaTiO_3$ multilayer ceramic capacitors. *J. Alloys Compd.* **2020**, *817*, 152804. [CrossRef]
14. Park, H.; Na, Y.; Choi, H.J.; Suh, S.; Baek, D.H.; Yoon, J.R. Electrical properties of $BaTiO_3$-based 0603/0.1 μF/0.3 mm ceramics decoupling capacitor for embedding in the PCB of 10G RF transceiver module. *J. Electr. Eng. Technol.* **2018**, *13*, 1637–1642.
15. Ianculescu, A.; Pintilie, I.; Vasilescu, C.A.; Botea, M.; Iuga, A.; Melinescu, A.; Dragan, N.; Pintilie, L. Intrinsic pyroelectric properties of thick, coarse grained $Ba_{1-x}Sr_xTiO_3$ ceramics. *Ceram. Int.* **2016**, *42*, 10338–10348. [CrossRef]
16. Wang, J.; Bai, J.; Jin, Z.; Chen, C.; Zhai, W. Temperature dependent magnetoelectric coupling in $BaTiO_3/La_{0.67}Sr_{0.33}MnO_3$ heterojunction. *J. Phys. D Appl. Phys.* **2018**, *51*, 135305. [CrossRef]
17. Hong, K.; Lee, T.H.; Suh, J.M.; Park, J.S.; Kwon, H.S.; Choi, J.; Jang, H.W. Direct observation of surface potential distribution in insulation resistance degraded acceptor-doped $BaTiO_3$ multilayered ceramic capacitors. *Electron. Mater. Lett.* **2018**, *14*, 629–635. [CrossRef]
18. Kim, J. Electrical and thermal properties of a $(Ba_{1-x-y}Sr_xCa_y)TiO_3$-based PTC thermistor for preheating light oil. *Funct. Mater. Lett.* **2018**, *11*, 1850076. [CrossRef]
19. Cernea, M.; Vasile, B.S.; Surdu, V.A.; Trusca, R.; Bartha, C.; Craciun, F.; Galassi, C. Electric and magnetic properties of ferromagnetic/piezoelectric bilayered composite. *J. Mater. Sci.* **2018**, *53*, 14160–14171. [CrossRef]
20. Wu, J.; Qin, N.; Yuan, B.; Lin, E.; Bao, D. Enhanced pyroelectric catalysis of $BaTiO_3$ nanowires for utilizing waste heat in pollution treatment. *ACS Appl. Mater. Interfaces* **2018**, *10*, 37963–37973. [CrossRef]
21. Jin, M.H.; Shin, E.; Jin, S.; Jo, H.; Ok, K.M.; Hong, J.; Jun, B.H.; Durrant, J.R. Solvothermal synthesis of ferroelectric $BaTiO_3$ nanoparticles and their application to dye-sensitized solar cells. *J. Korean Phys. Soc.* **2018**, *73*, 627–631. [CrossRef]
22. Schadli, G.N.; Buchel, R.; Pratsinis, S.E. Nanogenerator power output: Influence of particle size and crystallinity of $BaTiO_3$. *Nanotechnology* **2017**, *28*, 275705. [CrossRef] [PubMed]
23. Aminirastabi, H.; Xue, H.; Mitic, V.V.; Lazovic, G.; Ji, G.L.; Peng, D.L. Novel fractal analysis of nanograin growth in $BaTiO_3$ thin film. *Mater. Chem. Phys.* **2020**, *239*, 122261. [CrossRef]
24. Chen, Y.; Ye, H.H.; Wang, X.S.; Li, Y.X.; Yao, X. Grain size effects on the electric and mechanical properties of submicro $BaTiO_3$ ceramics. *J. Eur. Ceram. Soc.* **2020**, *40*, 391–400. [CrossRef]
25. Gu, L.L.; Li, T.; Xu, Y.J.; Sun, C.H.; Yang, Z.Y.; Zhu, D.L.; Chen, D.L. Effects of the Particle Size of $BaTiO_3$ Fillers on Fabrication and Dielectric Properties of $BaTiO_3$/Polymer/Al Films for Capacitor Energy-Storage Application. *Materials* **2019**, *12*, 439. [CrossRef]
26. Thanki, A.A.; Goyal, R.K. Study on effect of cubic- and tetragonal phased $BaTiO_3$ on the electrical and thermal properties of polymeric nanocomposites. *Mater. Chem. Phys.* **2016**, *183*, 447–456. [CrossRef]
27. Chen, T.; Meng, J.; Wu, S.; Pei, J.; Lin, Q.; Wei, X.; Li, J.; Zhang, Z. Room temperature synthesized $BaTiO_3$ for photocatalytic hydrogen evolution. *J. Alloys Compd.* **2018**, *754*, 184–189. [CrossRef]
28. Shmyt'ko, I.M.; Frolov, D.D.; Aronin, S.; Ganeeva, G.R.; Kedrov, V.V. Formation of new structural states in compressed $BaTiO_3$ nanopowders. *Phys. Solid State* **2017**, *59*, 1196–1205. [CrossRef]

29. Rotaru, R.; Peptu, C.; Samoila, P.; Harabagiu, V. Preparation of ferroelectric barium titanate through an energy effective solid state ultrasound assisted method. *J. Am. Ceram. Soc.* **2017**, *100*, 4511–4518. [CrossRef]
30. Sampaio, D.V.; Silva, M.S.; Souza, N.R.S.; Santos, J.C.A.; Rezende, M.V.S.; Silva, R.S. Electrical characterization of BaTiO$_3$ and Ba$_{0.77}$Ca$_{0.23}$TiO$_3$ ceramics It synthesized by the proteic sol-gel method. *Ceram. Int.* **2018**, *44*, 15526–15530. [CrossRef]
31. Zhang, X.T.; Cui, B.; Wang, J.; Jin, Q. The effect of a barium titanate xerogel precursor on the grain size and densification of fine-grained BaTiO$_3$ ceramics. *Ceram. Int.* **2019**, *45*, 10626–10632. [CrossRef]
32. Hongo, K.; Kurata, S.; Jomphoak, A.; Hayashi, K.; Maezono, R. Stabilization mechanism of the tetragonal structure in a hydrothermally synthesized BaTiO$_3$ nanocrystal. *Inorg. Chem.* **2018**, *57*, 5413–5419. [CrossRef] [PubMed]
33. Zhao, D.; Ying, D.; Hua, X.; Zhou, M. Preparation of size controllable BaTiO$_3$ nanoparticles in microemulsion at low temperature. *Adv. Mater. Res.* **2014**, *1004*, 63–68. [CrossRef]
34. Shut, V.N.; Mozzharov, S.E.; Bobrovskii, V.V. Effect of ultrasonic processing on the synthesis of barium titanyl oxalate and the characteristics of the BaTiO$_3$ powder prepared from it. *Inorg. Mater.* **2018**, *54*, 72–78. [CrossRef]
35. Li, J.; Inukai, K.; Takahashi, Y.; Tsuruta, A.; Shin, W. Formation mechanism and dispersion of pseudo-tetragonal BaTiO$_3$-PVP nanoparticles from different titanium precursors: TiCl$_4$ and TiO$_2$. *Materials* **2018**, *11*, 51. [CrossRef]
36. Grendal, O.G.; Blichfeld, A.B.; Skjaervo, S.L.; Bee, W.; Selbach, S.M.; Grande, T.; Einarsrud, M. Facile low temperature hydrothermal synthesis of BaTiO$_3$ nanoparticles studied by in situ X-ray diffraction. *Crystals* **2018**, *8*, 253. [CrossRef]
37. Zhao, P.; Wang, L.; Bian, L.; Xu, J.B.; Chang, A.; Xiong, X.; Xu, F.; Zhang, J. Growth mechanism, modified morphology and optical properties of coral-like BaTiO$_3$ architecture through CTAB assisted synthesis. *J. Mater. Sci. Technol.* **2015**, *31*, 223–228. [CrossRef]
38. Ozen, M.; Mertens, M.; Snijkers, F.; Cool, P. Hydrothermal synthesis and formation mechanism of tetragonal barium titanate in a highly concentrated alkaline solution. *Ceram. Int.* **2016**, *42*, 10967–10975. [CrossRef]
39. Gao, J.; Shi, H.; Yang, J.; Li, T.; Zhang, R.; Chen, D. Influencing factor investigation on dynamic hydrothermal growth of gapped hollow BaTiO$_3$ nanospheres. *Nanoscale Res. Lett.* **2015**, *10*, 1033. [CrossRef]
40. Xiong, X.; Tian, R.; Lin, X.; Chu, D.; Li, S. Formation and photocatalytic activity of BaTiO$_3$ nanocubes via hydrothermal process. *J. Nanomater.* **2015**, *2015*, 692182. [CrossRef]
41. Moghtada, A.; Moghadam, A.H.; Ashiri, R. Tetragonality enhancement in BaTiO$_3$ by mechanical activation of the starting BaCO$_3$ and TiO$_2$ powders: Characterization of the contribution of the mechanical activation and postmilling calcination phenomena. *Int. J. Appl. Ceram. Technol.* **2018**, *15*, 1518–1531. [CrossRef]
42. Katsuki, H.; Furuta, S.; Komarneni, S. Semi-continuous and fast synthesis of nanophase cubic BaTiO$_3$ using a single-mode home-built microwave reactor. *Mater. Lett.* **2012**, *83*, 8–10. [CrossRef]
43. Yasui, K.; Kato, K. Numerical simulations of sonochemical production and oriented aggregation of BaTiO$_3$ nanocrystals. *Ultrason. Sonochem.* **2017**, *35*, 673–680. [CrossRef] [PubMed]
44. Xie, Y.C.; Wang, J.; Yu, Y.; Jiang, W.R.; Zhang, Z.C. Enhancing breakdown strength and energy storage performance of PVDF-based nanocomposites by adding exfoliated boron nitride. *Appl. Surf. Sci.* **2018**, *440*, 1150–1158. [CrossRef]
45. He, D.L.; Wang, Y.; Chen, X.Q.; Deng, Y. Core-shell structured BaTiO$_3$@Al$_2$O$_3$ nanoparticles in polymer composites for dielectric loss suppression and breakdown strength enhancement. *Compos. Part A Appl. Sci. Manuf.* **2017**, *93*, 137–143. [CrossRef]
46. Huang, Q.; Luo, H.; Chen, C.; Zhou, K.C.; Zhang, D. Improved energy density and dielectric properties of P(VDF-HFP) composites with TiO$_2$ nanowire clusters. *J. Electroceram.* **2018**, *40*, 65–71. [CrossRef]
47. Huang, Q.; Luo, H.; Chen, C.; Zhou, X.; Zhou, K.C.; Zhang, D. Enhanced energy density in P(VDF-HFP) nanocomposites with gradient dielectric fillers and interfacial polarization. *J. Alloys Compd.* **2017**, *696*, 1220–1227. [CrossRef]
48. Wang, J.C.; Long, Y.C.; Sun, Y.; Zhang, X.Q.; Yang, H.; Lin, B. Enhanced energy density and thermostability in polyimide nanocomposites containing core-shell structured BaTiO$_3$@SiO$_2$ nanofibers. *Appl. Surf. Sci.* **2017**, *426*, 437–445. [CrossRef]

49. Wang, J.C.; Long, Y.C.; Sun, Y.; Zhang, X.Q.; Yang, H.; Lin, B. Fabrication and enhanced dielectric properties of polyimide matrix composites with core-shell structured $CaCu_3Ti_4O_{12}@TiO_2$ nanofibers. *J. Mater. Sci. Mater. Electron.* **2018**, *29*, 7842–7850. [CrossRef]
50. Dai, Y.; Zhu, X. Improved dielectric properties and energy density of PVDF composites using PVP engineered BaTiO3 nanoparticles. *Korean J. Chem. Eng.* **2018**, *35*, 1570–1576. [CrossRef]

© 2020 by the authors. Licensee MDPI, Basel, Switzerland. This article is an open access article distributed under the terms and conditions of the Creative Commons Attribution (CC BY) license (http://creativecommons.org/licenses/by/4.0/).

Article

Impact of the Surface Properties of Cellulose Nanocrystals on the Crystallization Kinetics of Poly(Butylene Succinate)

Hatem Abushammala [1,*] and Jia Mao [2]

1. Fraunhofer Institute for Wood Research (WKI), Bienroder Weg 54E, 38108 Braunschweig, Germany
2. Department of Mechanical Engineering, Al-Ghurair University, Dubai International Academic City, P.O. Box 37374 Dubai, UAE; jia.mao@agu.ac.ae
* Correspondence: hatem.abushammala@wki.fraunhofer.de; Tel.: +49-531-215-5409

Received: 17 February 2020; Accepted: 11 March 2020; Published: 13 March 2020

Abstract: The hydrophilicity of cellulose nanocrystals (CNCs) is a major challenge for their processing with hydrophobic polymers and matrices. As a result, many surface modifications have been proposed to hydrophobize CNCs. The authors showed in an earlier study that grafting alcohols of different chain lengths onto the surface of CNCs using toluene diisocyanate (TDI) as a linker can systematically hydrophobize CNCs to a water contact angle of up to 120° depending on the alcohol chain length. Then, the hydrophobized CNCs were used to mechanically reinforce poly(butylene succinate) (PBS), which is a hydrophobic polymer. As a result of hydrophobization, PBS/CNCs interfacial adhesion and the composite mechanical properties significantly improved with the increasing CNC contact angle. Continuing on these results, this paper investigates the impact of CNC surface properties on the crystallization behavior of PBS using differential scanning calorimetry (DSC). The results showed that the crystallization temperature of PBS increased from 74.7 °C to up to 86.6 °C as a result of CNC nucleation activity, and its value was proportionally dependent on the contact angle of the CNCs. In agreement, the nucleation activity factor (ϕ) estimated using Dobreva and Gutzow's method decreased with the increasing CNC contact angle. Despite the nucleation action of CNCs, the rate constant of PBS crystallization as estimated using the Avrami model decreased in general as a result of a prevailing impeding effect. This decrease was minimized with increasing the contact angle of the CNCs. The impeding effect also increased the average activation energy of crystallization, which was estimated using the Kissinger method. Moreover, the Avrami exponent (n) decreased because of CNC addition, implying a heterogeneous crystallization, which was also apparent in the crystallization thermograms. Overall, the CNC addition facilitated PBS nucleation but retarded its crystallization, and both processes were significantly affected by the surface properties of the CNCs.

Keywords: cellulose; nanocrystals; modification; poly(butylene succinate); crystallization; kinetics

1. Introduction

Cellulose nanocrystals (CNCs) are rod-shaped nanoparticles with a thickness of 3–10 nm and a length of few hundreds of nanometers [1]. They can be extracted from pulp fibers, microcrystalline cellulose, and wood using a variety of reagents including strong acids, bases, oxidizing agents, and ionic liquids [2–4]. CNCs have shown great potential in a wide range of applications due to their advantageous mechanical properties, high surface area, biocompatibility, and biodegradability [5,6]. Furthermore, the possibility to modify their surfaces has led to CNCs with a wide range of surface properties and functionalities [7,8]. Due to their mechanical strength, CNCs have often been used to reinforce synthetic and bio-based polymers [9–11]. One main obstacle is the poor interfacial adhesion between CNCs and their matrices, which results in weak interfaces and reduced mechanical reinforcement [12,13].

Most of the commonly used polymers such as polyolefins are hydrophobic, while CNCs are hydrophilic. Two main approaches have been used to overcome this issue. One approach relies on the use of compatibilizers or coupling agents such as maleic anhydride-grafted-polyethylene (MAPE) for nanocellulose/polyethylene composites [14,15]. The other approach reduces the hydrophilicity of CNCs by reacting the surface hydroxyls with a variety of hydrophobic chemicals through relatively simple reactions such as acetylation and carbanilation or through grafting bulkier chemicals and polymers [16–18]. In a previous report by the authors, the surface properties of CNCs were tailored by grafting alcohols of different chain lengths onto their whole surface (almost all available surface hydroxyls, i.e., a degree of surface modification of ca. 100%) using 2,4-toluene diisocyanate (TDI) as a linker (Figure 1) [19]. Four alcohols were explored: ethanol, 1-butanol, 1-hexanol, and 1-octanol. As a result of this surface modification, CNCs with a tailored water contact angle of up to 120° were prepared, which was a result of growing a hydrophobic shell around the CNCs. These modified CNCs were used to reinforce poly(butylene succinate) (PBS), a hydrophobic polymer. The resultant composite showed improved interfacial adhesion, which was dependent on the alcohol chain length as confirmed by microscopic and thermomechanical investigations. This approach was proved to be simple for tailoring the surface properties of CNCs by only varying the alcohol chain length. It is also expected to improve the interfacial adhesion between CNCs and other hydrophobic matrices.

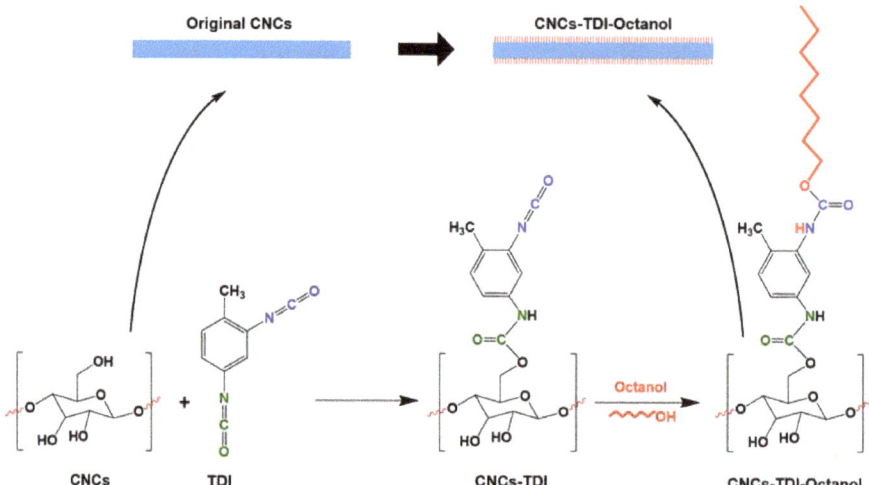

Figure 1. Tailoring the hydrophilicity of cellulose nanocrystals (CNCs) by grafting alcohols of different chain lengths on the CNC surface using toluene diisocyanate (TDI) as a linker.

PBS, which is produced by the polycondensation of succinic acid and butanediol, has shown great potential in many applications including automotive and packaging industries due to its similar properties to polyolefins and its advantageous biodegradability [20–23]. It still suffered some drawbacks in terms of its mechanical and gas barrier properties and its slow crystallization [24]. In order to modulate its properties, PBS has been processed with a variety of polymers, fibers, microparticles, and nanoparticles including CNCs, starch nanocrystals, silicon nitride, carbon nanotubes, calcium carbonate, etc. [25–27]. This resulted in significant changes in its mechanical, morphological, optical, and thermal properties [28–32], which were strongly influenced by the size and distribution of the generated spherulites upon PBS crystallization [33]. Therefore, the impact of many of these "modifiers" on the crystallization kinetics of PBS has been isothermally and non-isothermally investigated using

Avrami and Ozawa models and their combination [34,35]. In general, the modifiers or nanofillers acted as nucleating agents, and their nucleation activity was dependent on their amount, size, and morphology [36–38]. For instance, Filizgok et al. studied the influence of the shape of different carbon-based nanoparticles on the crystallization kinetics of PBS and showed that carbon nanotubes have a stronger nucleation activity than carbon black and fullerene [39].

This paper studies the crystallization kinetics of PBS upon the addition of CNCs of different surface properties to shed more light on the impact of interfacial adhesion on the nucleation activity of CNCs in PBS composites.

2. Materials and Methods

2.1. Materials

The CNC suspension with a solid content of 10.4% w/w was purchased from the University of Maine, which was prepared following a sulfuric acid-mediated procedure. Ethanol, 1-butanol, 1-hexanol, 1-octanol, acetone, toluene, triethylamine (TEA), and chloroform were purchased from VWR (Darmstadt, Germany) and stored over A3 or A4 molecular sieves, which were purchased from Carl Roth (Karlsruhe, Germany) and regenerated before use. 2,4-toluene diisocyanate (2,4-TDI) was purchased from TCI Chemicals (Eschborn, Germany) and bio-based poly(butylene succinate) was purchased from PTTMCC (Bangkok, Thailand) (commercial name is BioPBS^{TM}).

2.2. Hydrophobization of CNCs by Grafting Alcohols of Different Chain Lengths onto Their Surfaces

The CNCs were reacted with TDI following the method of Habibi and Dufresne after minor modifications [40] and the optimum conditions suggested by Abushammala [41,42]. First, 9.6 g of 10.4% CNC suspension (equivalent to 1.0 g of dried CNCs) was solvent-exchanged from water to anhydrous acetone using a washing/precipitation procedure (three times) and then to anhydrous toluene using the same procedure (twice). The precipitation was performed using a Sigma 3-16P centrifuge (5000 rpm for 30 min) (Sigma Laborzentrifugen, Osterode am Harz, Germany). After the final washing with anhydrous toluene, the precipitated CNCs were transferred to a 100 mL round-bottom flask using 46.3 mL of anhydrous toluene. To them, 3.3 g of 2,4-TDI and 3.0 mL of triethylamine (TEA) as catalyst were added. The reaction proceeded at 35 °C in a nitrogen environment. After 24 h, the reaction mixture was centrifuged to isolate the TDI-carbamated CNCs (CNCs-TDI) from the unreacted TDI and TEA. Then, the CNCs-TDI were washed three times with anhydrous toluene before transferred to 50 mL of anhydrous ethanol and stirred for 24 h at room temperature to allow a complete grafting of ethanol onto the CNC surface. Then, the CNCs were collected by centrifugation and dried at 60 °C under vacuum to a constant mass. The reaction was repeated to assure reproducibility. The reaction was also performed using 1-butanol, 1-hexanol, and 1-octanol instead of ethanol to produced CNCs with different surface properties. The produced CNCs are referred to in this paper as CNCs-TDI-Eth, CNCs-TDI-But, CNCs-TDI-Hex, and CNCs-TDI-Oct, respectively. Following this procedure, almost every CNC surface hydroxyl had an alcohol chain attached to it. This has been confirmed in a previous paper by the authors [19].

2.3. Water Contact Angle Measurements

Powdered samples of the neat and modified CNCs were pressed in an FT-IR mold to obtain discs (diameter: 1 cm) of smooth surfaces. The water contact angle was determined by placing a water droplet (volume of 15 μL) on each disc surface using OCA20 equipment (DataPhysics Instruments GmbH, Filderstadt, Germany). The standard tangent procedure was used to determine the contact angle. To assure reproducibility, the measurements were performed in triplicate.

2.4. Processing of the Modified CNCs with PBS

First, 0.25 g of the modified CNCs was dispersed in 20 mL of chloroform using the ultrasonicator UW2200 (power of 10% of 2200 W for 20 s) (Bandelin Electronic, Berlin, Germany). Then, 4.75 g of PBS was added and stirred to a complete dissolution. The homogeneous mixture was transferred to an aluminum dish and heated gradually to 135 °C for 24 h to evaporate the chloroform and melt the PBS. Then, the samples were cooled down to room temperature to obtain films containing 5% (w/w) of CNCs. They are referred to in the paper as PBS+CNCs-TDI-Eth, PBS+CNCs-TDI-But, PBS+CNCs-TDI-Hex, and PBS+CNCs-TDI-Oct. The dispersibility of the CNCs was confirmed using scanning electron microscopy and published somewhere else [19]. It is important to mention that it was not possible to prepare films using the unmodified CNCs because of their high hydrophilicity.

2.5. Crystallization and Melting Behavior of Neat and CNC-Reinforced PBS Using Differential Scanning Calorimetry (DSC)

The crystallization and melting behavior of the neat and CNC-reinforced PBS was investigated using the differential scanning calorimeter DSC 3+ (Mettler Toledo, Giessen, Germany). Around 10 mg of each sample was heated from −60 to 135 °C in a nitrogen environment to destroy the thermal history of the sample. Afterwards, it was cooled down to −60 °C then heated again to 135 °C using a heating/cooling rate of 10 °C/min. This procedure was repeated using different heating/cooling rates of 8, 6, and 4 °C/min, and the measurements were performed in duplicate to assure reproducibility. The crystallization temperature (T_c) and melting temperature (T_m) were determined as the peak of the crystallization and melting curves, respectively. The degree of crystallinity of PBS (X) was calculated from the enthalpy of melting (ΔH_m) of the sample in comparison to the enthalpy of melting of 100% crystalline PBS (110.3 J/g) [43]. For the CNC-reinforced PBS samples, the ΔH_m values were adjusted to account for CNC mass.

3. Results and Discussion

A previous work by the authors has shown that grafting alcohols of different chain lengths onto the surface of CNCs using TDI as a linker is a relatively simple process for tailoring CNC surface properties [19]. As a result of this grafting, the water contact angle of the modified CNCs increased to 34° ± 4, 52° ± 3, 104° ± 1, or 120° ± 5 using ethanol, 1-butanol, 1-hexanol, or 1-octanol, respectively. In comparison, the water contact angle of PBS is 77° ± 3. This systematic reduction in hydrophilicity allowed the dispersion of CNCs in hydrophobic matrices such as PBS, which is in general not possible for unmodified CNCs. Moreover, thermomechanical studies showed that the surface modification had a direct positive impact on the reinforcement capabilities of the CNCs, as it improved their interfacial adhesion with PBS. The improvement increased with increasing the chain length of the grafted alcohol. In this paper, we performed further studies to investigate the impact of interfacial adhesion on the crystallization kinetics of PBS using the modified CNCs as filler.

The non-isothermal crystallization and melting behaviors of the neat and CNC-reinforced PBS were studied using DSC at four cooling/heating rates (4, 6, 8, and 10 °C/min) (Figure 2 and Table 1). The results showed that crystallization took place at higher temperatures for the reinforced PBS compared to the neat one due to the nucleation activity of the modified CNCs. A similar behavior has been observed for PBS composites using other nanofillers such as carbon nanotubes [39]. It is interesting that the nucleation activity of the modified CNCs was dependent on the chain length of the grafted alcohol. Using a cooling rate of 10 °C/min, the crystallization temperature (T_c) of the reinforced PBS was 82.2, 83.4, 84.5, and 86.6 °C using CNCs-TDI-Eth, CNCs-TDI-But, CNCs-TDI-Hex, and CNCs-TDI-Oct compared to 74.7 °C for neat PBS. This means a stronger interaction between PBS molecular chains and the modified CNCs with higher degrees of hydrophobization. The homogeneity of the crystallization process was also affected by the addition of the CNCs. Compared to the typical single crystallization peak of the neat PBS, the crystallization peak of the reinforced PBS had shoulders on its both sides. These shoulders indicate a special interaction between the CNCs and PBS during nucleation (first shoulder) and secondary crystallization (second shoulder), which resulted in accelerating both of them. However, the shape of the primary crystallization peak does not look significantly different for the neat and reinforced PBS despite taking place at significantly higher temperatures. This implies that the interaction of the CNCs with PBS changes as crystallization progresses.

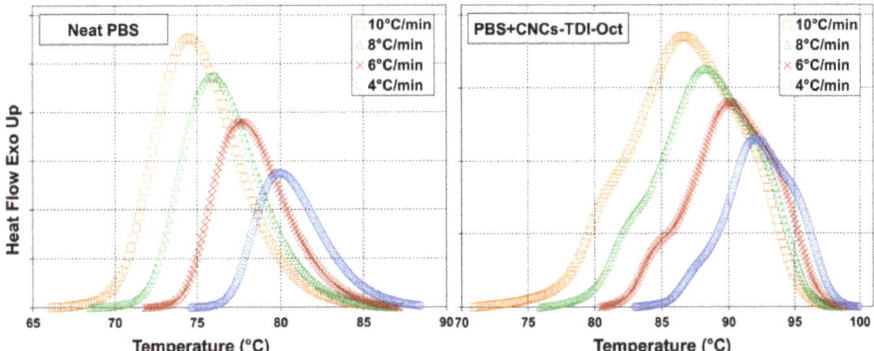

Figure 2. The crystallization exotherms of neat and reinforced poly(butylene succinate) (PBS (left)) (PBS+CNCs-TDI-Oct (right)) at different cooling rates (4, 6, 8, and 10 °C/min).

Table 1. The crystallization and melting parameters of neat and reinforced PBS at different cooling rates.

Sample	B (°C/min)	T_c (°C)	T_m (°C)	ΔH_m (J/g PBS)	X (%)
PBS	4	80.2 ± 0.1	115.7 ± 0.0	55.1 ± 0.4	49.9 ± 0.4
	6	78.0 ± 0.3	115.6 ± 0.0	56.2 ± 0.1	51.0 ± 0.1
	8	76.2 ± 0.1	115.5 ± 0.0	59.6 ± 0.2	54.0 ± 0.2
	10	74.7 ± 0.1	115.7 ± 0.1	62.3 ± 0.5	56.5 ± 0.5
PBS+CNCs-TDI-Eth	4	85.8 ± 0.1	115.6 ± 0.1	53.8 ± 0.6	48.8 ± 0.5
	6	84.6 ± 0.1	115.6 ± 0.0	55.6 ± 0.1	50.4 ± 0.1
	8	83.4 ± 0.2	115.7 ± 0.1	56.6 ± 0.7	51.3 ± 0.6
	10	82.2 ± 0.1	115.8 ± 0.0	58.9 ± 0.4	53.4 ± 0.3
PBS+CNCs-TDI-But	4	87.6 ± 0.0	115.7 ± 0.1	55.6 ± 0.9	50.4 ± 0.8
	6	85.8 ± 0.0	115.5 ± 0.0	57.0 ± 0.4	51.7 ± 0.3
	8	84.6 ± 0.3	115.6 ± 0.0	58.1 ± 0.5	52.6 ± 0.4
	10	83.4 ± 0.1	115.9 ± 0.1	58.8 ± 1.1	53.3 ± 1.0
PBS+CNCs-TDI-Hex	4	88.6 ± 0.1	115.5 ± 0.3	56.2 ± 0.5	50.9 ± 0.4
	6	87.1 ± 0.1	115.6 ± 0.3	59.3 ± 0.4	53.8 ± 0.3
	8	85.1 ± 0.2	115.7 ± 0.4	63.3 ± 0.9	57.4 ± 0.9
	10	84.5 ± 0.3	115.8 ± 0.5	65.3 ± 0.3	59.2 ± 0.3
PBS+CNCs-TDI-Oct	4	91.7 ± 0.4	115.7 ± 0.1	56.8 ± 0.5	51.5 ± 0.5
	6	89.5 ± 0.6	115.5 ± 0.0	59.7 ± 0.8	54.2 ± 0.7
	8	87.9 ± 0.4	115.6 ± 0.0	64.9 ± 1.4	58.8 ± 1.3
	10	86.6 ± 0.1	115.9 ± 0.1	67.3 ± 1.0	61.0 ± 0.9

In terms of PBS melting behavior, the melting temperature was not affected by the CNCs (around 115 °C for all samples). However, the crystallinity was slightly affected. The PBS samples reinforced by CNCs-TDI-Eth and CNCs-TDI-But showed a slightly lower crystallinity than neat PBS, while those reinforced by CNCs-TDI-Hex and CNCs-TDI-Oct showed higher crystallinity. It was dependent on the alcohol chain length of the modified CNCs, which could be a result of the higher nucleation activity of the modified CNCs with increasing the chain length of the grafted alcohol. Overall, the crystallinity values are in accordance with those reported in the literature [34].

To estimate the nucleation activity of nucleating agents, a simple method was proposed by Dobreva and Gutzow [44]. They proposed the nucleation activity factor φ, which varies from 0 to 1 and represents the ratio between the heterogeneous and homogenous nucleation parameters, B* and B, respectively (φ = B*/B). B* and B can be estimated by plotting lnβ versus the reciprocal of $(T_m - T_c)^2$ according to the following equation (C is a constant):

$$ln\beta = -\frac{B}{(T_m - T_c)^2} + C.$$

when applied on the crystallization of neat and CNC-reinforced PBS, straight lines were obtained (Figure 3), and the slope was either the B value for the neat PBS or the B* value for the reinforced PBS. Then, the nucleation activity factor was calculated (Table 2). According to the results, it is clear that PBS nucleation was accelerated in the presence of CNCs in general, as the B* values of the reinforced PBS were all lower than the B value of neat PBS. Moreover, the nucleation activity was dependent on the surface properties of the CNCs as it increased with increasing the chain length of the grafted alcohol. This implies that stronger PBS/CNCs interfacial adhesion results in stronger nucleation activity, as suggested earlier in Table 1.

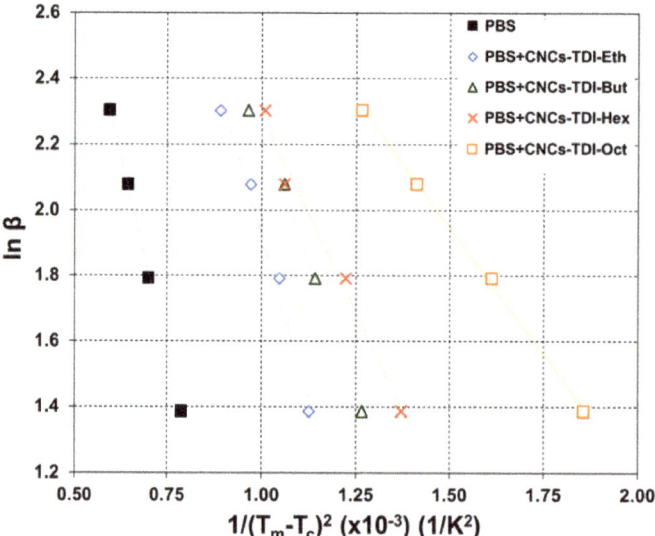

Figure 3. Nucleation activity determination for the neat and reinforced PBS using Dobreva and Gutzow's method.

Table 2. The estimated nucleation activity values for neat and reinforced PBS.

Sample	B or B*	φ
PBS	4638 ± 112	-
PBS+CNCs-TDI-Eth	3821 ± 62	0.82
PBS+CNCs-TDI-But	2907 ± 177	0.63
PBS+CNCs-TDI-Hex	2398 ± 16	0.52
PBS+CNCs-TDI-Oct	1605 ± 63	0.35

B value for the neat PBS and B* values for the reinforced PBS.

To shed more light on the interaction between the modified CNCs and PBS, the crystallization kinetics of the composite samples were studied using the Avrami model [45]. It is one of the most commonly used models to describe the crystallization process of semi-crystalline polymers. It expresses the relative crystallinity of a polymer ($X(t)$) as a function of time (t) according to the following equation (where Z is the crystallization rate constant and n is the Avrami exponent) [46]:

$$X(t) = 1 - \exp(-Zt^n).$$

The Avrami exponent describes the mechanism of the crystallization process, and it is a term of two components: the dimensionality of crystal growth (n_d) and the time dependence of nucleation (n_n). The value of n_d can be 1, 2, or 3 depending on if the crystal growth takes place in 1D, 2D, or 3D, respectively. The value of n_n is close to 1 when nucleation is homogenous and close to 0 when nucleation is heterogeneous. As a result, the value of n is in the range from 1 to 4 ($n_d + n_n$) [47]. The crystallization rate constant and Avrami exponent can be estimated from the slope and intercept of the logarithmic version of Avrami equation [48]:

$$\ln[-\ln(1 - X(t))] = n \ln t + \ln Z.$$

The crystallization profiles of PBS and reinforced PBS at different cooling rates were all sigmoidal similar to the profiles of many other semi-crystalline polymers (Figure 4) [49]. The sigmoidal shape represents the different stages of crystallization. At first, nucleation takes place (first 60 s) followed by a rapid crystal growth (next 60–120 s) and ends by secondary crystallization (the last 30–60 s) [50]. Expectedly, crystallization took a shorter time to completion using higher cooling rates. It was also observed that the heterogeneity of crystallization seen in Figure 2 for the reinforced PBS did not significantly affect the smoothness of its crystallization profile because the relative crystallinity is cumulative. During the nucleation stage (first 60 s), the reinforced PBS samples reached higher relative crystallinity compared to neat PBS as a result of the nucleation action of the CNCs. This may explain the right shoulder of the crystallization peak of the reinforced PBS samples (Figure 2). However, crystallization took a longer time as a result of the hindrance imposed on the motion of the PBS chains by the surrounding CNCs (impeding effect).

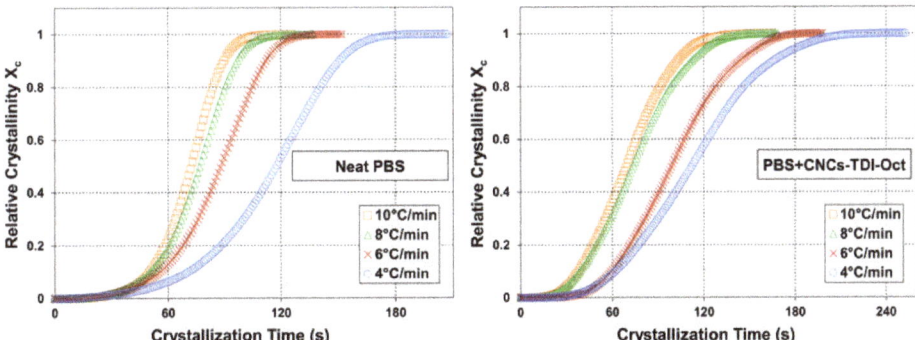

Figure 4. Plots of relative crystallinity with time at different cooling rates for neat PBS (left) and PBS+CNCs-TDI-Oct (right).

To quantitatively assess the impact of the CNCs on PBS crystallization, the Avrami kinetic parameters were estimated by plotting the logarithmic version of the Avrami equation (Figure 5 and Table 3). The crystallization of the reinforced PBS samples followed the Avrami model better than neat PBS. It deviated mainly in the beginning due to the slow nucleation of neat PBS (as also observed in Figure 2) and due to ignoring the role of secondary crystallization in the Avrami model [46]. In the reinforced PBS samples, the nucleation activity of the CNCs compensated for this deviation by accelerating PBS nucleation. The Avrami exponent for the neat PBS decreased upon CNC reinforcement, indicating increased crystallization heterogeneity. Avrami exponent values close to 4 imply a three-dimensional crystal growth and homogeneous nucleation (neat PBS), while values closer to 3 implies also three-dimensional growth but following a heterogeneous nucleation (reinforced PBS). This is in accordance with the crystallization curves of the neat and reinforced PBS in Figure 2. When it comes to the crystallization rate constant (Z), it was lower for the reinforced PBS samples compared to the neat PBS due to the hindrance imposed by the CNCs on the migration and diffusion of the PBS chains to the growing crystals [51]. This was also evident from the higher half-time of crystallization ($t_{\frac{1}{2}}$) values. This hindrance became less significant by reducing the hydrophilicity of the CNCs. It was the highest for the PBS sample reinforced by CNCs-TDI-Eth and diminished for the sample reinforced by CNCs-TDI-Oct. This may imply an optimum interfacial adhesion between PBS and CNCs-TDI-Oct, which supported a hindrance-free mobility of PBS chains during crystallization.

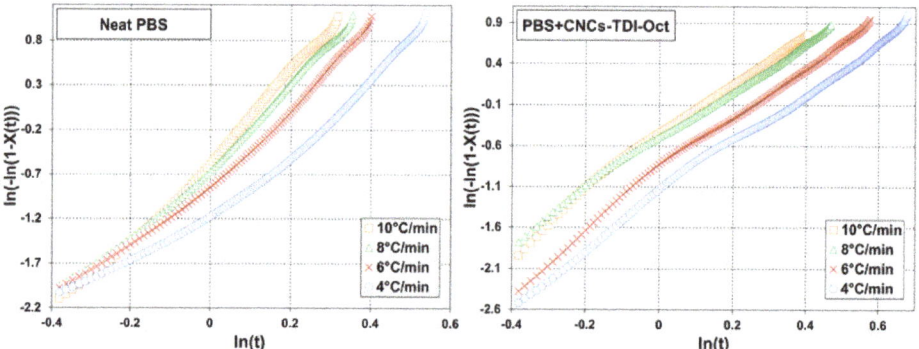

Figure 5. Avrami curves describing the crystallization of neat PBS (left) and PBS+CNCs-TDI-Oct (right) at different cooling rates.

Table 3. Avrami crystallization kinetics parameters of neat and reinforced PBS at different cooling rates.

Sample	B (°C/min)	n	Z	$t_{\frac{1}{2}}$ (min)
PBS	4	3.6 ± 0.1	0.05 ± 0.03	2.0 ± 0.1
	6	3.7 ± 0.0	0.19 ± 0.03	1.5 ± 0.1
	8	3.8 ± 0.1	0.30 ± 0.02	1.3 ± 0.1
	10	3.8 ± 0.0	0.39 ± 0.02	1.1 ± 0.1
PBS+CNCs-TDI-Eth	4	3.4 ± 0.0	0.03 ± 0.00	3.0 ± 0.0
	6	3.4 ± 0.0	0.11 ± 0.04	2.1 ± 0.0
	8	3.4 ± 0.0	0.12 ± 0.01	1.7 ± 0.1
	10	3.3 ± 0.0	0.29 ± 0.03	1.4 ± 0.0
PBS+CNCs-TDI-But	4	3.4 ± 0.1	0.03 ± 0.01	2.6 ± 0.2
	6	3.2 ± 0.2	0.08 ± 0.01	1.9 ± 0.1
	8	3.7 ± 0.2	0.20 ± 0.02	1.7 ± 0.0
	10	3.5 ± 0.2	0.32 ± 0.03	1.3 ± 0.0
PBS+CNCs-TDI-Hex	4	3.7 ± 0.1	0.07 ± 0.01	2.6 ± 0.0
	6	3.4 ± 0.1	0.10 ± 0.02	1.7 ± 0.0
	8	3.1 ± 0.2	0.31 ± 0.04	1.3 ± 0.1
	10	3.0 ± 0.1	0.33 ± 0.05	1.1 ± 0.0
PBS+CNCs-TDI-Oct	4	3.2 ± 0.1	0.03 ± 0.01	2.0 ± 0.2
	6	3.6 ± 0.3	0.14 ± 0.01	1.7 ± 0.0
	8	3.1 ± 0.0	0.30 ± 0.02	1.3 ± 0.0
	10	3.4 ± 0.1	0.40 ± 0.04	1.1 ± 0.1

It is possible to estimate the average activation energy (E_a) of crystallization for neat and reinforced PBS following Kissinger's method (R is the universal gas constant and β_o is the exponential factor):

$$ln\beta = -\frac{E_a}{R*T_c} + ln\beta_o.$$

The estimated crystallization activation energy of neat PBS was 169 ± 5 kJ/mol (Figure 6). This value is in accordance with the values reported in the literature [36]. It increased to 266 ± 16 kJ/mol when PBS was reinforced by CNCs-TDI-Eth. This implies that crystallization was hindered, although the addition of the modified CNCs in general facilitated PBS nucleation. However, the activation energy was dependent on the surface properties of the CNCs. It decreased with increasing the chain length of the grafted alcohol. The activation energy was 236 ± 8, 225 ± 5, and 196 ± 11 kJ/mol using CNCs-TDI-But, CNCs-TDI-Hex, and CNCs-TDI-Oct, respectively. These conclusions agree with those made based on the estimated Avrami kinetics parameters (Table 2).

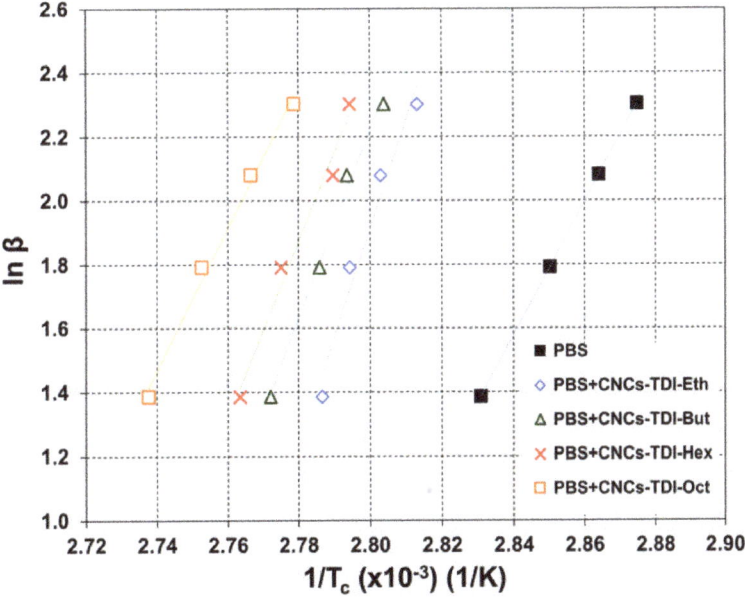

Figure 6. Kissinger plot to determine the activation energy of crystallization of neat and reinforced PBS.

In summary, the crystallization behavior of PBS was significantly affected by the addition of CNCs, which was dependent on their surface properties (Table 4). In general, CNC addition increased PBS crystallization temperature as a result of the nucleation activity of CNCs. The nucleation activity of the CNCs increased with the increase in their contact angle upon hydrophobization, which could be a result of the improved interaction between PBS and CNCs. Despite the significant improvement in PBS nucleation, its crystallization kinetics were slower as indicated by the drop of Z value and increase of $t_{\frac{1}{2}}$ and E_a. This outcome indicates that the impeding effect of the CNCs was stronger than their nucleation activity, which as a result hindered the molecular motion of PBS chains and decelerated overall crystallization. However, the impeding effect of CNCs became less significant with reducing their hydrophilicity. In terms of the PBS melting temperature and crystallinity, the CNCs did not have a significant impact on them.

Table 4. Summary of the impact of CNC surface properties on the crystallization of PBS (at 10 °C/min).

Sample	Contact Angle (°)	T_c (°C)	T_m (°C)	X (%)	n	Z	$t_{\frac{1}{2}}$ (min)	E_a (kJ/mol)	Φ
PBS	77	74.7	115.7	56.5	3.8	0.39	1.1	169	-
PBS+CNCs-TDI-Eth	34 *	82.2	115.8	53.4	3.3	0.29	1.4	266	0.82
PBS+CNCs-TDI-But	52 *	83.4	115.9	53.3	3.5	0.32	1.3	236	0.63
PBS+CNCs-TDI-Hex	104 *	84.5	115.8	59.2	3.0	0.33	1.1	225	0.52
PBS+CNCs-TDI-Oct	120 *	86.6	115.9	61.0	3.4	0.40	1.1	196	0.35

* Contact angle of the CNCs not the composite.

4. Conclusions

The impact of CNC surface properties on the crystallization behavior of PBS in CNC-reinforced PBS composites was investigated using differential scanning calorimetry. The surface properties of the CNCs were tailored by grafting alcohols of different chain lengths on their surface using TDI as a linker. The water contact angle of the modified CNCs increased to 34°, 52°, 104°, or 120° using ethanol, 1-butanol, 1-hexanol, or 1-octanol, respectively. Compared to neat PBS, the crystallization thermograms of the PBS composites showed a heterogeneous crystallization behavior with increased crystallization temperature as a result of CNC nucleation activity, which was proportionally dependent on the contact angle of the CNCs. However, the estimated Avrami kinetic parameters and Kissinger activation energy of crystallization showed that the impeding effect of CNCs is prevailing, resulting in retarded crystallization. The impeding effect was less significant for the CNCs with higher contact angles. Overall, the surface properties of CNCs had a crucial role in the crystallization behavior of PBS and can be advantageous to tailor its properties.

Author Contributions: Conceptualization, H.A. and J.M.; methodology, H.A. and J.M.; formal analysis, H.A. and J.M.; investigation, H.A. and J.M.; writing—original draft preparation, H.A. and J.M.; writing—review and editing, H.A. and J.M.; funding acquisition, H.A. All authors have read and agreed to the published version of the manuscript.

Funding: This research was funded by the Fraunhofer Institute for Wood Research (WKI) through the Wilhelm-Klauditz Fellowship.

Acknowledgments: Many thanks to Nadine Nöcker from the Fraunhofer Institute for Surface Engineering and Thin Films (IST) for the contact angle measurements and to Martin Eichler from the Fraunhofer Institute for Wood Research (WKI) for performing the DSC measurements.

Conflicts of Interest: The authors declare no conflict of interest.

References

1. ISO. *Standard Terms and Their Definition for Cellulose Nanomaterial*; International Organization for Standardization (ISO): Geneva, Swizerland, 2017.
2. Abushammala, H.; Mao, J. A review on the partial and complete dissolution and fractionation of wood and lignocelluloses using imidazolium ionic liquids. *Polymers* **2020**, *12*, 195. [CrossRef]
3. Mao, J.; Abushammala, H.; Brown, N.; Laborie, M.P. Comparative assessment of methods for producing cellulose I nanocrystals from cellulosic sources. In *Nanocelluloses: Their Preparation, Properties, and Applications, ACS Symposium Series*; ACS Publications: Washington, DC, USA, 2017; Volume 1251, pp. 19–53.
4. Mao, J.; Abushammala, H.; Hettegger, H.; Rosenau, T.; Laborie, M.P. Imidazole, a new tunable reagent for producing nanocellulose, part I: Xylan-Coated CNCs and CNFs. *Polymers* **2017**, *9*, 473. [CrossRef] [PubMed]
5. Moon, R.J.; Martini, A.; Nairn, J.; Simonsen, J.; Youngblood, J. Cellulose nanomaterials review: Structure, properties and nanocomposites. *Chem. Soc. Rev.* **2011**, *40*, 3941–3994. [CrossRef] [PubMed]
6. Eichhorn, S.J.; Dufresne, A.; Aranguren, M.; Marcovich, N.E.; Capadona, J.R.; Rowan, S.J.; Weder, C.; Thielemans, W.; Roman, M.; Renneckar, S.; et al. Review: Current international research into cellulose nanofibres and nanocomposites. *J. Mater. Sci.* **2010**, *45*, 1–33. [CrossRef]

7. Abushammala, H.; Mao, J. A review of the surface modification of cellulose and nanocellulose using aliphatic and aromatic mono- and di-isocyanates. *Molecules* **2019**, *24*, 2782. [CrossRef] [PubMed]
8. Eyley, S.; Thielemans, W. Surface modification of cellulose nanocrystals. *Nanoscale* **2014**, *6*, 7764–7779. [CrossRef] [PubMed]
9. Pirani, S.; Abushammala, H.M.; Hashaikeh, R. Preparation and characterization of electrospun PLA/nanocrystalline cellulose-based composites. *J. Appl. Polym. Sci.* **2013**, *130*, 3345–3354. [CrossRef]
10. Sapkota, J.; Natterodt, J.C.; Shirole, A.; Foster, E.J.; Weder, C. Fabrication and properties of polyethylene/cellulose nanocrystal composites. *Macromol. Mater. Eng.* **2017**, *302*, 1600300. [CrossRef]
11. Kargarzadeh, H.; Mariano, M.; Huang, J.; Lin, N.; Ahmad, I.; Dufresne, A.; Thomas, S. Recent developments on nanocellulose reinforced polymer nanocomposites: A review. *Polymer* **2017**, *132*, 368–393. [CrossRef]
12. Ferreira, F.V.; Dufresne, A.; Pinheiro, I.F.; Souza, D.H.S.; Gouveia, R.F.; Mei, L.H.I.; Lona, L.M.F. How do cellulose nanocrystals affect the overall properties of biodegradable polymer nanocomposites: A comprehensive review. *Eur. Polym. J.* **2018**, *108*, 274–285. [CrossRef]
13. Sinko, R.; Qin, X.; Keten, S. Interfacial mechanics of cellulose nanocrystals. *MRS Bull.* **2015**, *40*, 340–348. [CrossRef]
14. Inai, N.H.; Lewandowska, A.E.; Ghita, O.R.; Eichhorn, S.J. Interfaces in polyethylene oxide modified cellulose nanocrystal-polyethylene matrix composites. *Compos. Sci. Technol.* **2018**, *154*, 128–135. [CrossRef]
15. Lewandowska, A.; Eichhorn, S. Raman imaging as a tool for assessing the degree of mixing and the interface between polyethylene and cellulose nanocrystals. In Proceedings of the IOP Conference Series: Materials Science and Engineering 37th Risø International Symposium on Materials Science, Risø, Denmark, 5–8 September 2016.
16. Hu, F.; Lin, N.; Chang, P.R.; Huang, J. Reinforcement and nucleation of acetylated cellulose nanocrystals in foamed polyester composites. *Carbohydr. Polym.* **2015**, *129*, 208–215. [CrossRef] [PubMed]
17. Zhang, C.; Salick, M.R.; Cordie, T.M.; Ellingham, T.; Dan, Y.; Turng, L.S. Incorporation of poly(ethylene glycol) grafted cellulose nanocrystals in poly(lactic acid) electrospun nanocomposite fibers as potential scaffolds for bone tissue engineering. *Mater. Sci. Eng. C* **2015**, *49*, 463–471. [CrossRef] [PubMed]
18. Yu, H.Y.; Qin, Z.Y.; Yan, C.F.; Yao, J.M. Green nanocomposites based on functionalized cellulose nanocrystals: A study on the relationship between interfacial interaction and property enhancement. *ACS Sustain. Chem. Eng.* **2014**, *2*, 875–886. [CrossRef]
19. Abushammala, H. Nano-brushes of alcohols grafted onto cellulose nanocrystals for reinforcing poly(butylene succinate): Impact of alcohol chain length on interfacial adhesion. *Polymers* **2020**, *12*, 95. [CrossRef]
20. Ferreira, F.V.; Pinheiro, I.F.; de Souza, S.F.; Mei, L.H.; Lona, L.M. Polymer composites reinforced with natural fibers and nanocellulose in the automotive industry: A short review. *J. Compos. Sci.* **2019**, *3*, 51. [CrossRef]
21. Shi, J.; Xu, W.; Li, D.; Liao, R.; Zhang, L. The Innovation Research of Biodegradable Polymers for Sustainable Packaging. *DEStech Trans. Environ. Energy Earth Sci.* **2016**. [CrossRef]
22. Gigli, M.; Fabbri, M.; Lotti, N.; Gamberini, R.; Rimini, B.; Munari, A. Poly (butylene succinate)-based polyesters for biomedical applications: A review. *Eur. Polym. J.* **2016**, *75*, 431–460. [CrossRef]
23. Xu, J.; Guo, B.H. Poly (butylene succinate) and its copolymers: Research, development and industrialization. *Biotechnol. J.* **2010**, *5*, 1149–1163. [CrossRef]
24. Bin, T.; Qu, J.P.; Liu, L.M.; Feng, Y.H.; Hu, S.X.; Yin, X.C. Non-isothermal crystallization kinetics and dynamic mechanical thermal properties of poly (butylene succinate) composites reinforced with cotton stalk bast fibers. *Thermochim. Acta* **2011**, *525*, 141–149. [CrossRef]
25. Zhang, G.; Xie, W.; Wu, D. Selective localization of starch nanocrystals in the biodegradable nanocomposites probed by crystallization temperatures. *Carbohydr. Polym.* **2020**, *227*, 115341. [CrossRef] [PubMed]
26. Pramoda, K.P.; Linh, N.T.T.; Zhang, C.; Liu, T. Multiwalled carbon nanotube nucleated crystallization behavior of biodegradable poly(butylene succinate) nanocomposites. *J. Appl. Polym. Sci.* **2009**, *111*, 2938–2945. [CrossRef]
27. Bosq, N.; Aht-Ong, D. Nonisothermal crystallization behavior of poly(butylene succinate)/NaY zeolite nanocomposites. *Macromol. Res.* **2018**, *26*, 13–21. [CrossRef]

28. Li, Y.D.; Fu, Q.Q.; Wang, M.; Zeng, J.B. Morphology, crystallization and rheological behavior in poly(butylene succinate)/cellulose nanocrystal nanocomposites fabricated by solution coagulation. *Carbohydr. Polym.* **2017**, *164*, 75–82. [CrossRef] [PubMed]
29. Bao, L.; Chen, Y.; Zhou, W.; Wu, Y.; Huang, Y. Bamboo fibers@ poly (ethylene glycol)-reinforced poly (butylene succinate) biocomposites. *J. Appl. Polym. Sci.* **2011**, *122*, 2456–2466. [CrossRef]
30. Xu, J.; Manepalli, P.H.; Zhu, L.; Narayan-Sarathy, S.; Alavi, S. Morphological, barrier and mechanical properties of films from poly (butylene succinate) reinforced with nanocrystalline cellulose and chitin whiskers using melt extrusion. *J. Polym. Res.* **2019**, *26*, 188. [CrossRef]
31. Joy, J.; Jose, C.; Yu, X.; Mathew, L.; Thomas, S.; Pilla, S. The influence of nanocellulosic fiber, extracted from Helicteres isora, on thermal, wetting and viscoelastic properties of poly(butylene succinate) composites. *Cellulose* **2017**, *24*, 4313–4323. [CrossRef]
32. Phua, Y.; Pegoretti, A.; Araujo, T.M.; Ishak, Z.M. Mechanical and thermal properties of poly (butylene succinate)/poly (3-hydroxybutyrate-co-3-hydroxyvalerate) biodegradable blends. *J. Appl. Polym. Sci.* **2015**, *132*, 47. [CrossRef]
33. Chen, C.H. Effect of attapulgite on the crystallization behavior and mechanical properties of poly (butylene succinate) nanocomposites. *J. Phys. Chem. Solids* **2008**, *69*, 1411–1414. [CrossRef]
34. Liang, Z.; Pan, P.; Zhu, B.; Dong, T.; Inoue, Y. Mechanical and thermal properties of poly (butylene succinate)/plant fiber biodegradable composite. *J. Appl. Polym. Sci.* **2010**, *115*, 3559–3567. [CrossRef]
35. Han, H.; Wang, X.; Wu, D. Mechanical properties, morphology and crystallization kinetic studies of bio-based thermoplastic composites of poly (butylene succinate) with recycled carbon fiber. *J. Chem. Technol. Biotechnol.* **2013**, *88*, 1200–1211. [CrossRef]
36. Chen, G.X.; Yoon, J.S. Nonisothermal crystallization kinetics of poly (butylene succinate) composites with a twice functionalized organoclay. *J. Polym. Sci. Part B Polym. Phys.* **2005**, *43*, 817–826. [CrossRef]
37. Yarici, T.; Kodal, M.; Ozkoc, G. Non-isothermal crystallization kinetics of Poly (Butylene succinate)(PBS) nanocomposites with different modified carbon nanotubes. *Polymer* **2018**, *146*, 361–377. [CrossRef]
38. Bosq, N.; Aht-Ong, D. Isothermal and non-isothermal crystallization kinetics of poly (butylene succinate) with nanoprecipitated calcium carbonate as nucleating agent. *J. Therm. Anal. Calorim.* **2018**, *132*, 233–249. [CrossRef]
39. Filizgok, S.; Kodal, M.; Ozkoc, G. Non-isothermal crystallization kinetics and dynamic mechanical properties of poly (Butylene succinate) nanocomposites with different type of carbonaceous nanoparticles. *Polym. Compos.* **2018**, *39*, 2705–2721. [CrossRef]
40. Habibi, Y.; Dufresne, A. Highly filled bionanocomposites from functionalized polysaccharide nanocrystals. *Biomacromolecules* **2008**, *9*, 1974–1980. [CrossRef]
41. Abushammala, H. On the para/ortho reactivity of isocyanate groups during the carbamation of cellulose nanocrystals using 2,4-toluene diisocyanate. *Polymers* **2019**, *11*, 1164. [CrossRef]
42. Abushammala, H. A simple method for the quantification of free isocyanates on the surface of cellulose nanocrystals upon carbamation using toluene diisocyanate. *Surfaces* **2019**, *2*, 444–454. [CrossRef]
43. Correlo, V.M.; Boesel, L.F.; Bhattacharya, M.; Mano, J.F.; Neves, N.M.; Reis, R.L. Properties of melt processed chitosan and aliphatic polyester blends. *Mater. Sci. Eng. A* **2005**, *403*, 57–68. [CrossRef]
44. Dobreva, A.; Gutzow, I. Kinetics of Non-isothermal Overall Crystallization in Polymer Melts. *Cryst. Res. Technol.* **1991**, *26*, 863–874. [CrossRef]
45. Avrami, M. Kinetics of phase change. I General theory. *J. Chem. Phys.* **1939**, *7*, 1103–1112. [CrossRef]
46. Liu, Y.; Wang, L.; He, Y.; Fan, Z.; Li, S. Non-isothermal crystallization kinetics of poly (L-lactide). *Polym. Int.* **2010**, *59*, 1616–1621. [CrossRef]
47. Zhou, W.Y.; Duan, B.; Wang, M.; Cheung, W.L. Isothermal and non-isothermal crystallization kinetics of poly (L-Lactide)/carbonated hydroxyapatite nanocomposite microspheres. *Adv. Divers. Ind. Appl. Nanocompos.* **2011**, *1*, 231–260.
48. Ravari, F.; Mashak, A.; Nekoomanesh, M.; Mobedi, H. Non-isothermal cold crystallization behavior and kinetics of poly (l-lactide): Effect of l-lactide dimer. *Polym. Bull.* **2013**, *70*, 2569–2586. [CrossRef]
49. Papageorgiou, G.Z.; Achilias, D.S.; Bikiaris, D.N. Crystallization kinetics of biodegradable poly (butylene succinate) under isothermal and non-isothermal conditions. *Macromol. Chem. Phys.* **2007**, *208*, 1250–1264. [CrossRef]

50. Seven, K.M.; Cogen, J.M.; Gilchrist, J.F. Nucleating agents for high-density polyethylene—A review. *Polym. Eng. Sci.* **2016**, *56*, 541–554. [CrossRef]
51. Calabia, B.P.; Ninomiya, F.; Yagi, H.; Oishi, A.; Taguchi, K.; Kunioka, M.; Funabashi, M. Biodegradable poly (butylene succinate) composites reinforced by cotton fiber with silane coupling agent. *Polymers* **2013**, *5*, 128–141. [CrossRef]

© 2020 by the authors. Licensee MDPI, Basel, Switzerland. This article is an open access article distributed under the terms and conditions of the Creative Commons Attribution (CC BY) license (http://creativecommons.org/licenses/by/4.0/).

MDPI
St. Alban-Anlage 66
4052 Basel
Switzerland
Tel. +41 61 683 77 34
Fax +41 61 302 89 18
www.mdpi.com

Crystals Editorial Office
E-mail: crystals@mdpi.com
www.mdpi.com/journal/crystals

www.ingramcontent.com/pod-product-compliance
Lightning Source LLC
LaVergne TN
LVHW070448100526
838202LV00014B/1686